# Numerical Methods for the Navier-Stokes Equations

Edited by
F.-K. Hebeker,
R. Rannacher and
G. Wittum

# Notes on Numerical Fluid Mechanics (NNFM)  Volume 47

Series Editors: Ernst Heinrich Hirschel, München (General Editor)
Kozo Fujii, Tokyo
Bram van Leer, Ann Arbor
Keith William Morton, Oxford
Maurizio Pandolfi, Torino
Arthur Rizzi, Stockholm
Bernard Roux, Marseille



Volumes 1 to 25 are out of print.
The addresses of the Editors are listed at the end of the book.

# Numerical Methods for the Navier-Stokes Equations

Proceedings of the International Workshop held at Heidelberg, October 25–28, 1993

Edited by
Friedrich-Karl Hebeker,
Rolf Rannacher and
Gabriel Wittum

Die Deutsche Bibliothek – CIP-Einheitsaufnahme

**Numerical methods for the Navier-Stokes equations:**
proceedings of the international workshop held at Heidelberg,
October 25–28, 1993 / ed. by Friedrich-Karl Hebeker ... –

(Notes on numerical fluid mechanics; Vol. 47)
  ISBN 978-3-528-07647-4     ISBN 978-3-663-14007-8 (eBook)
  DOI 10.1007/978-3-663-14007-8
NE: Hebeker, Friedrich-Karl [Hrsg.]; GT

Produced by Langelüddecke, Braunschweig
Printed on acid-free paper

ISSN 0179-9614
ISBN 978-3-528-07647-4

# PREFACE

The Sonderforschungsbereich "Reactive Flow, Diffusion and Transport" (SFB 359) at Heidelberg University and the IBM Scientific Center Heidelberg have jointly organized a workshop on

"Numerical Methods for the Navier-Stokes Equations".

This workshop took place from October 25-28, 1993, at the IBM Scientific Center and was attended by 113 scientists from 13 countries. The scientific program consisted of 12 invited and 34 contributed lectures which dealt with various aspects of the numerical solution of the Navier-Stokes equations describing compressible as well as incompressible flows. The main topics were stable and higher-order discretization schemes, discretizations based on non-standard variational formulations, operator splitting methods, multilevel and domain decomposition techniques, a posteriori error control and adaptivity, and implementation issues on parallel computers. These proceedings contain 29 of the contributions to the workshop in alphabetical order.

The editors thank the Deutsche Forschungsgemeinschaft (DFG) for its financial support through the SFB 359. They also like to express their gratitude to all persons involved in the organization of the workshop and the preparation of these proceedings.

April 1994

F. K. Hebeker  
R. Rannacher  
G. Wittum

# CONTENTS

# CONTENTS (continued)

# RELIABLE FINITE VOLUME METHODS FOR NAVIER STOKES EQUATIONS.


M. Berzins and J.M. Ware

Centre for the Development of CFD and School of Computer Studies,
The University of Leeds, Leeds LS2 9JT, UK.


## SUMMARY


The use of adaptive mesh spatial discretisation methods, coupled spatial and temporal error control and domain decomposition methods make it possible to construct efficient automatic methods for the numerical solution of time-dependent Navier Stokes problems. This paper describes the unstructured triangular mesh spatial discretisation method being used in a prototype package for compressible flows. The scheme is a cell-centred, second-order finite volume scheme that uses a ten triangle stencil. Previous work has concentrated on algorithms and error estimates for convection dominated problems. In this paper the algorithm is extended to include a new treatment of the diffusion terms. The prototype software uses an adaptive time error control and space remeshing strategy is used to attempt to control the numerical error in the solution.


## TRIANGULAR MESH SPATIAL DISCRETISATION METHOD

Although finite element and finite volume schemes based on unstructured triangular meshes have been used for many years, only recently have a number of high-order cell-centred finite volume schemes been developed, [5, 11, 8] . This paper is concerned with the Ware and Berzins [11, 2, 3] method. Although this method has been developed for systems of equations, for ease of exposition, consider the class of scalar p.d.e.s:

$$\frac{\partial u}{\partial t} + \frac{\partial f}{\partial x} + \frac{\partial g}{\partial y} = 0 \tag{1}$$

where $f = f(x,y,u,\frac{\partial u}{\partial x},\frac{\partial u}{\partial y})$ and $g = g(x,y,u,\frac{\partial u}{\partial x},\frac{\partial u}{\partial y})$ are the flux functions in $x$ and $y$ respectively and with appropriate boundary and initial conditions. The cell-centred finite volume scheme described here uses triangular elements as the control volumes over which the divergence theorem is applied. The solution values are deemed to be associated with the centroids of the triangles. In Figure 1, for example, the solution at the centroid of triangle $i$ is $U_i$ , the solutions at the centroids of the triangles surrounding triangle $i$ are $U_l$, $U_j$ and $U_k$ and the next level of centroid values used by the discretisation method on the ith triangle are: $U_m, U_n, U_p, U_q, U_r$ and $U_s$. The mesh point at which a solution value, say $U_s$, is defined is denoted by $(x_s, y_s)$ . Integration of equation (1) on the ith triangle, which has area $A_i$, and use of the divergence theorem gives:

$$A_i \frac{\partial U_i}{\partial t} = -\oint_{C_i} (f.\underline{n}_x + g.\underline{n}_y)dS,$$

where $C_i$ is the circumference of triangle $i$. The line integral along each edge is approximated by using the midpoint quadrature rule. The numerical flux is evaluated at the midpoint of the edge:

$$\frac{\partial u}{\partial t} = -\frac{1}{A_i}(f_{ik}\Delta y_{0,1} - g_{ik}\Delta x_{0,1} + f_{ij}\Delta y_{1,2} - g_{ij}\Delta x_{1,2} + f_{il}\Delta y_{2,0} - g_{il}\Delta x_{2,0}), \tag{2}$$

where $\Delta x_{i,j} = x_j - x_i$ , $\Delta y_{i,j} = y_j - y_i$. The fluxes $f_{ij}$ and $g_{ij}$ in the $x$ and $y$ directions respectively are evaluated at the midpoint of the triangle edge separating the triangles associated with $U_i$ and $U_j$. The convective parts of these fluxes are evaluated by using approximate Riemann solvers $f_{Rm}$ and $g_{Rm}$ respectively with the *left* solution value being defined as that internal to triangle $i$ and the *right* solution value being defined as that external to triangle $i$:

$$
\begin{aligned}
\frac{\partial u}{\partial t} = \frac{-1}{A_i}( \ &f_{Rm}(U_{ik}^l, U_{ik}^r, (U_{ik})_x, (U_{ik})_y)\Delta y_{0,1} \ &- \ &g_{Rm}(U_{ik}^l, U_{ik}^r, (U_{ik})_x, (U_{ik})_y)\Delta x_{0,1} \ &+ \\
&f_{Rm}(U_{ij}^l, U_{ij}^r, (U_{ij})_x, (U_{ij})_y)\Delta y_{1,2} \ &- \ &g_{Rm}(U_{ij}^l, U_{ij}^r, (U_{ij})_x, (U_{ij})_y)\Delta x_{1,2} \ &+ \\
&f_{Rm}(U_{il}^l, U_{il}^r, (U_{il})_x, (U_{il})_y)\Delta y_{2,0} \ &- \ &g_{Rm}(U_{il}^l, U_{il}^r, (U_{il})_x, (U_{il})_y)\Delta x_{2,0} \ &),
\end{aligned}
$$

$$\tag{3}$$

where $U_{ij}^l$ is the internal solution, with respect to triangle $i$, at the midpoint of the edge between $U_i$ and $U_j$ and $U_{ij}^r$ is the external solution, with respect to triangle $i$, on edge $j$. Note that $U_{i,j}^r = U_{j,i}^l$ as a consequence of this notation. Standard approximate Riemann solvers such as those of Osher and Roe are used to define the convective fluxes. The left and right values for the Riemann solver are created using limited linear upwind values. The internal and external values at cell interface of two triangular elements, $U_{ij}^l$ and $U_{ij}^r$ in equation (3) are replaced with the limited linearly interpolated values defined by

$$
U_{ij}^l = U_i + \Phi(r_{ij}^l)\left(U_{ij}^L - U_i\right) \quad \text{and} \quad U_{ij}^r = U_j + \Phi(r_{ij}^r)\left(U_{ij}^R - U_j\right), \tag{4}
$$

where $U_{ij}^L$ is the internal linear upwind value, $U_{ij}^R$ is the external linear upwind value, $r_{ij}^l$ is the internal upwind bias ratio of gradients and $r_{ij}^r$ is the external upwind bias ratio of gradients. The internal and external ratio of linear gradients are defined by

$$
r_{ij}^l = \frac{U_{ij}^C - U_i}{U_{ij}^L - U_i} \quad \text{and} \quad r_{ij}^r = \frac{U_{ij}^C - U_j}{U_{ij}^R - U_j}. \tag{5}
$$

$U_{ij}^C$ is the linear centred value at the cell interface. The choice of limiter function $\Phi(.)$ is left open at this point although it should be noted that a zero limiter gives a first-order method. Equations (4) and (5) depend on the as yet undefined, interpolated and extrapolated values: $U_{ij}^L$, $U_{ij}^R$ and $U_{ij}^C$.

The value $U_{ij}^L$ is constructed by using linear extrapolation based on the solution value $U_i$ and an intermediate solution value (again calculated by linear interpolation) $U_{lk}$ which lies on the line joining the centroids at which $U_l$ and $U_k$ are defined (see Figure 1) i.e.

$$
U_{ij}^L = U_i + d_{ij,i}\frac{U_i - U_{lk}}{d_{i,lk}}, \tag{6}
$$

where the term $d_{a,b}$ denotes the positive distance between points $a$ and $b$, so for example $d_{ij,i}$ denotes the positive distance between points $ij$ and $i$, see Figure 1, as defined by

$$
d_{i,ij} = \sqrt{(x_i - x_{ij})^2 + (y_i - y_{ij})^2} \ , \tag{7}
$$

where $(x_{ij}, y_{ij})$ are the co-ordinates of $U_{ij}$ . The value $U_{ij}^R$ is defined in a similar way using linear extrapolation based on the solution value $U_j$ and an intermediate solution value (itself calculated by linear interpolation) $U_{rs}$ which lies on the line joining the centroids at which $U_r$ and $U_s$ are defined, see Figure 1. In the case when the three

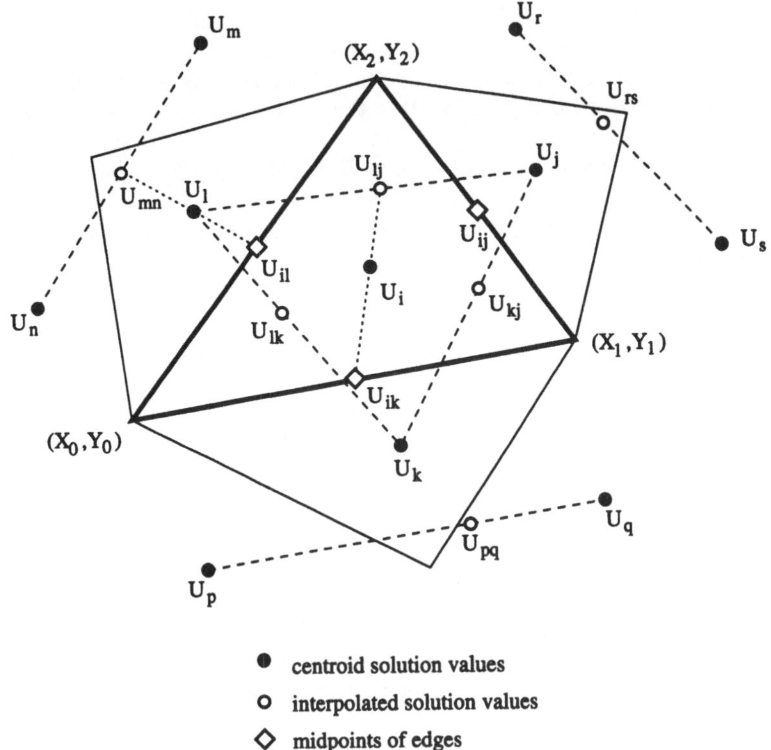


● centroid solution values

○ interpolated solution values

◇ midpoints of edges

Figure 1: Construction of Interpolants

centroid points are collinear it is not possible to define a linear interpolant and so the immediate upwind centroid value will be used: internally $U_i$ or externally $U_j$.

Assuming that all the centroid values are exact then the interpolation errors associated with the linear interpolants defined above may be determined by standard Taylor's series analysis. which shows that both interpolation errors are second order in the mesh spacing distances $d_{**}$, [3].

The centered value, $U_{ij}^C$, is constructed from the six values: $U_i$, $U_j$, $U_k$, $U_l$, $U_s$ and $U_r$ by a series of one-dimensional linear interpolations. Three linear interpolations onto the edge being considered are performed using *opposing* pairs of centroid values, see Figure 1. $U_{lr}$, $U_{ij}$ and $U_{ks}$ are found using the pairs $U_l$ and $U_r$, $U_i$ and $U_j$ and $U_k$ and $U_s$ respectively. If the midpoint of the edge lies between $U_{ks}$ and $U_{ij}$ then the centred value is found by linear interpolation using these two values. Otherwise the values $U_{lr}$ and $U_{ij}$ are used to compute the centred value at the midpoint by using linear interpolation.

## APPROXIMATION OF DIFFUSIVE FLUXES

In order to compute the diffusive flux contributions at mid-points of edges it is necessary to estimate the derivatives $\partial u/\partial x$ and $\partial u/\partial y$ at these points. Consider the mid-point $(x_{il}, y_{il})$ which lies inside the triangle formed by the centroids $i, l$ and $k$ . Durlofsky et. al. [5] construct first-order derivative approximations by differentiating the linear interpolant defined by the solution values at these points. An alternative is to use the six centroid values $U_i, U_l, U_k, U_m, U_n$ and $U_j$ to form a quadratic interpolant and then to differentiate this. Hyman et. al. , [6], show that this is not possible for an arbitrary set of points.

An alternative to this is to use the four points $U_i, U_l, U_k$, and $U_n$ to form a bilinear interpolant and then to differentiate this. For ease of notation suppose that the edge mid-point $il$ is the origin and assume that all derivatives are evaluated there. Standard Taylor's series expansions then yield:

$$\Delta U_{li} = \Delta x_{li}\frac{\partial u}{\partial x} + \Delta y_{li}\frac{\partial u}{\partial y} + \Delta x y_{li}\frac{\partial^2 u}{\partial x \partial y} + h.o.t. \tag{8}$$

where $\Delta U_{li} = U_l - U_i$ , $\Delta x_{li} = x_l - x_i$ and $\Delta x y_{li} = x_l x_i - y_l y_i$ .

Similar equations for $\Delta U_{lk}$ and $\Delta U_{in}$ may be written using matrix notation as

$$\begin{bmatrix} \Delta U_{lk} \\ \Delta U_{in} \end{bmatrix} = M_{in}^{lk} \begin{bmatrix} \partial u/\partial x \\ \partial u/\partial y \end{bmatrix} + \frac{\partial^2 u}{\partial x \partial y} \begin{bmatrix} \Delta x y_{lk} \\ \Delta x y_{in} \end{bmatrix} + h.o.t. \tag{9}$$

$$\text{where } M_{in}^{lk} = \begin{bmatrix} \Delta x_{lk} & , & \Delta y_{lk} \\ \Delta x_{in} & , & \Delta y_{in} \end{bmatrix}$$

which is an invertible matrix. Applying the inverse of this matrix to equation (9) and then substituting for $\partial u/\partial x$ and $\partial u/\partial y$ in equation (8) gives:

$$\alpha \frac{\partial^2 u}{\partial x \partial y} = \Delta U_{li} - \begin{bmatrix} \Delta x_{li} & , & \Delta y_{li} \end{bmatrix} \begin{bmatrix} M_{in}^{lk} \end{bmatrix}^{-1} \begin{bmatrix} \Delta U_{lk} \\ \Delta U_{in} \end{bmatrix} + h.o.t. \tag{10}$$

$$\text{where } \alpha = \Delta x y_{li} - \begin{bmatrix} \Delta x_{li} & , & \Delta y_{li} \end{bmatrix} \begin{bmatrix} M_{in}^{lk} \end{bmatrix}^{-1} \begin{bmatrix} \Delta x y_{lk} \\ \Delta x y_{in} \end{bmatrix} \tag{11}$$

An approximation to the required derivatives is then given by

$$\begin{bmatrix} \partial u/\partial x \\ \partial u/\partial y \end{bmatrix} = \begin{bmatrix} M_{in}^{lk} \end{bmatrix}^{-1} \begin{bmatrix} \Delta U_{lk} & - & \Delta x y_{lk}\frac{\partial^2 u}{\partial x \partial y} \\ \Delta U_{in} & - & \Delta x y_{in}\frac{\partial^2 u}{\partial x \partial y} \end{bmatrix} + h.o.t. \tag{12}$$

In order to avoid the case $\alpha = 0$ it necessary to take care when choosing the local co-ordinate system in which to calculate the bilinear approximation. The Durlofsky et. al. approach is still used at the boundaries and gives a fallback position should $\alpha$ be zero.


## PROPERTIES OF SPATIAL DISCRETIZATION METHOD

Berzins and Ware [3] have considered whether or not the new scheme has the properties of *linearity preservation* and *positivity*, as proposed by Struijs et. al., [9]. The definition of positivity requires that every new value at a particular time can be written as a convex combination of old values at the previous time step. Berzins and Ware considered different flow paths through the triangle in Figure 1 and showed that three sufficient conditions for positivity are:

**1.** For every upwind interpolant the centroid value nearest the edge at whose midpoint the upwind value is being calculated is the maximum or minimum of the three values used to form the interpolant.

**2.** The centred interpolant must be bounded by the centroid values on either side i.e.

$$U_{il}^C = \alpha U_l + (1 - \alpha)U_i , \quad \text{for} \ \ 0 \le \alpha \le 1. \tag{13}$$

**3.** The limiter $\Phi(.)$ must be positive and $\Phi(S)/S \le 1$ . This last condition is satisfied, for example, by a modified van Leer limiter defined by

$$\Phi(S) = \frac{S + |S|}{1 + v} \quad \text{where} \ \ v = \text{Max}(1, |S|) \ . \tag{14}$$

A linearity-preserving spatial discretization method is defined by [9], as one which preserves the exact steady state solution whenever this is a linear function of the space coordinates $x$ and $y$, for any arbitrary triangulation of the domain. This is equivalent to second order accuracy on regular meshes, see [9]. Berzins and Ware were able to show that the method in its unlimited form is linearity preserving but that in some cases condition 2 above may force linearity preservation to be violated.

The previous results on interpolation errors may be combined with standard results for the effect of quadrature errors, see [7], to show how the errors at the mid-points of edges accumulate in the truncation error. Consider the spatial truncation error in the approximation of the Laplacian $c \left[ \frac{\partial^2 u}{\partial x^2} + \frac{\partial^2 u}{\partial y^2} \right]$ on the $i$th triangle, as denoted by $TE_i$. This is, after ignoring the second order quadrature error, see Jeng and Chen [7], a combination of the derivative errors at the mid-points of the edges i.e.

$$\begin{aligned} TE_i = \frac{-c}{A_i}[ \ &(E_{ik})_x \Delta y_{0,1} \ + \ (E_{ij})_x \Delta y_{1,2} \ + \ (E_{il})_x \Delta y_{2,0} \ - \\ &(E_{ik})_y \Delta x_{0,1} \ - \ (E_{ij})_y \Delta x_{1,2} \ - \ (E_{il})_y \Delta x_{2,0} \ ]. \end{aligned} \tag{15}$$

where the individual errors in the derivative approximations are defined such that $(E_{ij})_x$ is the error in $\partial u/\partial x$ at the $ik$th midpoint for example. Assuming from the definitions of the differentiation approximations that it is possible to extract a constant factor, say $d_{min}$, depending on the minimum of the distances, $d_{ab}$, as defined in equation (7), from each of the errors in this equation and assuming still further that the individual errors all have the form

$$(E_{ik})_x = \ d_{min} \ (e_{ik})_x \ \text{and} \ (E_{ik})_y = \ d_{min} \ (e_{ik})_y,$$

the expression for the truncation error may be rewritten as:

$$\begin{aligned} TE_i = -\frac{c \ d_{min}}{A_i}[ \ &(e_{ik})_x \Delta y_{0,1} \ + \ (e_{ij})_x \Delta y_{1,2} \ + \ (e_{il})_x \Delta y_{2,0} \ - \\ &(e_{ik})_y \Delta x_{0,1} \ - \ (e_{ij})_y \Delta x_{1,2} \ - \ (e_{il})_y \Delta x_{2,0} \ ]. \end{aligned} \tag{16}$$

It is now possible to define two linear functions on the $i$th triangle $E_f(x, y)$ and $E_g(x, y)$ such that $E_f(x, y)$ has values $(e_{ik})_x, (e_{ij})_x$ and $(e_{il})_x$ at the midpoints $ik, ij$ and $il$ , $E_g(x, y)$ has values $(e_{ik})_y, (e_{ij})_y$ and $(e_{il})_y$ at the midpoints $ik, ij$ and $il$ . From the linearity of these functions and the divergence theorem it follows that

$$\frac{\partial E_f}{\partial x} = \frac{1}{A_i}[(e_{ik})_x \Delta y_{0,1} + (e_{ij})_x \Delta y_{1,2} + (e_{il})_x \Delta y_{2,0}] \tag{17}$$

and

$$\frac{\partial E_g}{\partial y} = -\frac{1}{A_i}[(e_{ik})_y \Delta x_{0,1} + (e_{ij})_y \Delta x_{1,2} + (e_{il})_y \Delta x_{2,0}]. \tag{18}$$

Hence the truncation error (ignoring the quadrature error due to the use of the mid-point rule) may be written as

$$TE_i = -c\,d_{min}\left[\frac{\partial E_f}{\partial x} + \frac{\partial E_g}{\partial y}\right]. \tag{19}$$

The error due to the use of the quadrature rule is derived by Jeng and Chen [7].


## TIME INTEGRATION AND ERROR CONTROL

The above spatial discretization scheme results in a system of differential equations, which can be written as the initial value problem:

$$\underline{\dot{U}} = \underline{F}_N\ (\ t,\ \underline{U}(t)\ )\ ,\ \underline{U}(0)\ \text{given}\ , \tag{20}$$

where the vector, $\underline{U}(t)$, is defined by $\underline{U}(t) = [U(x_1, y_1, t), U(x_2, y_2, t), ..., U(x_N, y_N, t)\ ]^T$. The point $x_i, y_i$ is the centre of the $i$ th cell and $U_i(t)$ is a numerical approximation to $u(x_i, y_i, t)$. Numerical integration of equation (20) provides the approximation, $\underline{V}(t)$, to the vector of exact p.d.e. solution values at the mesh points, $\underline{u}(t)$. The global error in the numerical solution can be expressed as the sum of the spatial discretization error, $\underline{e}(t) = \underline{u}(t) - \underline{U}(t)$, and the global time error, $\underline{g}(t) = \underline{U}(t) - \underline{V}(t)$. That is,

$$\begin{aligned}\underline{E}(t) = \underline{u}(t) - \underline{V}(t) &= (\underline{u}(t) - \underline{U}(t)) + (\underline{U}(t) - \underline{V}(t)) \\ &= \underline{e}(t) + \underline{g}(t).\end{aligned} \tag{21}$$

Efficient time integration requires that the spatial and temporal are roughly the same order of magnitude. The need for spatial error estimates unpolluted by temporal error requires the spatial error to be the larger of the two errors.

The Theta method code, see [1] used here defines the numerical solution at $t_{n+1} = t_n + k$, where $k$ is the time step size, as denoted by $\underline{V}(t_{n+1})$, by

$$\underline{V}(t_{n+1}) = \underline{V}(t_n) + (1 - \theta)k\,\underline{\dot{V}}(t_n) + \theta\,k\,\underline{F}_N(t_{n+1}, \underline{V}(t_{n+1})),\ \theta = 0.55, \tag{22}$$

in which $\underline{V}(t_n)$ and $\underline{\dot{V}}(t_n)$ are the numerical solution and its time derivative at the previous time $t_n$. Berzins and Ware [3] show that the method will preserve positivity if a CFL-like condition is satisfied. Although such a condition is often used to choose a stable timestep it may be imprecise as an accuracy control. In contrast when a standard local error $\underline{l}_{n+1}(t_{n+1})$ control i.e. $\|\ \underline{l}_{n+1}(t_{n+1})\ \| < TOL$ is used it is difficult to establish a relationship between the accuracy tolerance, $TOL$, and the global time error.

An alternative approach is described by Berzins [1] who balances the spatial and temporal errors by controling the local time error to be a fraction of the local growth in the spatial discretization error. The local-in-time spatial error, $\underline{\hat{e}}(t_{n+1})$, for the timestep from $t_n$ to $t_{n+1}$ is defined as the spatial error at time $t_{n+1}$ given the assumption that the spatial error, $\underline{e}(t_n)$, at time $t_n$ is zero. A local error balancing approach is then:

$$\|\ \underline{l}_{n+1}(t_{n+1})\ \| < \epsilon\ \|\ \underline{\hat{e}}(t_{n+1})\ \|,\ 0 < \epsilon < 1. \tag{23}$$

The error $\hat{\underline{e}}(t_{n+1})$ is estimated by the difference between the computed solution and the first-order solution which satisfies a modified o.d.e. system denoted by

$$\dot{\underline{v}}_{n+1}(t) = \underline{G}_N(t, \underline{v}_{n+1}(t)),\qquad(24)$$

where $\underline{v}_{n+1}(t_n) = \underline{V}(t_n)$ , $\dot{\underline{v}}_{n+1}(t_n) = \underline{G}_N(t, \underline{V}(t_n))$ and where $\underline{G}_N(.,.)$ is obtained simply by setting the limiter function in the space discretisation to zero and by using the first order space derivative approximations. The local-in-time space error is then given by

$$\hat{\underline{e}}(t_{n+1}) = \underline{V}(t_{n+1}) - \underline{v}_{n+1}(t_{n+1})\qquad(25)$$

and is computed by applying the $\theta$ method with one functional iteration to equation (24). Equations (22) and (25) combined with the conditions on $\underline{v}_{n+1}(t_n)$ then give, [1],

$$\begin{aligned}\hat{\underline{e}}(t_{n+1}) = \theta \quad k \quad &[\underline{F}_N(t_{n+1}, \underline{V}(t_{n+1})) - \underline{G}_N(t_{n+1}, \underline{V}(t_{n+1}))] \ + \\ (1-\theta) \quad k \quad &[\underline{F}_N(t_n, \underline{V}(t_n)) - \underline{G}_N(t_n, \underline{V}(t_n))].\end{aligned}\qquad(26)$$

## NUMERICAL EXAMPLES

The properties of the diffusive approximation may be illustrated by two example problems with analytic solutions on $(0,1) \times (0,1)$. Problem A is a simple Poisson equation with an analytic solution $u(x,y) = 3e^{x+y}(x - x^2)(y - y^2)$. Problem B is the system of two p.d.e.s used by de Goede and Boonkamp [4] and is similar to the Navier Stokes equations while still having an exact solution. The equations are modified to be in conservative form, a Reynolds number of 100 is used so that accuracy in the diffusive part is important and integration halted at $t = 2.5$. The L1 norms obtained by discretization on the unit square are given in Table 1. On regular meshes both the Durlofsky and the bilinear approximation are second order accurate . The fixed meshes were varied so as to be irregular and have between 136 and 8704 triangles.

Table 1: L1 Error Norms

| Method | Prob. | No. of Triangles | | | |
|---|---|---|---|---|---|
| | | 136 | 544 | 2176 | 8704 |
| Durlo. | A | 6.9617e-3 | 2.0171e-3 | 7.2810e-4 | 2.1478e-4 |
| Bilin. | A | 6.5321e-3 | 1.7258e-3 | 5.5671e-4 | 1.5385e-4 |
| Durlo. | B | 9.6889e-3 | 3.0749e-3 | 7.5211e-4 | 2.1395e-4 |
| Bilin. | B | 9.5294e-3 | 2.7675e-3 | 6.8262e-4 | 1.8787e-4 |

## CONCLUSIONS

The improved accuracy of the new bilinear derivative on fixed meshes is demonstrated by Table 1. Preliminary results suggest that this improvement carries across to adaptive unstructured meshes.

The prototype adaptive software based on this discretisation method is being used to solve a variety of problems using fully automatic mesh generation and mesh adaptation software. The adaptivity tracks features in the solution automatically whilst using large elements away from these features to increase the efficiency. The spatial error estimate is used successfully in these examples to control the error through mesh adaptivity. Time integration is performed in such a way that the spatial error dominates, see [1]. The selection of appropriate times to adapt the spatial mesh is made by using a combination of estimated errors and predicted future errors, [2] . The prototype package also can be used on both shared and distributed memory computers as the flux calculation used in the residual is designed to operate in parallel. The mapping of unstructured meshes onto distributed memory processors is achieved by using graph-theoretic techniques, [10]; this ensures good speed-ups on both shared and distributed memory parallel computers.

**Acknowledgement.** The authors would like to thank Shell Research Ltd. for funding.

## REFERENCES


[1] M. BERZINS. *Temporal error control for convection-dominated equations in two space dimensions.*, SIAM Journal of Scientific Computing (to appear), 199x.

[2] M. BERZINS, J.M. WARE AND J. LAWSON. *Spatial and temporal error control in the adaptive solution of systems of conservation laws.* Advances in Computer Methods for Partial Differential Equations: IMACS PDE VII. IMACS, 1992.

[3] M. BERZINS AND J.M. WARE. *Positive discretization methods for hyperbolic equations on irregular meshes.* Submitted to Applied Numerical Mathematics, 1994.

[5] L. J. DURLOFSKY, B. ENQUIST AND S. OSHER. *Triangle based adaptive stencils for the solution of hyperbolic conservation laws.* Jour.Of Comp. Phys., 98:64–73, 1992.

[4] E. DE GOEDE AND M. BOONKAMP .*Vectorisation of the Odd/Even Hopscotch Scheme and the ADI scheme for the Two Dimensional Burgers' Equation.*, SIAM Jour. of Sci. Comp., 11:354–367, 1990.

[6] J. M. HYMAN, R. J. KNAPP, AND J. E. SCOVEL. *High order finite volume approximations of differential operators on nonuniform grids.* Physica D. , 60:112–138, 1992.

[7] Y.N. JENG AND J.L. CHEN. *Truncation error analysis of the finite volume method for a model steady convective equation.* Jour. of Comp. Phys., 100:64–76, 1992.

[8] S.Y. LIN, T.M. WU AND Y.S. CHIN. *Upwind finite-volume method with a triangular mesh for conservation laws.* Jour. of Comp. Phys., 107:324–337, 1993.

[9] R. STRUIJS, H. DECONINCK AND P. L. ROE. *Fluctuation splitting schemes for the 2D Euler equations.* Technical report, von Karman Institute for Fluid Dynamics, Chaussee de Waterloo, 72, B-1640 Rhode Saint Genese - Belgium, 1991.

[10] C.H. WALSHAW AND M. BERZINS. *Enhanced dynamic load-balancing of adaptive unstructured meshes.* In R.F. Sincovec et. al., editor, Parallel Processing for Scientific Computing, pages 971–978. SIAM, 1993.

[11] J.M. WARE AND M. BERZINS. *Finite volume techniques for time-dependent fluid-flow problems.* In Advances in Computer Methods for Partial Differential Equations: IMACS PDE VII. IMACS, 1992.

# PARALLEL COMPUTING AND MULTIGRID SOLUTION ON ADAPTIVE UNSTRUCTURED MESHES


S. Bikker, H. Greza and W. Koschel
Institute for Jet Propulsion and Turbomachinery
Aachen University of Technology, 52056 Aachen, Germany


## SUMMARY


An unstructured Finite Element method for the transient solution of the compressible Navier-Stokes equations using triangular respectively tetrahedral elements is presented. For convergence acceleration a Multigrid solution method has been implemented. A Domain Splitting scheme improves the performance of the solver mainly in case of time-dependent flows. An efficient solution method for parallel computations is proposed including two decomposition methods. The automatic mesh generation scheme allows for the application to arbitrarily shaped boundaries. A linear behaviour of the generation time in dependence on the mesh size has been achieved. An adaptive remeshing procedure enables the directional refinement and coarsening. The mesh quality is improved by an adaptive smoothing process.


## 1. INTRODUCTION

The employment of unstructured grids in conjunction with a powerful mesh generation scheme enables both the use of complex boundaries and the incorporation of an effective adaptation procedure by completely regenerating the mesh [1]. The application of unstructured grids inevitably results in a limited efficiency of the flow solver. To eliminate this disadvantage various methods for the improvement of the convergence behaviour and for the implementation on parallel computers have been investigated. The efficiency of the described methods is demonstrated by several applications including the turbulent flow-field of a horseshoe-vortex.

## 2. NUMERICAL SCHEME

The three-dimensional Navier-Stokes equations governing the transient flow of a viscous, compressible fluid completed by the assumption of perfect gas are considered. In case of turbulent simulations a Baldwin-Lomax or a $k$-$\varepsilon$-turbulence model is incorporated. The Navier-Stokes equations are integrated by a MacCormack two-step scheme with a weighted residual Finite Element approach using linear shape functions. The cell-centered formulation of the predictor step leads to a piecewise constant approximation of the state variables at an intermediate time level. The corrector step computes a piecewise linear continuous solution at the advanced time level with use of the Galerkin weighted residual method and represents a node-centered scheme. The global equation system arising from the assembly of all nodal element contributions can be solved iteratively by introducing the lumped mass matrix without the necessity of storing the global consistent mass matrix. For relaxing the solution towards a steady state local time steps are used and one iteration is sufficient. In order to stabilize the scheme for shock capturing artificial viscosity is added using a second order difference term controlled by a pressure switch. For compressible high-speed flow computations a Flux-Corrected Transport scheme enables high resolution for capturing discontinuous flow effects. Furthermore a five-stage stage Runge-Kutta time-stepping scheme is available employing the spatial scheme already described for the corrector step. The artificial damping terms are recomputed only in the first two steps and are constructed as a combination of the same second-order operator as used for the two-step scheme and a fourth-order operator for damping out high frequencies. After the removal of recurrences the algorithms were vectorized for suitable machines yielding to a vectorization rate of about 99% corresponding to a program speed-up factor of about 40 on a Siemens S600. For a more detailed information about the solution method see [2] and [3].

## 3. MESH GENERATION SCHEME

The automatic mesh generation scheme utilizes the concept of the generalized advancing front method [1], which allows for the application to arbitrarily shaped boundaries with embedded interior boundaries. The shape of the triangles is controlled by the mesh parameters element size, element stretching factor and stretching direction (Fig. 1). A spatial distribution of these parameters is provided by a background grid, which consists of linear triangular elements completely covering the computational domain (Fig. 2). The generation starts with the construction of the initial front which is made of the assembly of orientated line segments discretised into sides. In order to define the geometry of these line segments, linear, arched, parabolic and cubic Spline segments are available. During the generation new elements and points are simultaneously introduced permitting significant changes in the local mesh structure (Fig. 3). After the generation of each element the front is updated. A stretching of the elements is achieved by a local transformation to an unstretched space. Detailed information may be taken from [3].
In principle the method of the three-dimensional mesh generation follows the two-dimensional scheme with the front now containing the faces of tetrahedrons instead of sides. Apart from the element size the mesh parameters consist of two stretching factors and two stretching directions. In order to define the domain to be gridded the boundaries are subdivided into surface segments, which again are bounded by common line segments (Fig. 4). The orientation of the surface segments indicates on which side of the surface the domain to be gridded is situated. At present for the surface definition plane, cylindrical and triangular parabolic surface segments are available (Fig. 5). The assembling of the boundary begins by discretising all line segments into sides. Afterwards all the surface segments in turn are transformed into a quasi two-dimensional domain with the corresponding line segments acting as the initial front for the subsequent triangulation. After the application of the two-dimensional procedures the generated surface mesh is transfered back to physical space.

## 4. ADAPTIVE REMESHING AND MESH SMOOTHING

The adaptive remeshing method considered here offers the ability to improve the solution quality in a computationally efficient manner. During the analysis of a certain flow problem the adaptation process is achieved by repeatedly regenerating the complete computational mesh allowing for a directional refinement and coarsening independent of the previous grid. The initial computational mesh is now acting as a background grid providing the spatial distribution of the mesh parameters. A one-dimensional estimation of the interpolation error leads to an indicator based upon the second derivatives of a certain scalar variable (usually density or Mach number).

For multi-dimensional problems the mean values and principle directions of the matrix of the derivatives give some indication of the error magnitude and direction. The condition of a uniformly distributed error indicator within the domain leads to the optimal nodal values of the mesh parameters [1]. Fig. 6 shows the adaptation for a inviscid shock reflection at Ma=2 for the two- and three-dimensional case [3].

It turns out that ill-distorted elements may affect the solution quality and the convergence rate of the flow solver. For this reason an adaptive smoothing process is applied after the mesh generation. First the element connectivity is optimized by swapping the element sides. The mesh is relaxed to the optimal case with each node being surrounded by six elements. Subsequently the nodes are moved with regard to the prescribed distribution of the mesh parameters using a spring system analogy. Instead of a remeshing the adaptive smoothing process can be used for the purpose of mesh adaptation too. The improvement of the mesh structure after the application of the smoothing procedures becomes evident in Fig. 7. The enhanced shape of the triangles is obvious and the flow lines of the mesh are more apparent. Up to the present for the three-dimensional case only the surface meshes are smoothed.

## 5. GENERATION TIMING

When using unstructured grids the different search operations involved during the mesh generation and the Multigrid data transfer naturally lead to a quadratic development of the computational time in dependence on the mesh size. By the incorporation of fast search algorithms and specially suited data structures like connectivity lists and multidimensional tree structures (binary list, Quad-, Oct-, 4D- and 6D-Tree) a linear behaviour has been achieved (Fig. 8).

## 6. MULTIGRID SOLUTION METHOD

The presented direct Multigrid method was implemented to accelerate the convergence to steady state. It enables the solution of the compressible Euler equations in two dimensions using the (5,2)-Runge-Kutta scheme (Section 2). The algorithm operates on a sequence of fine and coarse grids allowing of an arbitrary number of meshes. Because of the unstructured generation of the elements and the points (Section 3) the meshes employed are completely unrelated. The advantages of the time-stepping on the coarse meshes are on the one hand larger elements permitting a larger time-step and on the other hand less grid points causing less computational work. The algorithm uses a V-cycle performing one time-step per level when proceeding from the fine to the coarse meshes and no time-step when interpolating back the corrections. To maintain a conservative data transfer the solution and the residual vector are transfered by interpolation respectively distribution using linear shape functions. Furthermore a fine-to-coarse—defect correction is added in order to maintain the solution quality of the finest mesh on the coarse grids [4]. The four adapted meshes employed for the Multigrid computation of a subsonic flow past a cylinder together with the number of elements is shown in Fig. 9. The convergence history of the Multigrid and the fine grid solution in dependence on work units is depicted in Fig. 10 where the improvements become obvious. The convergence is measured by the $L_2$-norm of the density residuals and one work unit means the CPU-time normalized by the time which is needed for a single time-step on the finest grid only. The solution quality is determined by the finest mesh solely (Fig. 11).

# 7. PARALLEL COMPUTATION

The following section describes the implementation of the explicit Taylor-Galerkin two-step algorithm on a MIMD machine. The algorithm can efficiently be parallelized by load balancing of the elements. An additional amount of CPU-time will result from both double calculations on the boundary points, and communication between neighbouring processors. In case of steady state computations local time-stepping and local communication structures in the corrector step and in the smoothening process are used. Under PARIX-Fortran two different communication models, namely synchronous and loosely asynchronous tasks, are available.

Additional work has to be carried out when partitioning is made for unstructured triangular or tetrahedral meshes. In order to achieve a good efficiency the boundaries of the decomposed domains have to be small. In the following two different approaches are presented, namely a simple geometrical method (ORS) and an algorithm based on a spectral decomposition (ERS) [5]. A simple way to scatter the elements on the processors is a subdivision along predefined directions, like the orthogonal coordinate axes. Fig. 12 shows an example for a mesh already shown in Fig. 7. The boundaries between the processors are straight yielding an overall number of 157 (39%) communication points. Within the ERS method the second eigenpair of the Laplacian of the dual graph is used in order to find accurate partitions with small boundaries [6]. It can be found that the second eigenvector itself gives some directional information and differences of the vector components describe distances in the graph. Fig. 12b) depicts a decomposition with curved boundaries and an amount of 132 (32.8 %) communication points. In order to judge the efficiency of the MIMD machine (Parsytec SC256), speed-ups for a flow simulation on the mesh of Fig. 10 are shown in Fig. 13. The major acceleration can be achieved with asynchronous tasks. In this case the ERS method proves to be superior to ORS, but needs about ten times more CPU-time for the partitioning. It is dependant on the problem size, which method is to be preferred.

# 8. DOMAIN-SPLITTING

In contrast to the pure spatial decomposition described in the previous section this algorithm splits the domain according to the allowed time-step coupled to the mesh size by the CFL criterion. Usually the mesh size varies by orders of magnitudes in the discretized domain. This leads to some disadvantages, because very small Courant numbers are imposed on the large elements, yielding to an increased truncation error. The idea of the algorithm is to advance the solution on different time levels with larger Courant numbers reducing simultaneously computational costs. Following a suggestion of Morgan et al. [7] the numerical implementation is carried out with an overlap of the subdomains of at least two elements. For validation the two-dimensional time dependant viscous flow behind a circular cylinder resulting in the well-known Karman vortex-street is shown. The Reynolds number used for this example was Re=200 and a Mach number of the incoming flow of Ma=0.3 was chosen. Fig. 14a) and b) depict the Finite Element mesh and the resulting distribution of the domains, where the overlap regions are visible. In Fig. 14c) the Mach number isolines are displayed. Fig. 14d) shows, that the frequency respectively the Strouhal number meets well the measured data from Schlichting. Compared to the single-domain run a gain in CPU-time by a factor of 1.4 has been obtained.

# 9. HORSESHOE-VORTEX FLOW

As an example of a viscous flow simulation the three-dimensional formation of a horseshoe-vortex has been computed. A special version of the Baldwin-Lomax turbulence model [8] is used on structured subgrids. All results are compared with experimental data from Eckerle/Langston [9]. The oncoming flow rolls up in front of the cylinder and results in a vortex (Fig. 15a)). In Fig. 15b) the surface elements of the mesh with a dense region in the vicinity of the vortex is shown. Fig. 15c) depicts some computed stream-ribbons around the cylinder detecting the location of the horseshoe-vortex. In Fig. 15d) the dimensionless static pressure coefficient $c_p$ on the endwall in the plane of symmetry versus the radius $r/D$ is given. The computed pressure profile reflects well the measured one. Fig. 15e) and f) display measured and computed pressure isolines. In the next figures the endwall streaklines are compared. The experimental flow vectors have been determined by an ink painting (Fig. 15g)). In addition the vectors obtained from the calculation are depicted in Fig. 15h). The saddle point and the separation line are in good agreement with the measured data. For the computation of the horseshoe vortex the speed-up with 256 processors amounts 198, leading to an efficiency of 77%.

# 10. CONCLUSIONS

It has been demonstrated that unstructured grids may be employed advantageously for the accurate simulation of both geometrically as well as physically complex flow fields in a computationally efficient manner.

# ACKNOWLEDGEMENTS

Work on CFD-parallelization is funded by the Deutsche Forschungsgemeinschaft (DFG) within the Priority Research Programme 'Flow Simulation on High-Performance Computers'. The development of the automatic generation methods of unstructured adaptive grids is funded by the German Minister of Science and Technology (BMFT) within the program AG TURBO in connection with the Daimler Benz AG.

# REFERENCES

[1] J. Peraire, M. Vahdati, K. Morgan, O.C. Zienkiewicz: "Adaptive Remeshing for Compressible Flow Computations", J. Comp. Phys. 72, pp. 449-466, 1987.
[2] W. Rick, H. Greza, W. Koschel: "FCT-Solution on Adapted Unstructured Meshes for Compressible High Speed Flow Computations", Notes on Numerical Fluid Mechanics, Volume 38, pp. 334-348, Vieweg, Braunschweig 1993.
[3] H. Greza: "Automatische Neugenerierung lösungsadaptierter unstrukturierter Rechennetze zur numerischen Strömungssimulation nach der Methode der Finiten Elemente", DGLR Jahrbuch 1991, Bd. I, pp. 389-399, 1991.
[4] D.J. Mavriplis: "Multigrid Solution of the Two-Dimensional Euler Equations on Unstructured Triangular Meshes", AIAA Journal, Vol. 26, No. 7, pp. 824-831, 1988
[5] A. Pothen, H.D. Simon, K. Liou: "Partitioning Sparse Matrices with Eigenvectors of Graphs", SIAM J. Matrix Anal. Appl., Vol. 11, No. 3, pp. 430-452, 1990.
[6] H.D. Simon: "Partitioning of Unstructured Problems for Parallel Processing", Comp. Systems in Engineering, Vol. 2, No. 2/3, pp. 135-148, 1991.
[7] R. Löhner, K. Morgan, O.C. Zienkiewicz: "The Use of Domain-Splitting with an Explicit Hyperbolic Solver", Comp. Meth. Appl. Mech. Eng. 45, pp. 313-329, 1984.
[8] H. Zimmermann: "Berechnung von zwei- und dreidimensionalen Strömungen in einem transsonischen Turbinengitter unter besonderer Berücksichtigung der Verluste" , Dissertation, TU Braunschweig, 1990.
[9] W.A. Eckerle, L.S. Langston: "Horseshoe Vortex Formation around a Cylinder", ASME Journal of Turbomachinery, Vol. 109, pp. 278-285, 1987.

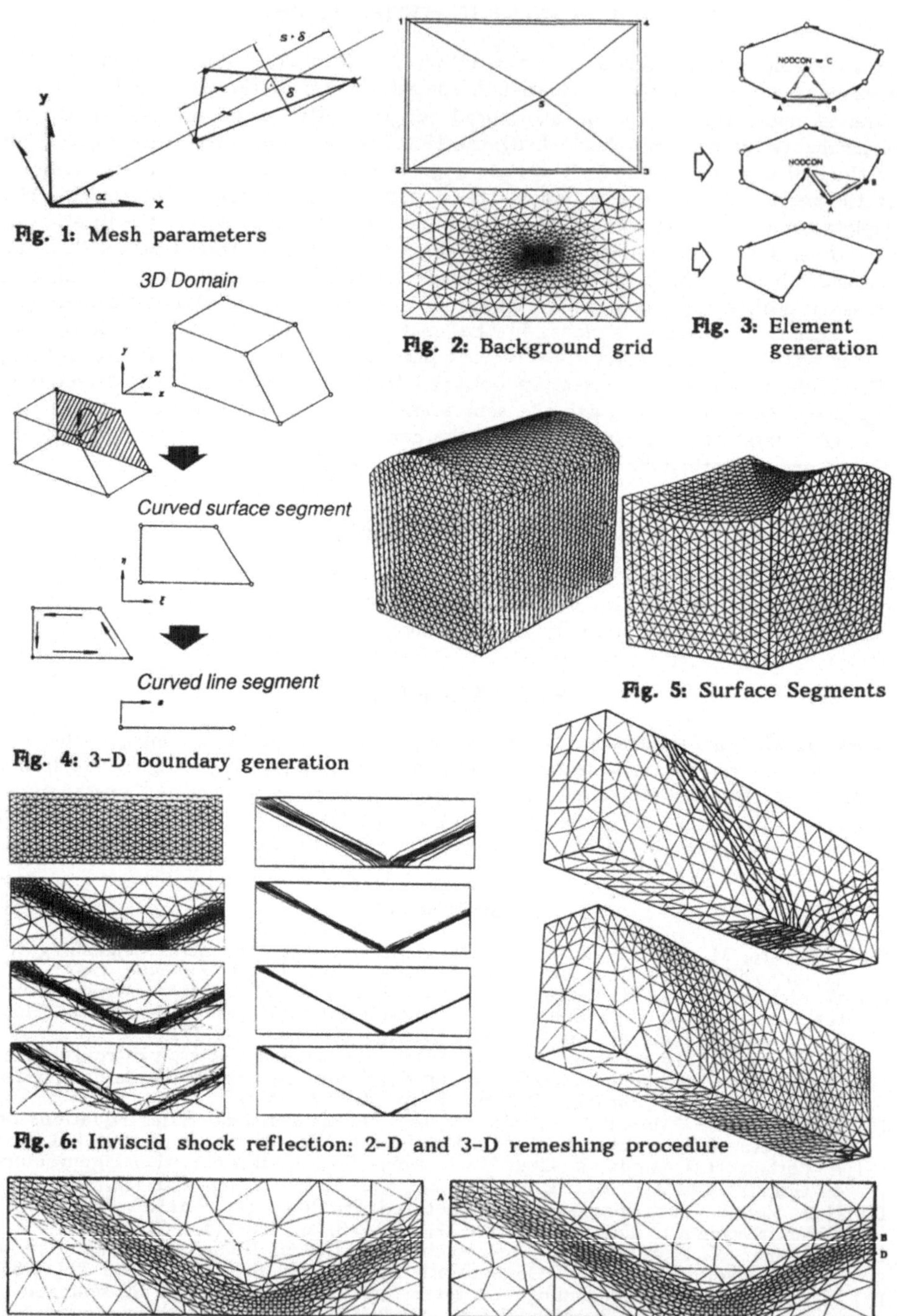

**Fig. 1:** Mesh parameters

**Fig. 2:** Background grid

**Fig. 3:** Element generation

**Fig. 4:** 3-D boundary generation

**Fig. 5:** Surface Segments

**Fig. 6:** Inviscid shock reflection: 2-D and 3-D remeshing procedure

**Fig. 7:** Mesh before and after the application of adaptive smoothing methods
(Lines A–B and C–D: examples of flow lines of the mesh)

14

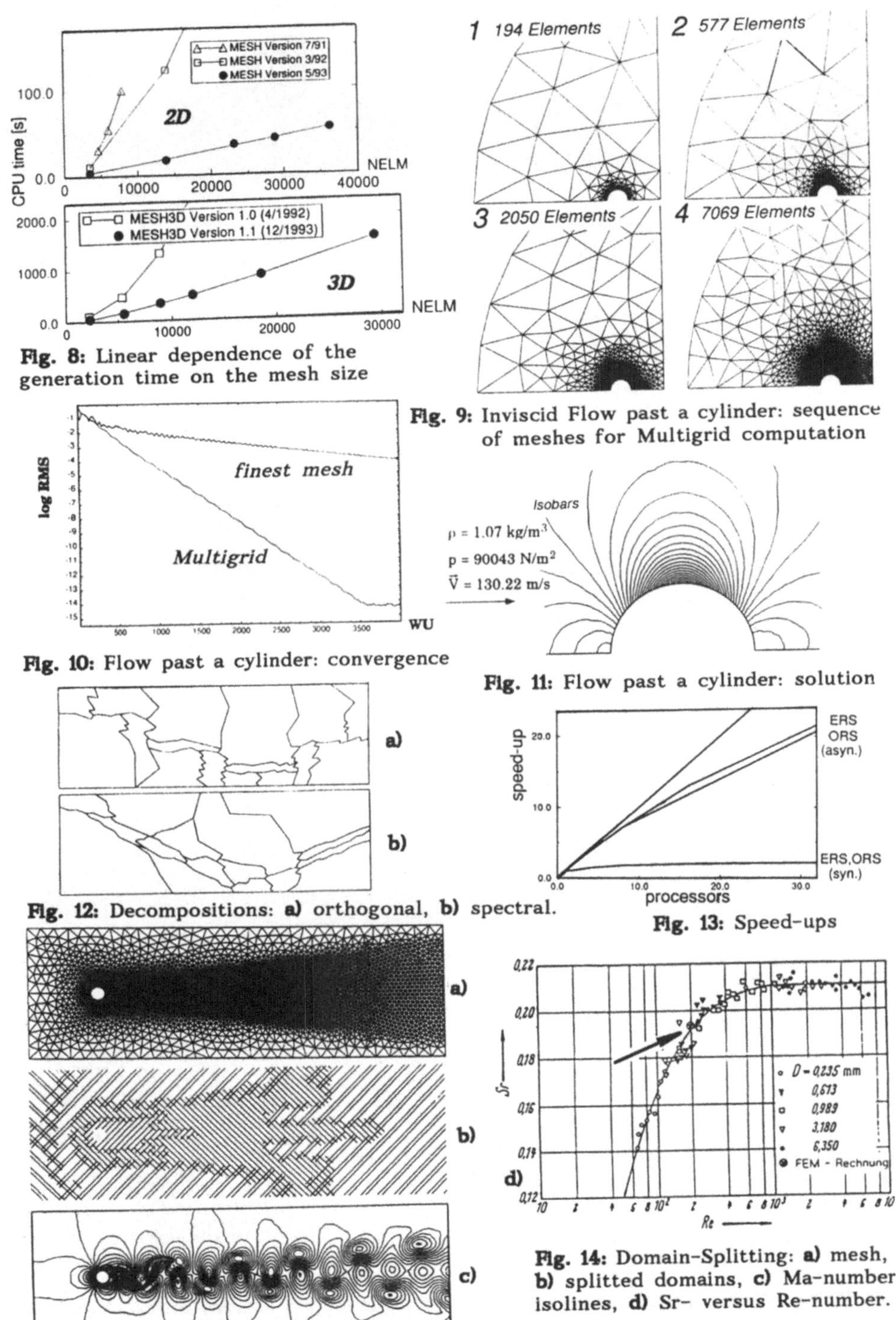



**Fig. 8:** Linear dependence of the generation time on the mesh size

**Fig. 9:** Inviscid Flow past a cylinder: sequence of meshes for Multigrid computation

**Fig. 10:** Flow past a cylinder: convergence

**Fig. 11:** Flow past a cylinder: solution

**Fig. 12:** Decompositions: **a)** orthogonal, **b)** spectral.

**Fig. 13:** Speed-ups

**Fig. 14:** Domain-Splitting: **a)** mesh, **b)** splitted domains, **c)** Ma-number isolines, **d)** Sr- versus Re-number.

15

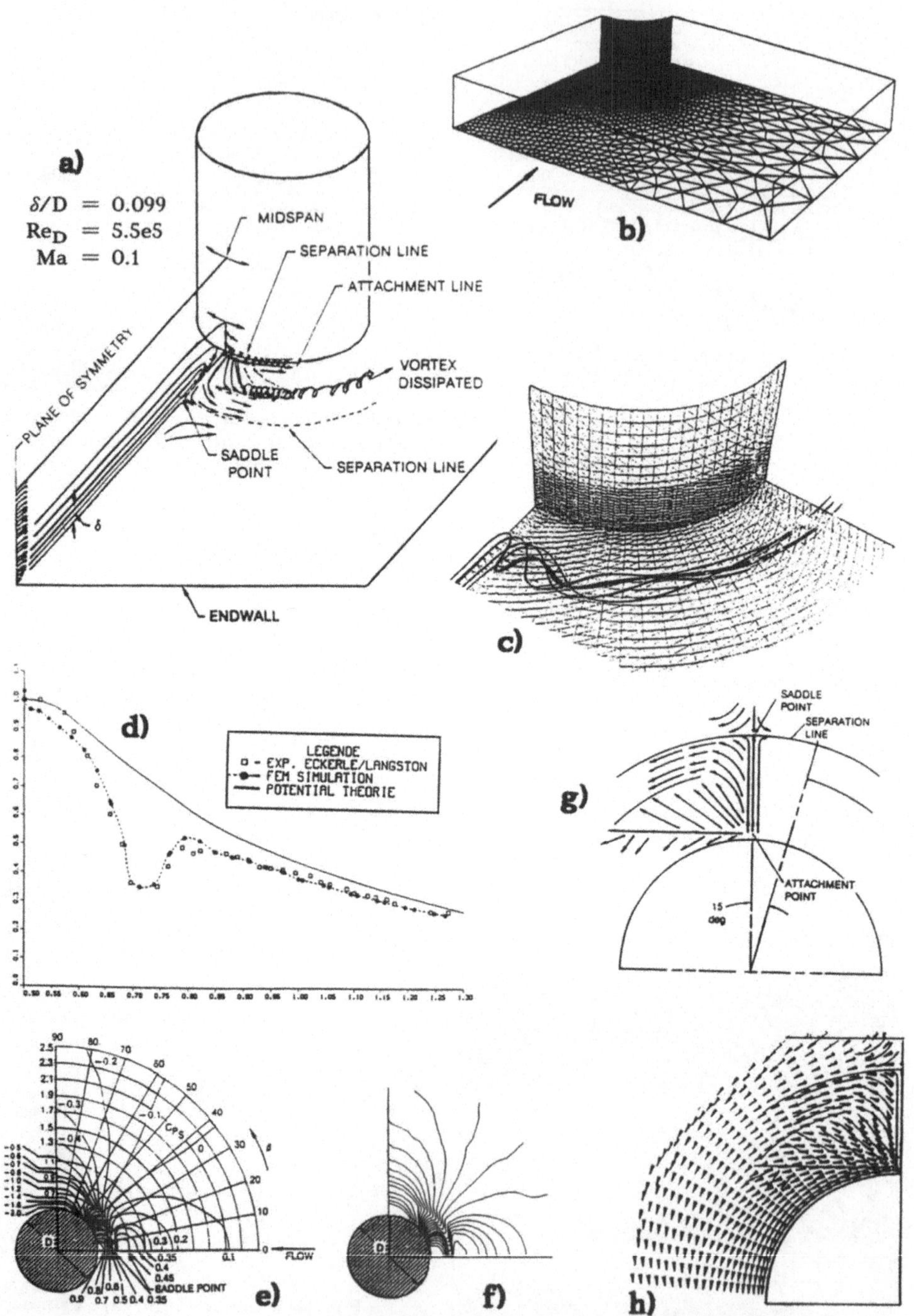


**Fig. 15:** Horseshoe-vortex computation: **a)** physical model, **b)** surface mesh, **c)** stream ribbons, **d)** static pressure distribution, **e)** measured pressure isolines, **f)** computed pressure isolines, **g)** vector plot from experiment, **h)** computed flow vectors.

16

# NEWTON-KRYLOV-SCHWARZ METHODS IN CFD


**X.-C. Cai**
Department of Computer Science
University of Colorado, Boulder, CO 80309 USA

**W. D. Gropp**
Mathematics and Computer Science Division
Argonne National Laboratory, Argonne, IL 60439 USA

**D. E. Keyes**
Institute for Computer Applications in Science and Engineering
NASA-LaRC, Hampton, VA 23681 USA,
Department of Computer Science
Old Dominion University, Norfolk, VA 23529 USA, and
Department of Mechanical Engineering
Yale University, New Haven, CT 06520 USA

**M. D. Tidriri**
Institute for Computer Applications in Science and Engineering
NASA-LaRC, Hampton, VA 23681 USA



## Summary

Newton-Krylov methods are potentially well suited for the implicit solution of non-linear problems whenever it is unreasonable to compute or store a true Jacobian. Krylov-Schwarz iterative methods are well suited for the parallel implicit solution of multidimensional systems of boundary value problems that arise in CFD. They provide good data locality so that even a high-latency workstation network can be employed as a parallel machine. We call the combination of these two methods Newton-Krylov-Schwarz and report numerical experiments on some algorithmic and implementation aspects: the use of mixed discretization schemes in the (implicitly defined) Jacobian and its preconditioner, the selection of the differencing parameter in the formation of the action of the Jacobian, the use of a coarse grid in additive Schwarz preconditioning, and workstation network implementation. Three model problems are considered: a convection-diffusion problem, the full potential equation, and the Euler equations.


## 1. Introduction

Newton-like methods, together with fully implicit linear solvers, in principle allow a more rapid asymptotic approach to steady states, $f(u) = 0$, than do time-explicit methods or semi-implicit methods based on defect correction. Strict Newton methods have the disadvantage of requiring solutions of linear systems of equations based on the Jacobian, $f_u(u)$, of the true steady nonlinear residual and are often impractical in several respects:

1. Their quadratic convergence properties are realized only asymptotically. In early

stages of the nonlinear iteration, continuation or regularization is typically required in order to prevent divergence.

2. Some popular discretizations (e.g., using limiters) of $f(u)$ are nondifferentiable, leaving the Jacobian undefined in a continuous sense.

3. Even if $f_u(u)$ exists, it is often inconvenient or expensive to form either analytically or numerically, and may be inconvenient to store.

4. Even if the true Jacobian is easily formed and stored, it may have a bad condition number.

5. The most popular family of preconditioners for large sparse Jacobians on structured or unstructured two- or three-dimensional grids, incomplete factorization, is difficult to parallelize efficiently.

In this paper we examine how points (3) through (5) may be addressed through Newton-Krylov-Schwarz methods. Our point of view with respect to (1) is that there will usually be an asymptotic regime in which the power of Newton's method is desirable if the storage overhead is not too great. To connect the opening iterations to the asymptotic regime, polyalgorithmic linear solvers for the Newton corrections were shown to be desirable in, for instance, [8]. Regarding (2), we refer to [19] for recent developments. The last three considerations are the most important with respect to parallel CFD. For a variety of reasons, industrial CFD groups are inclining towards the distributed network computing environment characterized by coarse to medium granularity, large memory per node, and very high latency. The all-to-all data dependencies between the unknown fields in a fully implicit method have led to a resurgence of interest in less rapidly convergent methods in high-latency parallel environments. Resisting, we present related investigations that lie along the route to parallel implicit CFD. Sections §2 and §3 briefly review Newton-Krylov and Krylov-Schwarz domain decomposition methods, respectively. Numerical results on three model problems, each focusing on different parts of the overall development of parallel Newton-Krylov-Schwarz methods, are then presented in §4 through §6. It is our intention to bring these developments together in a Navier-Stokes code, as described in the conclusions.

## 2. Newton-Krylov Methods

High-accuracy evaluation of the discrete residuals of $d$-dimensional flow formulations may require a large number of arithmetic operations. (For instance, a $(d+2)$-dimensional eigendecomposition may be required at each grid point in an Euler code.) Their Jacobians, though block-sparse, have dense blocks and are usually an order of magnitude even more complex to evaluate, whether by analytical or numerical means. Hence, matrix-free Newton-Krylov methods, in which the action of the Jacobian is required only on a set of given vectors, instead of all possible vectors, are natural in this context. To solve the nonlinear system $f(u) = 0$, given $u^0$, let $u^{l+1} = u^l + \lambda^l \delta u^l$, for $l = 0, 1, \ldots$, until the residual is sufficiently small, where $\delta u^l$ approximately solves the Newton correction equation $J(u^l)\delta u^l = -f(u^l)$, and parameter $\lambda^l$ is selected by some line search or trust region algorithm [6]. Krylov methods, such as the method of conjugate gradients for symmetric positive definite systems or GMRES for general nonsingular systems, find the best approximation of the solution in a relatively small-dimensional subspace that is built up from successive powers of the Jacobian on the initial residual. The Krylov solver used throughout this paper is GMRES [15], because of previous comparisons [10] with other modern Krylov solvers on the same problem class that showed CPU cost differences to be small and unsystematic when well-enough preconditioned that any of the methods were practical.

The action of Jacobian $J$ on an arbitrary Krylov vector $w$ can be approximated by

$$J(u^l)w \approx \frac{1}{\epsilon}\left[f(u^l + \epsilon w) - f(u^l)\right].$$

Finite-differencing with $\epsilon$ makes such matrix-free methods potentially much more susceptible to finite word-length effects than ordinary Krylov methods [13]. Steady aerodynamics applications require the solution of linear systems that lack strong diagonal dominance, so it is important to verify that properly-scaled matrix-free methods can be employed in this context.

An approximation to the Jacobian can be used to precondition the Krylov process. Examples are:

1. the Jacobian of a lower-order discretization,
2. the Jacobian of a related discretization that allows economical analytical evaluation of elements,
3. a finite-differenced Jacobian computed with lagged values for expensive terms, and
4. domain decomposition-parallel preconditioners composed of Jacobian blocks on subdomains of the full problem domain.

We consider case (1) in §4, case (2) in §6, and case (4) in §5 and §6. Case (4) can be combined with any of the split-discretization techniques (cases (1)–(3)), in principle.

Left preconditioning of the Jacobian with an operator $B^{-1}$ can be accommodated via

$$B^{-1}J(u^l)w \approx \frac{1}{\epsilon}\left[B^{-1}f((u^l + \epsilon w)) - \tilde{f}(u^l)\right],$$

where $\tilde{f}(u^l) = B^{-1}f(u^l)$ is stored once, and right preconditioning via

$$J(u^l)B^{-1}w \approx \frac{1}{\epsilon}\left[f((u^l + \epsilon B^{-1}w)) - f(u^l)\right].$$

Right preconditioning is preferable when the focus is on comparing different preconditioners, since the residual norm measured as a by-product in GMRES and used in the termination test is independent of any right preconditioning. On the other hand, any left preconditioning changes the by-product residual norm in GMRES. Left preconditioning may be preferable when GMRES is applied in practice as the solver for an inexact Newton method. When the preconditioning $B^{-1}$ is of high quality, the left-preconditioned residual serves as an estimate of the error in the Newton update vector. This leads to a useful termination condition when Newton step acceptance tests are based on $\|\delta u\|$.

## 3. Krylov-Schwarz Methods

A variety of parallel preconditioners, whose inverse action we denote by $B^{-1}$, can be induced by decomposing the domain of the underlying PDE, finding an approximate representation of $J$ on each subdomain, inverting locally, and combining the results. Generically, we seek to approximate the inverse of $J$ by a sum of local inverses:

$$B^{-1} = R_0^T J_{0,u^l}^{-1} R_0 + \sum_{k=1}^{K} R_k^T J_{k,u^l}^{-1} R_k \ , \quad \text{where } J_{k,u^l} = \left\{ \frac{\partial f_i(u^l)}{\partial u_j} \right\}$$

is the Jacobian of $f(u)$ for $i$ and $j$ in subdomain $k$ ($k > 0$), subscript "0" corresponds to a possible coarse grid, and where $R_k$ is a restriction operator that takes vectors spanning the entire space into the smaller dimensional subspace in which $J_k$ is defined. We use the term "Krylov-Schwarz" to distinguish these methods within the general class of domain decomposition methods. In the parallel computing literature the latter term is now used as a synonym for "data parallelism," whereas in the computational engineering literature it has come to be associated with any algorithm based on traversing a "multiblock" data structure. Meanwhile, in the applied mathematics literature, domain decomposition has become associated with the process of identifying the subdomains in which different dominant balances between terms of the governing equations hold, in the sense of asymptotic analysis.

The simplest of the Schwarz preconditioners is block Jacobi, which can be regarded as a zero-overlap form of additive Schwarz [7]. The convergence rate of block Jacobi can be improved, at higher cost per iteration, with subdomain overlap and (for many problems) by solving an additional judiciously chosen coarse grid system. It is demonstrated numerically in [5] for a variety of nonselfadjoint scalar elliptic problems that additive Schwarz with a nested coarse grid, containing one degree of freedom per subdomain, provides an "optimal" preconditioning, in the sense that the number of iterations required to attain a fixed reduction in residual is bounded by a constant as either the mesh spacing $h$ or the diameter of the subdomains $H$ is indefinitely refined. Multiplicative Schwarz methods improve on additive methods as block Gauss-Seidel improves upon block Jacobi, by roughly a factor of two, with the same serialization penalty. In a situation in which there are more subdomains than processors, hybrid multicolored multiplicative/additive Schwarz is recommended for optimal convergence at a given parallel granularity [2].

Parallelism is not the sole motivation for Schwarz methods. We remark that, given a preconditioner for the global domain, a Krylov-Schwarz method in which the same preconditioner is applied locally on each subdomain may provide a better serial algorithm than Krylov acceleration of the original global preconditioner. Given a problem of size $N$ and a preconditioner with arithmetic complexity $c \cdot N^\alpha$, partition the problem into $P$ subproblems of size $N/P$. The complexity of applying the solver independently to the set of subproblems is $P \cdot c \cdot (N/P)^\alpha$. Even in serial, $P^{\alpha-1}$ sets of subdomain iterations iterations can be afforded to coordinate the solutions of the subproblems per single global iteration, while breaking even in total complexity. If $\alpha > 1$, there is "headroom" for the domain-decomposed approach, depending upon the overall spectral properties of the global and multidomain preconditioners. There may still be parallel headroom even if $\alpha = 1$, since the global method may involve too much communication to parallelize efficiently. In addition, a hierarchical data structure is often natural for modeling or implementation reasons; and memory requirements, cache thrashing, or I/O costs on large problems may demand decomposition anyway.

## 4. A Convection-Diffusion Problem

The academic nonlinear convection-diffusion Dirichlet problem

$$\frac{\partial u}{\partial t} + u\frac{\partial u}{\partial x} + \frac{\partial u}{\partial y} - \nu\nabla^2 u = 0$$

from [17] is employed for tests of the Newton-Krylov method because, under the assumption of backward Euler time-differencing, $\frac{\partial u}{\partial t} \approx (u^{n+1} - u^n)/\Delta t$, an exact semi-discrete solution can be constructed for $u^{n+1}$. Specifically, if we set $u^n = (x^2 + y^2 + 1) + \Delta t \cdot (2x(x^2 + y^2 + 1) + 2y - 4\nu)$, then $u^{n+1} = x^2 + y^2 + 1$, and Dirichlet boundary values are set accordingly. This problem is discretized on a Courant-triangulated unit square using a hybrid finite-volume/finite-element first-order approximation [17]. It is extended to second-order upwinding for the convective terms using MUSCL-type approach [22].

The discretization is general enough to accommodate unstructured triangulated grids in two-dimensions; however, for easy visualization of the effects of inconsistent discretization of the true Jacobian and its preconditioner, four pairs of inconsistent preconditioner/Jacobian discretizations were first applied to the one-dimensional constant coefficient steady submodel,

$$a\frac{\partial u}{\partial x} - \nu\frac{\partial^2 u}{\partial x^2} = 0,$$

with $a > 0$ for a range of Peclet numbers, $\mathrm{Pe} \equiv a\Delta x/\nu$. In this model, the second-derivative term is always approximated by the standard central difference formula $-\frac{\partial^2 u}{\partial x^2}|_i \approx \frac{1}{h^2}(-u_{i-1}+2u_i-u_{i+1}) \equiv Du_i$ wherever it appears in the Jacobian, and whenever the diffusive terms are made a part of the preconditioner. The first-derivative term was variously approximated by either of the second-order formulae, $\frac{\partial u}{\partial x}|_i \approx \frac{1}{4h}(u_{i-2}-5u_{i-1}+3u_i+u_{i+1}) \equiv C_{U2}u_i$ or $\frac{\partial u}{\partial x}|_i \approx \frac{1}{2h}(u_{i-2} - 4u_{i-1} + 3u_i) \equiv C_{U2,FD}u_i$, or by the first-order formula, $\frac{\partial u}{\partial x}|_i \approx \frac{1}{h}(-u_{i-1} + u_i) \equiv C_{U1}u_i$. The last of these, the only diagonally dominant formula, and the formula with the most compact stencil, is used whenever a convective term appears in the preconditioner. Of the first two, the four-point formula, with one stencil point on the downwind side, corresponds to the second-order upwind extension of the cell-upwind scheme developed in [22] and considered further in the two-dimensional nonlinear convection-diffusion cases described below. The three-point formula, with all stencil points on the upwind side is a commonly used finite-difference form, which appears further below only for one-dimensional comparison purposes. We do not consider second-order central discretization of the convective term in the Jacobian, as is sometimes allowed in stationary defect correction methods, since, as was shown in [12], the preconditioned operator has an eigenvalue that tends to zero in the limit of large Peclet number for this choice.

Figure 1, generated with MATLAB, shows the spectra for four Jacobian/preconditioner pairs at each of three cell Peclet numbers: 0.1, 1.0, and 10.0. All Jacobian/preconditioner pairs considered have spectra that stay bounded away from the origin over the full range of Peclet number from zero to infinity. However, preconditioning with the convective

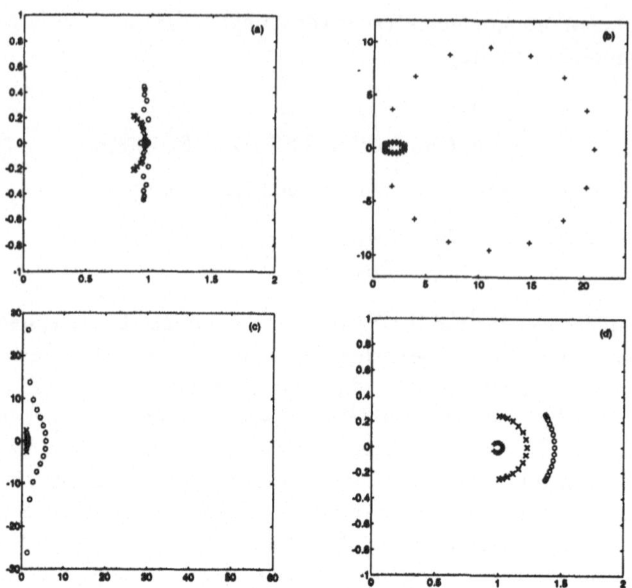

FiG. 1. *Spectra, in the right-half of the complex plane, of inconsistently preconditioned convection-diffusion operators in one-dimension, for Peclet numbers of 0.1 ('+'), 1.0 ('×'), and 10.0 ('○'). (a) $C_{U2} + D$ preconditioned by $C_{U1} + D$. (b) $C_{U2} + D$ preconditioned by $C_{U1}$. (c) $C_{U2} + D$ preconditioned by $D$. (d) $C_{U2,FD} + D$ preconditioned by $C_{U1} + D$.* Note differences in scales.

operator alone obviously fails as Pe $\rightarrow$ 0 and preconditioning with the diffusive operator alone fails as Pe $\rightarrow$ $\infty$, in the sense that their spectra contain some elements that become arbitrarily large in the respective limits while others are clustered near unity, resulting in intolerably large condition numbers. In spite of the first of these unsuitable limits, upwind Euler discretizations for the left-hand side are sometimes applied to full Navier-Stokes residuals on the right-hand side in stationary defect correction methods. Both of the inconsistent Jacobian/preconditioner pairs that contain first-order upwind convection and standard diffusion together in the preconditioner lead to spectra that stay bounded in a small region of the complex plane near unity as either Pe $\rightarrow$ 0 or Pe $\rightarrow$ $\infty$. In fact, in case (d), when the convective term in the Jacobian contains no nonzero coefficients on the downwind side, the preconditioned spectrum clusters at a single point $(\frac{3}{2}, 0)$ in the infinite Peclet limit. (Both (a) and (d) have preconditioned spectra cluster that at a single point $(1, 0)$ in the zero Peclet limit.) Therefore, it appears possible to accelerate an inconsistent discretization pair and obtain good conditioning. It should be noted that since the preconditioned operators are generally non-normal, no conclusions should be drawn about the performance of Krylov-accelerated version of these methods on the basis of the exact spectra alone. It has been shown [18] that, in the presence of finite-precision arithmetic, the spectrum itself may be misleading for non-normal operators, and the *pseudo*-spectrum is more revealing. For present purposes, we merely show the spectra, and rely on the actual iteration counts as evidence of successful application of the inconsistent discretization of Fig. 1(a) in its two-dimensional, variable-coefficient generalization.

Before presenting the results of inconsistently preconditioned convection-diffusion Jacobians, we explore another axis of discretization parameter space, namely that of matrix-free approximation of the Jacobian-vector product, as described in §2. Our aim is to validate the existence of a range of the differencing parameter $\epsilon$ in the approximation of the Jacobian-vector product in which $\epsilon$ is simultaneously small enough for the Jacobian-vector product to be accurately estimated by just the first two terms of the Taylor series

$$f(u^l + \epsilon v) \approx f(u^l) + \epsilon J(u^l)v,$$

and large enough to avoid catastrophic cancellation for moderately ill-conditioned $J$.

Two techniques for choosing the scalar $\epsilon$ were investigated:

$$\epsilon = \sqrt{\varepsilon_{mach}} \cdot (\|u^l\|_2 + 1), \text{ and}$$
$$\epsilon = \sqrt{\varepsilon_{mach}} \cdot \frac{|(u^l, v)|}{\|v\|_2^2}.$$

The first choice is simply the square root of the machine epsilon (or unit roundoff) multiplied by the norm of the current solution vector, so that the size of an assumed order-unity norm perturbation vector is not buried by a potentially large $u^l$. Conversely, by adding unity to $\|u^l\|$, the perturbation vector is kept large enough for there to be significance in the difference of the two residual vectors in the limit as $\|u^l\|$ becomes small. The second technique takes into account the magnitude of $v$, the Krylov vector, as well as that of $u^l$. The second was considered preferable to the first. We note here that GMRES may have an advantage over other Krylov methods in the matrix-free context in that the vectors $v$ that arise in GMRES have unit two-norm, but may have widely varying scale in other Krylov methods for nonsymmetric systems. Right preconditioning spoils the perfect unit two-norm, however, and the second technique retains an advantage in this context. For an extended discussion of matrix-free applications of the Jacobian in the Krylov context, see [21].

TABLE 1

*Iteration counts for unpreconditioned solution via GMRES of the nonlinear convection-diffusion problem, and the discrete two-norm of the overall error between the algebraic and exact solution for the finest grid case, without preconditioning.*

| $h^{-1}$ | $\Delta t = 10^{-2}$ | | $\Delta t = 10^{-1}$ | |
|---|---|---|---|---|
| | $A_{U2}^{ex}$ | $A_{U2}^{mf}$ | $A_{U2}^{ex}$ | $A_{U2}^{mf}$ |
| 4 | 4 | 4 | 5 | 5 |
| 8 | 4 | 4 | 8 | 11 |
| 16 | 5 | 5 | 14 | 20 |
| 32 | 7 | 7 | 23 | 36 |
| 64 | 10 | 10 | 38 | 64 |
| 128 | 14 | 14 | 68 | 113 |
| Error | 1.12(-8) | 1.88(-8) | 3.20(-5) | 3.55(-5) |

Table 1 shows the iteration count for a $10^{-5}$ reduction in residual of the unpreconditioned Newton correction equation, in both explicit (superscript $ex$) and matrix-free (superscript $mf$) implementations. Two different time-step sizes are considered, with the larger ($\Delta t = 0.1$) corresponding to a worse conditioned system. Deterioration of the convergence of the matrix-free method without preconditioning is evident on the finer grids; however, the error in the converged solution does not much suffer.

TABLE 2

Iteration counts for ILU(0)/GMRES preconditioned solution of the nonlinear convection-diffusion problem, and the discrete two-norm of the overall error between the algebraic and exact solution for the finest grid case, with inconsistent preconditioning.

| $h^{-1}$ | $\Delta t = 10^{-2}$ | | $\Delta t = 10^{-1}$ | |
| --- | --- | --- | --- | --- |
| | $(A_{U2}^{ex})(A_{U1}^{ex})^{-1}$ | $(A_{U2}^{mf})(A_{U1}^{ex})^{-1}$ | $(A_{U2}^{ex})(A_{U1}^{ex})^{-1}$ | $(A_{U2}^{mf})(A_{U1}^{ex})^{-1}$ |
| 4 | 1 | 2 | 1 | 4 |
| 8 | 1 | 2 | 1 | 4 |
| 16 | 1 | 2 | 2 | 4 |
| 32 | 2 | 2 | 3 | 5 |
| 64 | 2 | 2 | 3 | 5 |
| 128 | 2 | 2 | 3 | 5 |
| Error | 5.42(-11) | 2.03(-9) | 2.96(-5) | 3.35(-5) |

Table 2 shows the iteration count for the same cases in the presence of inconsistent preconditioning. For these tests, a global ILU(0) preconditioning was created from the convection-diffusion operator with first-order upwind convection. (The diagonal dominance of this system protects the ILU(0) factorization from breakdown.) The iteration count for the matrix-free method with preconditioning still deteriorates relative to the explicit Jacobian case. However, the explicit preconditioner is created from the explicitly available matrix, which is second-order accurate. The absolute extent of this deterioration is only a couple of extra Jacobian-vector products. In applications, in which one is faced with a choice between many extra stationary defect correction steps at limited CFL versus a few more costly accelerated Newton correction steps at high CFL, the matrix-free Newton method may prevail.

The numerical experiments in this section considered only a global preconditioner based on approximate factorization. In the next two sections, Schwarz-like domain decomposition preconditioners are considered, instead.


## 5. A Full Potential Problem

The full potential equation for the velocity potential, $\Phi$, is

$$\nabla \cdot \big(\rho(\|\nabla\Phi\|)\nabla\Phi\big) = 0,$$

where the density is given in terms of the potential by

$$\rho = \rho_\infty \left(1 + \frac{\gamma - 1}{2} M_\infty^2 (1 - \frac{q^2}{q_\infty^2})\right)^{\frac{1}{(\gamma-1)}},$$

where $q = \|\nabla\Phi\|$ and $M_\infty = q_\infty/a_\infty$. Here, $a$ is the sound speed, $q$ the flow speed, and $\infty$ refers to the freestream. When the flow is everywhere subsonic the full potential formulation fits within the monotone nonlinear elliptic framework of additive Schwarz methods [4]. For a simple non-lifting model problem of an airfoil lying along the symmetry axis $y = 0$, we choose boundary conditions as follows:

- Upstream and Freestream: $\Phi = q_\infty x$ (zero angle of attack),
- Downstream: $\Phi_{,n} = q_\infty$,
- Symmetry: $\Phi_{,n} = 0$,
- On the parameterized airfoil with shape $y = f(x)$: $\Phi_{,n} = -q_\infty f'(x)$.

The farfield boundary conditions lead to inaccuracies if applied too near the airfoil, but our interest is in algebraic convergence rates.

TABLE 3

*Average number of GMRES steps per Newton step for full potential Newton-Krylov-Schwarz solver with varying coarse grid size.*

| Coarse Grid | $0 \times 0$ | $4 \times 5$ | $8 \times 9$ | $12 \times 13$ | $16 \times 17$ | $20 \times 21$ |
|---|---|---|---|---|---|---|
| Analytical | 177 | 35 | 28 | 27 | 24 | 21 |
| Matrix-free | 183 | 41 | 28 | 27 | 25 | 23 |

A uniform fine grid of mesh size $h$ and a uniform coarse grid of mesh size $H_c$ are defined over the entire domain. The coarse grid is not necessarily nested within the fine grid, and its elements are not necessarily vertices of the subdomains, which have sides of size $H_s$, [3]. Bilinear rectilinear elements are used for both coarse and fine grids, and bilinear interpolation for intergrid transfers. An overlap $2h$ is employed on each of the subdomains, and problems posed on the subdomains as part of the Schwarz preconditioning are solved inexactly by ILU(0) with a drop tolerance 0.01. $M_\infty$ is 0.1 and the airfoil is the scaled upper surface of a NACA0012. Nonlinear convergence is declared following a $10^{-3}$ relative reduction in the steady-state residual, which requires only three Newton steps independent of inner linear method. Inner iteration convergence is a relative residual reduction of $10^{-4}$. We restart GMRES every 20 iterations. Table 3 shows convergence performance for a fixed-size problem of $128 \times 128$ uniform cells with a fixed number of subdomains in an $8 \times 8$ array as the density of the unnested uniform coarse grid varies. Key observations from this example are: (1) even a modest coarse grid makes a significant improvement in an additive Schwarz preconditioner; (2) a law of diminishing returns sets in at roughly one point per subdomain; and (3) matrix-free "matvecs" degrade convergence as much as 15-20% in the less well-conditioned cases.

## 6. An Euler Problem

Our Euler example is a two-dimensional transonic airfoil flow modeled using an EAGLE-derivative code [14] that employs a finite volume discretization over a body-fitted coordinate grid. The Euler equations for dependent variable vector $Q \equiv [\rho, \rho u, \rho v, e]^T$ are expressed in strong conservation curvilinear coordinate form as

$$(1) \qquad \tilde{Q}_\tau + (\tilde{F})_\xi + (\tilde{G})_\eta = 0,$$

where $\tilde{Q}$ and the contravariant flux vectors, $\tilde{F}$ and $\tilde{G}$, are defined in terms of the Cartesian fluxes and the Jacobian determinant of the coordinate system transformation, $J = x_\xi y_\eta -$

$y_\xi x_\eta$, through

$$\begin{aligned}
\tilde{Q} &= J^{-1}Q \\
\tilde{F} &= J^{-1}\left(\xi_t Q + \xi_x F + \xi_y G\right) \\
\tilde{G} &= J^{-1}\left(\eta_t Q + \eta_x F + \eta_y G\right).
\end{aligned}$$

C-grids of $128 \times 16$ or $128 \times 32$ cells (from [10]) around a NACA0012 airfoil at an angle of attack of $1.25°$ and an $M_\infty$ of 0.8 are considered. To obtain a representative matrix/RHS pair on which to test the behavior of Euler Jacobians under Krylov-Schwarz, we first ran a demonstration case from [14] partway to convergence and linearized about the resulting flow state.

The discrete equations take the form

$$\left[I + \Delta\tau(\delta_\xi A_\bullet^+ + \delta_\eta B_\bullet^+ + \delta_\xi A_\bullet^- + \delta_\eta B_\bullet^-)\right]\Delta Q^l = -\Delta\tau f^l,$$

where

$$A = \frac{\partial \tilde{F}^{\text{impl}}}{\partial \tilde{Q}} \quad \text{and} \quad B = \frac{\partial \tilde{G}^{\text{impl}}}{\partial \tilde{Q}},$$

the eigenvalues of $A$ and $B$, respectively, are the components of the characteristic velocities in the $\xi$ and $\eta$ directions, $\delta$ is the first-order spatial difference operator, superscripts $\pm$ denote the characteristic (upwind) direction in which the differencing occurs, and the bullets signify that each spatial differencing is carried out on the entire product to the right, for example, $\delta_\xi$ on $(A^+\Delta Q^l)$. Following the defect correction practice of [16], a flux vector split scheme of Van Leer type is employed for the implicit operators, $A$ and $B$, and the vector of steady-state residuals, $f(u)$, is discretized by a Roe-type flux difference split scheme. Characteristic variable boundary conditions are employed at farfield boundaries using an explicit, first-order accurate formulation.

TABLE 4

*Iteration counts for transonic Euler flow Jacobian with 8192 degrees of freedom discretized at local CFL numbers of 1 and 100, for various preconditioners and decomposition into 4, 16, or 64 subdomains.*

| Precond. | Block Jacobi | | Add. Schwarz | | Mult. Schwarz | |
|---|---|---|---|---|---|---|
| CFL No. | 1 | $10^2$ | 1 | $10^2$ | 1 | $10^2$ |
| $1 \times 1$ | 1 | 1 | 1 | 1 | 1 | 1 |
| $2 \times 2$ | 4 | 14 | 7 | 14 | 2 | 7 |
| $4 \times 4$ | 4 | 18 | 7 | 17 | 3 | 8 |
| $8 \times 8$ | 5 | 28 | 10 | 23 | 3 | 8 |

For a given granularity of decomposition, curvilinear "box" decompositions are generally better than curvilinear "strip" decompositions for this problem [10]. Table 4 shows that the zero-overlap results are only slightly less convergent than the corresponding $h$-overlapped additive Schwarz results at high Courant-Friedrichs-Lewy (CFL) number, and that $h$-overlapped multiplicative Schwarz is significantly better, though the latter is a less

TABLE 5

*Iteration counts and serial execution time for a transonic Euler flow Jacobian with 16384 degrees of freedom discretized at a local Courant number of 100, for Block Jacobi run in serial on a Sparc10.*

| Decomp. | Iterations | Serial Time |
|---------|------------|-------------|
| $1 \times 1$ | 1 | 38.1 |
| $2 \times 2$ | 12 | 37.9 |
| $4 \times 4$ | 14 | 21.0 |
| $8 \times 8$ | 22 | 20.3 |

parallel algorithm. Though we have not yet experimented with a coarse grid in the Euler context, [20] shows that even a piecewise constant coarse grid operator substantially improves Krylov-Schwarz convergence rates in unstructured problems.

The serial benefit of the Schwarz preconditioning is illustrated in Table 5. Here, a direct solve with a nested dissection ordering is found inferior to an additive Schwarz-GMRES iteration with nested dissection applied to successively smaller subdomains.

The Euler code has been executed on an ethernet network of workstations using a package of distributed sparse linear system routines developed at Argonne National Laboratory by Gropp and Smith [11], with p4 [9] as the data exchange layer.

Table 6 shows CPU times for each of three Schwarz preconditioners with different parallel granularity, for a fixed number of iterations on a fixed size problem. When exact solvers are used on each subdomain, speedups on a per iteration basis are seen on up to 16 processors. This advantage may be artificial in the sense that global incomplete LU is superior to a Schwarz method using exact subdomain solvers in serial, and ILU should replace nested dissection as the subdomain solver. In recent experiments [1] on Jacobians from the TRANAIR code, a threshhold drop tolerance form of ILU was employed as the subdomain solver in a Schwarz preconditioner, and the drop tolerance was varied from zero (exact nested dissection factorization) to a value that filtered out all but the largest entries in the approximate factors. The optimal drop tolerance in terms of number of operations per gridpoint varied with problem size and required converged precision, but the optimal average number of nonzeros per row hovered around 20 to 40, which is much sparser than a full nested dissection.

To test the nonlinear matrix-free approach in a situation with four differently scaled components per gridpoint, we approached the steady solution via a pseudo-transient continuation with a local adaptation of CFL number. Starting from a small initial CFL

TABLE 6

*Execution time, maximized over the processor gang, for an equal number (ten) of Block Jacobi preconditioned iterations run in parallel over a network of Sparc10's and SparcELC's, for a transonic Euler flow Jacobian with 16384 degrees of freedom discretized at a local Courant number of 100.*

| Decomp. | Parallel Time |
|---------|---------------|
| $2 \times 2$ | 20.3 |
| $4 \times 2$ | 11.5 |
| $4 \times 4$ | 8.3 |

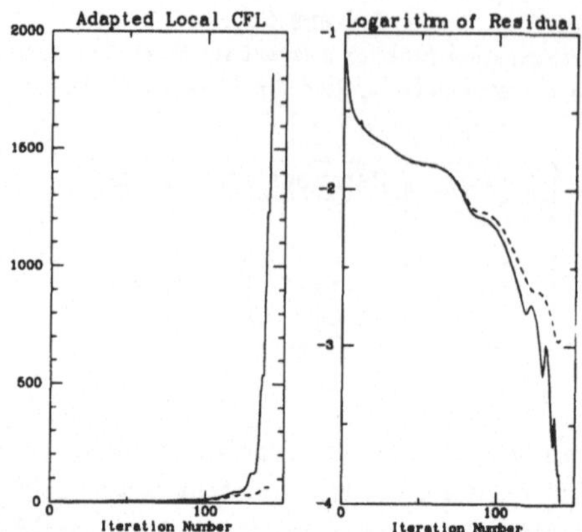


FIG. 2. *CFL and steady-state residual versus iteration count for defect correction and Newton-Krylov solvers.*

number (10), CFL may be adaptively advanced according to:

$$\mathrm{CFL}^{l+1} = \mathrm{CFL}^{l} \cdot \frac{\|f(u)\|^{l-1}}{\|f(u)\|^{l}} \ .$$

This was found preferable to an alternative local strategy:

$$\mathrm{CFL}_{ij}^{l+1} = \mathrm{CFL}_{ij}^{l} \cdot \frac{|f_{ij}(u)|^{l-1}}{|f_{ij}(u)|^{l}}$$

and also to higher powers of the ratio of successive norms.

Use of the baseline approximate factorization defect correction algorithm produces the dashed curves in Fig. 2. To obtain the solid curves, the explicitly available (Van Leer) flux vector split Jacobian ($J_{VL}$) is used to precondition the implicitly defined (Roe) flux difference split Jacobian ($J_R$) at each implicit time step. In matrix terms, the corrections $u$ are obtained as the approximate solutions of, respectively,

$$J_{VL}u = -f_R \ \ \text{and} \ \ (J_{VL})^{-1}J_R u = -(J_{VL})^{-1}f_R.$$

Unfortunately, in the retrofit of the existing code, transition to a full Newton method (CFL number approaching infinity) is precluded by explicit boundary conditions, but CFL number can be advanced, as shown in the figure, to $\mathcal{O}(10^3)$ with advantage.

The Schwarzian theory relies on the dominance of the elliptic operator. However, as in this example, multidimensional hyperbolic problems are often discretized in a defect-correction manner, with an artificially diffusive left-hand side operator. Some of the benefits of elliptic algorithms will be realized when these methods are accelerated with a Krylov method.

## 7. Conclusions

Several aspects of Newton-Krylov-Schwarz methods that are expected to be of value in a practical parallelized Navier-Stokes code are validated in model contexts. Simultaneous demands for higher-order discretizations in the Jacobian proper and diagonally dominant, easily invertible preconditioners are satisfied by mixed (inconsistent) discretizations. As in defect correction methods, the Jacobian proper need not be computed explicitly. However, by accelerating the stationary defect correction process, arbitrarily good convergence to the true Newton direction can be obtained, providing an asymptotic advantage. It is shown to be possible to find a differencing parameter for the matrix-free Newton method that is simultaneously small enough for good truncation error in approximating the action of the Jacobian and large enough to avoid the pitfalls of double precision round-off, at least for problems of $\mathcal{O}(10^2)$ mesh cells in each dimension, and in the presence of significant grid stretching. This problem is expected to grow more acute for high-aspect ratio Navier-Stokes grids. The use of a non-nested coarse grid provides a continuously adjustable "knob" with significant leverage in reducing the condition number of an elliptically dominated problem. The coarse grid entails a high communication cost in parallel implementations, but in additive algorithms it can be solved simultaneously with the subdomain fine grids operators, permitting overlap of its excess communication with useful subdomain computation. Finally, preliminary experience with treating Unix workstations spread throughout a building and connected only by a single ethernet is enouraging in that even for a modest fixed-size problem, wall clock execution time per iteration improves on up to sixteen workstations. When parallelism is exploited in its more advantageous scaling of fixed subdomain size per processor, we expect yet more encouraging results.

REFERENCES


[1] D. P. Young, C. C. Ashcraft, R. G. Melvin, M. B. Bieterman, W. P. Huffman, T. F. Johnson, C. L. Hilmes, and J. E. Bussoletti, *Ordering and Incomplete Factorization Issues for Matrices Arising from the TRANAIR CFD Code*, Boeing Computer Services TR BCSTECH-93-025, Seattle, 1993.

[2] X.-C. Cai, *An Optimal Two-level Overlapping Domain Decomposition Method for Elliptic Problems in Two and Three Dimensions*, SIAM J. Sci. Comp. (1993) 239–247.

[3] X.-C. Cai, *A Non-nested Coarse Space for Schwarz Type Domain Decomposition Methods*, Tech. Rep. CU-CS-705-94, Dept. of Comp. Sci., Univ. of Colorado at Boulder, 1994.

[4] X.-C. Cai and M. Dryja, *Domain Decomposition Methods for Monotone Nonlinear Elliptic Problems*, in D. E. Keyes and J. Xu, eds., Seventh International Conference on Domain Decomposition Methods in Scientific Computing (Penn. State Univ.), AMS, Providence, 1994, to appear.

[5] X.- C. Cai, W. D. Gropp, and D. E. Keyes, *A Comparison of Some Domain Decomposition and ILU Preconditioned Iterative Methods for Nonsymmetric Elliptic Problems*, J. Numer. Lin. Alg. Applics. (1994), to appear.

[6] J. E. Dennis, Jr. and R. B. Schnabel, *Numerical Methods for Unconstrained Optimization and Nonlinear Equations*, Prentice-Hall, Englewood Cliffs, NJ, 1983.

[7] M. Dryja and O. B. Widlund, *An Additive Variant of the Alternating Method for the Case of Many Subregions*, Courant Institute, NYU, TR 339, 1987.

[8] A. Ern, V. Giovangigli, D. E. Keyes and M. D. Smooke, *Towards Polyalgorithmic Linear System Solvers for Nonlinear Elliptic Problems*, SIAM J. Sci. Comp. 15(1994), to appear.

[9] Ralph Butler and Ewing Lusk, *Monitors, Messages, and Clusters: The p4 Parallel Programming System*, Argonne National Laboratory MCS Div. preprint P362-0493, and J. Parallel Comput., to appear.

[10] W. D. Gropp, D. E. Keyes and J. S. Mounts, *Implicit Domain Decomposition Algorithms for Steady, Compressible Aerodynamics*, in A. Quarteroni et al., eds., Sixth International Symposium on Domain Decomposition Methods for Partial Differential Equations (Como, Italy), AMS, Providence, 1994.

[11] W. D. Gropp and B. F. Smith, *Simplified Linear Equations Solvers Users Manual*, ANL-93/8, Argonne National Laboratory, 1993.

[12] D. E. Keyes and W. D. Gropp, *Domain-Decomposable Preconditioners for Second-order Upwind*

*Discretizations of Multicomponent Systems*, in R. Glowinski et al., eds., Fourth International Symposium on Domain Decomposition Methods for Partial Differential Equations (Moscow), SIAM, Philadelphia, 1991, pp. 129–139.

[13] P. R. McHugh and D. A. Knoll, *Inexact Newton's Method Solutions to the Incompressible Navier-Stokes and Energy Equations Using Standard and Matrix-Free Implementations*, AIAA Paper, 1993.

[14] J. S. Mounts, D. M. Belk and D. L. Whitfield, *Program EAGLE User's Manual, Vol. IV – Multiblock Implicit, Steady-state Euler Code*, Air Force Armament Laboratory TR-88-117, Vol. IV, September 1988.

[15] Y. Saad and M. H. Schultz, *GMRES: A Generalized Minimal Residual Algorithm for Solving Nonsymmetric Linear Systems*, SIAM J. Sci. Stat. Comp. 7(1986), 865–869.

[16] J. L. Steger and R. F. Warming, *Flux Vector Splitting of the Inviscid Gasdynamics Equations with Applications to Finite-Difference Methods*, J. Comp. Phys. 40(1981), 263–293.

[17] M. D. Tidriri, *PhD thesis*, Univ. of Paris XI, May 1992.

[18] L. N. Trefethen, *Approximation Theory and Numerical Linear Algebra*, in J. C. Mason and M. G. Cox, Algorithms for Approximation, Chapman (1990), pp. 336–360.

[19] V. Venkatakrishnan, *Convergence to Steady State Solutions of the Euler Equations on Unstructured Grids with Limiters*, J. Comp. Phys. (1994), submitted.

[20] V. Venkatakrishnan, *Parallel Implicit Unstructured Grid Euler Solvers*, AIAA Paper 94-0759, Reno, Nevada, January 1994.

[21] P. N. Brown and Y. Saad, *Hybrid Krylov Methods for Nonlinear Systems of Equations*, SIAM J. Sci. Stat. Comp. 11(1990), 450–481.

[22] B. Van Leer, *Towards the Ultimate Conservative Difference Scheme, II. Monotonicity and conservation combined in a Second Order Scheme*, J. Comp. Phys. 14(1974), 361–370.

# PASTIS-3D - A PARALLEL FINITE ELEMENT PROJECTION CODE FOR THE TIME- DEPENDENT INCOMPRESSIBLE NAVIER-STOKES EQUATIONS


Helmut Daniels and Alexander Peters
IBM Scientific Center, Vangerowstr. 18, 69115 Heidelberg, Germany


## 1. SUMMARY


PASTIS-3D (Projection Algorithm Solver for Time-dependent Incompressible flow Simulations in 3 Dimensions) is a computer program for the solution of the incompressible time-dependent Navier-Stokes equations. It uses the implicit projection-2 algorithm [9], which is 2nd order accurate [16] for the velocities. It works with a data parallel variant of the conjugate gradient method [13], [10] to run efficiently on parallel systems such as workstation clusters and the IBM Scalable POWERparallel Systems (9076 SPx). The parallel data structure is familiar from domain decomposition. But, we use a parallel implementation of a *sequential* algorithm and hence obtain the desired *sequential* results.


## 2. INTRODUCTION

More than twenty years ago Emmons [8] reviewed the possibilities for numerical modeling of fluid dynamics and concluded: 'Straightforward calculation of turbulent flows -necessarily three-dimensional and nonsteady- requires a number of numerical operations too great for the forseeable future.' With the advent of vector processors in the seventies and shared memory multi-processors in the eighties, the computer industry has provided more than four orders of magnitude improvement in computational capability ever since. In addition, an important number of new, much faster and more robust numerical and implementation techniques for computational fluid dynamics (CFD) has been developed.

Despite the progress done so far, the direct numerical simulation (DNS) of turbulence, i.e. the direct solution of the Navier-Stokes equations for large Reynolds numbers, remains beyond the reach. Recently Karniadakis and Orszag [12] noted: 'There is a broad consensus that major discoveries in key applications of turbulent flows would be within grasp if computers 1000 times faster than today's conventional supercomputers were available, assuming equal progress in algorithms and software....' There is also broad consensus that the teraflop speeds required by DNS of turbulence can be achieved only using massively parallel computer architectures.

Traditional supercomputers employed high-cost bipolar circuits and packing to satisfy the increasing demand of performance. A more recent design approach focuses on the reduction of the price to performance ratio. The effective utilization of a large number of low-cost powerful microprocessors is optimized to obtain highly competitive peak performances. Moreover, the number of processors participating in any parallel job can be scaled in accordance with the user's request.

The first practical realization of this *scalable parallel* approach was the workstations *cluster*, i.e. a collection of microcomputers interconnected by an existing network and able to work in parallel with the help of support software. Recently IBM announced the Scalable POWERparallel Systems (*9076 SP1*) -a natural, further development of this approach. Up to sixtyfour RS/6000 processors are connected via a high-speed switch. Together they are able to produce a peak performance of 8 GFlops.

The scalable parallel approach poses a series of tough questions to the designers of CFD codes. Key issues that must be addressed in achieving the effective utilization of a large number of microcomputers with distributed memory include the minimization of the inter-processor communication and the balancing of the memory.

In this paper we present the implementation of a state-of-the-art scheme for three-dimensional, time-dependent, incompressible Navier-Stokes simulations on a RS/6000 workstation cluster and the 9076 SP1. Present hardware limitations still do not allow us to perform high Reynolds number simulations without employing turbulence models. Consequently, this work should be seen as another contribution to the continuing effort of providing new and more efficent tools for large fluid dynamics computations.

The outline of this paper is as follows. Section 2 reviews the basic features of the *projection-2* method [9] which we apply to the solution of the time-dependent, incompressible Navier-Stokes equations. Section 3 presents the conjugate gradient (CG)-like data parallel model [13], [10] underlying our implementation. Section 4 illustrates benchmark results.

# 3. THE PROJECTION 2 METHOD

## 3.1. The Navier-Stokes Equations

We consider the Navier-Stokes equations in divergence notation for the primitive variables velocity ($\mathbf{u} = (u, v, w)^T$) and kinematic presure ($P = p/\rho_0$):

$$\frac{\partial \mathbf{u}}{\partial t} + \mathbf{u} \cdot \nabla \mathbf{u} = -\nabla P + \nu \nabla^2 \mathbf{u} + \mathbf{f}, \tag{3.1}$$

$$\nabla \cdot \mathbf{u} = 0. \tag{3.2}$$

The parameter $\nu$ denotes the kinematic viscosity. The vector $\mathbf{f}$ contains given body forces like buoyancy and surface forces. Typically, the system of equations (3.1) and (3.2) is defined over a domain $\Omega$ with Dirichlet (specified velocities) and Neumann (pseudo-traction) conditions on the boundary $\partial\Omega = \Gamma_1 \cap \Gamma_2$ :

$$\mathbf{u} = \mathbf{w} \qquad \text{on } \Gamma_1 \tag{3.3}$$

$$-P + \nu \frac{\partial u_\eta}{\partial \eta} = F_\eta \quad \text{and} \quad \nu \frac{\partial u_\tau}{\partial \eta} = F_\tau \qquad \text{on } \Gamma_2. \tag{3.4}$$

Here $u_\eta$ and $u_\tau$ are the projections of the velocity onto the tangential plane ($u_\tau = \mathbf{u} \cdot \tau$) and normal direction ($u_\eta = \mathbf{u} \cdot \eta$) to the boundary. Similarly, $F_\eta$ and $F_\tau$ are the normal and tangential components of specified boundary traction.

As a prerequisite for the existance of a solution the initial condition for velocities

$$\mathbf{u} = \mathbf{u}_0 \text{ in } \Omega \tag{3.5}$$

must fulfill the continuity equation (3.2) and the normal boundary condition (3.3) on $\Gamma_1$:

$$\nabla \cdot \mathbf{u}_0 \;=\; 0 \qquad\qquad \text{in } \Omega \tag{3.6}$$

$$\eta \cdot \mathbf{u}_0 \;=\; \eta \cdot \mathbf{w}(x_1, 0) \qquad \text{on } \Gamma_1. \tag{3.7}$$

Furthermore, if $\Gamma_1 = \partial\Omega$ (velocities specified on the intire boundary), the global mass conservation is another solvability constraint.


## 3.2. Basic Steps

The projection-2 method belongs to the more general class of correction methods [11], [3],[17],[18],[9],[1]). Rather than solve the Navier-Stokes equations simultaneously, these approximate methods decouple and solve the momentum equations (3.1) and the continuity equation (3.2) independently, to save memory and computer time at the expense of losses in accuracy. However, the projection-2 method does not loose much of the accuracy of a couple solution as [16] showed theoretically and we have observed numerically. Therefore, it is our method of choice for large problems. .

The most important steps of the projection-2 method are [9]:

1) Given the initial conditions $\mathbf{u}_0$ with $\nabla \cdot \mathbf{u}_0 = 0$ and $P_0$, solve the following system of equations

$$\frac{\partial\tilde{\mathbf{u}}}{\partial t} + \tilde{\mathbf{u}} \cdot \nabla\tilde{\mathbf{u}} - \nu\nabla^2\tilde{\mathbf{u}} \;=\; \mathbf{f} - \nabla P_0 \qquad \text{in } \Omega, \tag{3.8}$$

$$\tilde{\mathbf{u}} \;=\; \mathbf{w} \qquad\qquad \text{on } \Gamma_1, \tag{3.9}$$

$$\nu\frac{\partial\tilde{u}_\eta}{\partial\eta} = F_\eta(t) + P_0 \quad \text{and} \quad \nu\frac{\partial\tilde{u}_\tau}{\partial\tau} = F_\tau(t) \quad \text{on } \Gamma_2. \tag{3.10}$$

for the intermediate velocities $\tilde{\mathbf{u}}$, with $\tilde{\mathbf{u}}_0 = \mathbf{u}_0$. An essential observation is that the intermediate velocities $\tilde{\mathbf{u}}$ are not divergence free.

2) Therefore, solve the system of equations

$$(\mathbf{v} - \tilde{\mathbf{u}}) + \nabla\varphi \;=\; 0 \tag{3.11}$$

$$\nabla \cdot \mathbf{v} \;=\; 0 \qquad \text{in } \Omega. \tag{3.12}$$

for the unknowns $\mathbf{v}$ and $\varphi$, where the divergence free velocities $\mathbf{v}$ approximate the solution vector $\mathbf{u}$ at the projection time $\hat{t}$. Finally, $\varphi$ can be used to obtain a good estimate of the pressure $P(\hat{t})$ (see eq.(3.22)).

We recall that the system of equations (3.11) and (3.12) is obtained by solving the following least squares problem: Minimize with respect to $\mathbf{v}$ and $\varphi$ the integral with the boundary conditions

$$F(\mathbf{v}, \varphi) \;=\; \int_\Omega \frac{1}{2}(\mathbf{v} - \tilde{\mathbf{u}})^T(\mathbf{v} - \tilde{\mathbf{u}}) - \varphi^T\nabla \cdot \mathbf{v}, \tag{3.13}$$

$$\mathbf{n} \cdot \mathbf{v} \;=\; \mathbf{n} \cdot \tilde{\mathbf{u}} \qquad\qquad \text{on } \Gamma_1, \tag{3.14}$$

$$\varphi \;=\; -\frac{\hat{t}}{2}[F_n(\hat{t}) + P_0] \qquad\qquad \text{on } \Gamma_2. \tag{3.15}$$

## 3.3. Finite Element Formulation

Gresho [9] proposed the projecton 2 method in conjunction with finite elements. Daniels [4] implemented it on vector computers using three dimensional Q1-P0 finite elements in space and weighted finite differences in time. Q1-P0 elements employ trilinear ($C^0$-continuous) approximation functions for the velocities and elementwise constant ($C^{-1}$-discontinuous) approximation functions for the pressure. The parallel version presented in this paper is the further development of Daniels' work [4].

The finite element discretization of the Navier-Stokes equations (3.1) and (3.2) yields the semi-discrete matrix equations

$$M\dot{\mathbf{u}}_i + (D + V)\mathbf{u}_i \ = \ -C_i\mathbf{P} + \mathbf{f}_i, \tag{3.16}$$
$$C_i^T\mathbf{u}_i \ = \ \mathbf{g}, \tag{3.17}$$

where, the unknowns are the components of the velocity $\mathbf{u}_i \in R^n$ and the piece-wise constant element presures $\mathbf{P} \in R^m$. The index $i$ denotes the spacial dimension of the problem, $n$ is the number of grid nodes and $m$ the number of finite elements. The matrices $M$ (mass), $D$ (diffusion), and $V$ (convection) are $n$-by-$n$ and $C_i \in R^{m \times n}$ is the divergence matrix. The vectors $\mathbf{f}_i \in R^n$ and $\mathbf{g} \in R^m$ represent the discrete boundary traction and the divergence, respectively. More details about the assembly of the matrix equations (3.16) and (3.17) are given in [4], [5].

The basic steps of the projection-2 method (eqs.(3.8) to (3.12)) are applied to the system of equations (3.16) and (3.17). The following representation of the method (eqs.(3.18) to (3.22)) employs the Crank-Nicolson scheme for time differentiation. Other schemes can be derived easily [4]. When computing the presure in eq.(3.20), we replace the consistent mass matrix $M$ by the lumped form $L \in R^{n \times n}$, to reduce the computational complexity. Recall that matrix $L$ has diagonal structure and thus it can be inverted with ease. For the velocities, the consistent mass matrix is employed because it yields much more accurate phase speeds and its use here does not add much extra expense to the computational work.

The steps of the discrete formulation are [9]:

1) Given the nodal velocity vectors $\mathbf{u}_{0i}$ satisfying eq.(3.17) and the element presure vector $\mathbf{P}_0$ at the begining of a time step, solve the momentum equations independently for the intermediate velocity vectors $\tilde{\mathbf{u}}_i$ with $\tilde{\mathbf{u}}_{0i} = \mathbf{u}_{0i}$

$$\left[\frac{2}{\Delta t}M + D + V\right]\tilde{\mathbf{u}}_i \ = \ \left[\frac{2}{\Delta t}M - D - V\right]\tilde{\mathbf{u}}_{i0} - \tag{3.18}$$
$$ML^{-1}\left(2C_i\mathbf{P}_0 - \tilde{\mathbf{f}}_i - \mathbf{f}_{i0}\right).$$

Because $\mathbf{P} = \mathbf{P}_0$ body forces $\tilde{\mathbf{f}}_i = \mathbf{f}_{i0}$ are legitimate, too. The systems of linear equations (3.18) are solved in combination with a full set of Navier-Stokes boundary conditions (3.3) and (3.4). We refer to [4] and [5] for details on the discrete formulation of the boundary conditions.

2) Project the intermediate velocity vectors $\tilde{\mathbf{u}}_i$ onto the subspace of discretely divergence-free velocities $\mathbf{v}_i$, i.e. solve the discrete approximation of (3.11) and (3.12), i.e.

$$
\begin{bmatrix} L_1 & 0 & 0 & C_1 \\ 0 & L_2 & 0 & C_2 \\ 0 & 0 & L_3 & C_3 \\ C_1 & C_2 & C_3 & 0 \end{bmatrix} \cdot \begin{bmatrix} \mathbf{v}_1 \\ \mathbf{v}_2 \\ \mathbf{v}_3 \\ \varphi \end{bmatrix} = \begin{bmatrix} L_1\tilde{\mathbf{u}}_1 \\ L_2\tilde{\mathbf{u}}_2 \\ L_3\tilde{\mathbf{u}}_3 \\ 0 \end{bmatrix} \tag{3.19}
$$

where, $\varphi \in R^m$. The indices associated with the lumped mass matrix $L$ suggest that the Dirichlet boundary conditions (3.14) have been built into the system. The system of linear equations (3.19) is rewritten in terms of the Shur complement and solved for $\varphi$

$$
\left[ C_1^T L_1^{-1} C_1 + C_2^T L_2^{-1} C_2 + C_3^T L_3^{-1} C_3 \right] \varphi = C_1^T \tilde{\mathbf{u}}_1 + C_2^T \tilde{\mathbf{u}}_2 + C_3^T \tilde{\mathbf{u}}_3 . \tag{3.20}
$$

The approximate velocity components are obtained from the expressions

$$
\mathbf{v}_1 = \tilde{\mathbf{u}}_1 - L_1^{-1} C_1 \varphi, \ \mathbf{v}_2 = \tilde{\mathbf{u}}_2 - L_2^{-1} C_2 \varphi, \ \mathbf{v}_3 = \tilde{\mathbf{u}}_3 - L_3^{-1} C_3 \varphi . \tag{3.21}
$$

3)  Extrapolate a pressure at time $t + \Delta t$ from:

$$
\mathbf{P} = \mathbf{P}_0 + 2\varphi/\Delta t , \tag{3.22}
$$

set $\mathbf{P}_0 = \mathbf{P}, \mathbf{u}_{0t} = \mathbf{v}_t$, advance the time $t = t + \Delta t$ and go back to step 1).

## 4. THE PARALLEL DATA MODEL

We introduce a parallel data model, which allows coarse grain parallelism of the intire projection-2 process as in [4]. Preconditioned conjugate gradient (PCG)-like methods [14] are combined with domain decomposition techniques. Following [6], [13] and [10] we use a parallel implementation of the sequential PCG method. It is suitable for MIMD, distributed memory, loosely coupled systems which communicate by message passing.

First, the total mesh $\Omega$ of $m$ finite elements and $n$ nodes is partitioned into $s$ submeshes $\Omega_i$ consisting of $m_i$ elements and $n_t$ nodes. Adjacent meshes overlap at one line of nodes, the interface of two submeshes $\Omega_i \cap \Omega_j$. Elements do not overlap. We have:

$$
\Omega < \sum_{i=1}^{s} \Omega_t \qquad n < \sum_{t=1}^{s} n_i \qquad m \equiv \sum_{i=1}^{s} m_i. \tag{4.1}
$$

Then, we assemble the left hand side matrices $A_i$, size $R^{n_t} \times R^{n_t}$, and the right hand side vectors $b_i$, size $R^{n_t}$, for the local submeshes $\Omega_i, i = 1, s$ on $s$ distributed processors. The matrices $A_i$ and vectors $b_t$ are incomplete in those rows and colums, which correspond to nodes on interfaces $\Omega_i \cap \Omega_j$, because contributions to interface nodes from finite elements in the adjacent meshes are not assembled into the local matrices $A_i$ and $b_i$. However, $A_i$ and $A_j$ overlap at interface nodes and the sum $A_t \cap A_j$ is complete. It is the same with the right hand side vectors. Since $A_t$ and $A_j$ on distributed processors are additive, we can use a parallel algorithm and work on the serial problem. After initialization the PCG-like loop in a distributed host-node message passing model is:

- For $0 < k < limit$ Step 1 Until CONVERGE, Do:

$$
\tilde{q}_t^k = A_t p_i^k \tag{4.2}
$$

$$pap_i^k = \tilde{q}_i^{kT} p_i^k \tag{4.3}$$

$$pap^k = \sum_{i=1}^{s} \left[ pap_i^k \right] \tag{4.4}$$

$$\alpha^k = \frac{\rho^k}{pap^k} \tag{4.5}$$

$$x_i^{k+1} = x_i^k + \alpha^k p_i^k \tag{4.6}$$

$$\tilde{r}_i^{k+1} = \tilde{r}_i^k - \alpha^k \tilde{q}_i^k \tag{4.7}$$

$$\tilde{s}_i^{k+1} = (diagA)_i^{-1} \tilde{r}_i^{k+1} \tag{4.8}$$

$$s_i^{k+1} = SWITCH_i^{j=1,na_i}(\tilde{s}_i^{k+1,j}) \tag{4.9}$$

$$\rho_i^{k+1} = \tilde{r}_i^{k+1,T} s_i^{k+1} \tag{4.10}$$

$$\rho^{k+1} = \sum_{i=1}^{s} \left[ \rho_i^{k+1} \right] \tag{4.11}$$

$$\beta^k = \frac{\rho^{k+1}}{\rho^k} \tag{4.12}$$

$$\text{if} \quad \rho^{k+1} \leq epscg \quad \text{then} \quad \text{CONVERGED} \tag{4.13}$$

$$p_i^{k+1} = s_i^{k+1} + \beta^k p_i^k \tag{4.14}$$

- End For

In the above algorithm vectors like $\tilde{r}_i^k, \tilde{s}_i^{k,j}, \tilde{q}_i^k$ are incomplete on $\Omega_i$ at the interfaces $\Omega_i \cap \Omega_j$. These vectors are additive. Vectors like $p_i^k, s_i^k, x_i^k$ are complete on $\Omega_i$ and not additive. A $SWITCH$ adds into the vector $s_i^{k+1}$ on $\Omega_i$ all contributions from *connected* neighbors $\Omega_j$. The number of connected neighbors is $na_i$.

The computationally intensive steps (i.e. one incomplete matrix $\times$ vector product (4.2), two inner products (4.3, 4.10), three SAXPYs (4.6, 4.7, 4.14), one vector multiply (4.8) and a $SWITCH$ (4.9) are executed in parallel on $s$ distributed processors for submeshes $\Omega_i$. The global scalar operations (4.4, 4.5, 4.11 to 4.13) are completed on a HOST process. Communication between one processor (submesh $\Omega_i$) and its $1 < j < na_i$ neighbors and between $\Omega_i$ and the HOST process is of complexity:

$$2 \times \text{SEND a single scalar value to the HOST}$$
$$2 \times \text{RECEIVE a single scalar value from the HOST}$$
$$na_i \times \text{SEND an interface vector to neighbor } \Omega_j \ .$$

Obviously, the HOST-NODE communication is negligible, except for start up times for a SEND operation in the selected message passing library. Also, it scales linearly with the number of submeshes $\Omega_i$ involved in the computation. The bulk communication occurs in the $SWITCH$ (4.9). Minimal submesh connectivity and minimal size of the interfaces:

$$\sum_{i=1}^{s} [na_i] \to min \qquad \text{and} \qquad \sum_{i=1}^{s} \left[ \sum_{k=1}^{na_i} (\Omega_i \cap \Omega_k) \right] \to min \tag{4.15}$$

exploit the power of distributed systems with moderate communication bandwidth.

The data parallel PCG-like solver for element unknowns like the pressure can be derived similarly. However, there is no overlap between the matrices and vectors of the submeshes $\Omega_i$, because elements touch but do not overlap in the domain decomposed parallel data model. Hence, the matrices are not just additive, but even topologically incomplete after the serial matrix assembly process on $\Omega_i$ as shown in [4]. Missing pieces must be constructed explicitly on $\Omega_i$ in order to retrieve the sequential algorithm on distributed systems. The algorithm is given in [5] with all implementation details.

Preconditioning is a serious issue with conjugate gradient methods and we have made efforts in parallel preconditioners for the above algorithm. For more information see [15],[5].

# 5. Applications

The first example is a time dependent calculation of three-dimensional flow and transport around an object in a numerical wind tunnel. The example is small and can be simulated on a workstation in a couple of hours. The mesh consists of 28,960 3D hexahedral finite elements. The boundary conditions for this time dependent flow and heat transport problem are steady, the initial conditions (IC's) constant. They are:

|  | Flow BC's | Temperature BC's |
|---|---|---|
| Tunnel | slippery ($u_n = 0$ and $u_\tau =$ natural) | adiabatic ($\nabla T = 0 =$ natural) |
| Inflow | prescribed ($u_n = 1$ and $u_\tau = 0$) | prescribed ($T = 1$ ) |
| Outflow | open ($u, v, w =$ natural, $\nabla u_i = 0$) | open ($\nabla T = 0$) |
| Object | no slip on surface ($u, v, w = 0$) | prescribed ($T = 0$) |
| IC's: | $\mathbf{u}_0 : u = 0, v = 1, w = 0$ | $T_0 : T = 0$ |

A startup procedure [4] insures that the corrected initial conditions for the flow field fullfil a) the boundary conditions and b) the continuity equation $\nabla \mathbf{u}_0 = 0$. It corrects the ad hoc initial velocities to almost potential flow and produces the corresponding initial pressure. The model parameters, viscosity $\nu$ and thermal conductivity $\frac{\lambda}{\rho c}$, are chosen such that the Reynolds number is $Re = 5,000$ and the Peclet number is $Pe = 2,500$, based on the heights of the object. Fig. 1 shows the flow field in a horizontal and a vertical cut through the mesh at dimensionless times $t = 0.5$ and $t = 5.0$. Also the solid surface temperature of the object ($T = 0.0$) and the $T = 0.7$ dimensionless temperature isosurface (transparent) are indicated.

Runs on RS/6000 clusters and the IBM 9076 SP1 for mesh partitions of 1 up to 8 subdivisions showed reasonable speed ups with 2 and 3 subdomains when the mesh was devided into strips and the public domain PVM [7] was used with Ethernet. The algorithm did not scale further, because the ratio between (low) communication bandwith plus network latency and the (very high) compute power of the processors was not balanced for more than 3 parallel processes and this small example.

Much better scalability can be achieved for a large real world application. One of the benchmarks is an application to 3D time-dependent flow and contaminant transport in a drinking water reservoir. The project and data as well as the finite element mesh with 241,557 3D hexahedra was provided by the Institut für Wasserbau of the Aachen University of Technology [2]. The benchmark results for mesh subdivisions into 3,4,6,7 and 8 subdomains are shown in tab.1 below. Obviously we achieve high scalability for the large problem even though public domain PVM and Ethernet were used for the communication and the current implementation is not yet optimal in the pressure part. Also, the load balancing was not perfect, because the mesh is unstructured.

Fig. 1: Flow fields at times t=0.5 and t=5.0 for numerical wind tunnel

Tab.1: Benchmark for reservoir with PVM 2.4.2 and Ethernet on IBM 9076 SP1
*(8 nodes RS/6000 Model 370 with 256 MB each and 121.1 MFLOPS each for SPECfp92)*

| $N_{mesh}$ | 3 | 4 | 6 | 7 | 8 |
|---|---|---|---|---|---|
| $CPU/\Delta t[s]$ | 464-484 | 426-468 | 284-310 | 185-214 | 161-208 |
| $elapsed/\Delta t[s]$ | 505-554 | 502-525 | 345-375 | 242-263 | 238-258 |
| $efficiency$ | 98.3% | 93.3% | 88.3% | 85.6% | 87.4% |

A very important feature of the parallel data model is the scalability of the feasible problem size (mesh resolution) with the growth of the distributed memory. One workstation cannot handle the reservoir job, because the local memory (256 MB per node) is too small. Obviously, a distributed parallel environment allows bigger jobs. With regard to further development we have the potential to boost the scalability by more than one order of magnitude by a) using the IBM High Performance Switch in the SP1 with IBM-PVMe or MPI instead of Ethernet and public domain PVM b) optimizing the current implementation of the parallel PCG solver with regard to better preconditioning and communication reduction.

# References


[1] Bell, J.B. and D.L. Marcus, *A Second-Order Projection Method For Variable-Density Flows*, **UCRL-JC-104123**, LLNL, Livermore, CA, 1990.

[2] Bergen, O., *Numerische Simulation der Strömung und des Transports wasserlöslicher Stoffe in Seen und Talsperren am Beispiel des Vorbeckens der Möhnetalsperre*, Thesis, Inst. f. Wasserbau, RWTH Aachen, 1993.

[3] Chorin, A.J., *Numerical Solution of the Navier-Stokes Equations*, Math. Comp., **22**, 1968, 745-763.

[4] Daniels, H., *PASTIS-3D Finite Element Projection Algorithm Solver for Transient Incompressible Flow Simulations - Manual*, **UCRL-MA-111833**, LLNL, Livermore, CA, 1992.

[5] Daniels, H. and A. Peters, *Solving Large Incompressible Time-Dependent Flow Problems on Scalable Parallel Systems*, prep. for Int. J. Num. Meth. Fluids, IBM TR **75.94**, 1994.

[6] Dryja, M., *A finite element capacitance method for elliptic problems on regions partitioned into subdomains*, Numer. Math., **44**, 1984, 153-168.

[7] Geist, A., A. Beguelin, J. Dongarra, W. Jiang, R. Manchek and V. Sunderam, *PVM 3 User's Guide and Manual*, Report No. **ORNL/TM-12187** , Eng. Phys. and Math. Div., ORNL, Oak Ridge, TN, 1993.

[8] Emmons H.W., Annu. Rev. Fluid Mech., 2, 15, 1970.

[9] Gresho, P.M., *On the theory of semi-implicit projection methods for viscous incompressible flow and its implementation via a finite element method that also introduces a nearly-consistent mass matrix, Part 1: Theory*, Int. J. Num. Meth. Fluids, **11**, 1990, 587-620.

[10] Haase, G. and U. Langer, *Parallelisierung und Vorkonditionierung des CG-Verfahrens durch Gebietszerlegung*, Num. Algebra auf Transputersystemen, Teubner, May 1993.

[11] Harlow, F.H. and J.E. Welch, *Numerical Calculation of Time-Dependent Viscous Incompressible Flow of Fluids with Free Surface*, Physics of Fluids, **8**, No. 12, 1965, 2182-2189.

[12] Karniadakis G.E. and S.A. Orszag, *Nodes, Modes, and Flow Codes*, Physics Today, **46**, No. 3, 1993, 34-42.

[13] Keyes, D.E. and W.D. Gropp, *A comparison of domain decompositions techniques for elliptic partial differential equations and their parallel implementation*, SIAM J. SCI. STAT. COMPUT., **8**, No. 2, 1987, 166-202.

[14] Peters, A., *Non-symmetric CG-like schemes and the finite element solution of the advection-dispersion equation*, Int. J. Num. Meth. Fluids, **17**, 1993, 955-974.

[15] Schmidt, P., *Vorkonditioniertes paralleles Verfahren der konjugierten Gradienten für elliptische Differentialgleichungen* , Thesis, Prakt. Math., Karlsruhe Univ., IBM, 1993.

[16] Shin, J., *On Error Estimates of Some Higher Order Projection and Penalty-Projection Methods for Navier-Stokes Equations* , Report No. **A1190**, Dept. of Math., Penn State, subm. Num. Mathematik, 1991.

[17] Témam, R., *Sur l'approximation de la solution des équation de Navier-Stokes par la méthode des pas fractionnaires (I)*, Arch. Rat. Mech. Anal., **32**, 1969, 135-153.

[18] Van Kan, J., *A Second-Order Accurate Pressure Correction Scheme for Viscous Incompressible Flow*, SIAM J. Sci. Comp., **7**, No. 3, 1986, 870-891.

# COUPLED SOLUTION OF THE STEADY COMPRESSIBLE NAVIER-STOKES EQUATIONS AND THE K-$\mathcal{E}$ TURBULENCE EQUATIONS WITH A RELAXATION METHOD.


E. DICK and J. STEELANT

Department of mechanical and thermal engineering, Universiteit Gent
Sint-Pietersnieuwstraat 41, B-9000 Gent, Belgium


## SUMMARY


A relaxation method for the steady turbulent compressible Navier-Stokes equations is developed. The flow equations are fully coupled to the turbulence equations. The method is illustrated for three different k-$\epsilon$ models on a transitional flat plate test case. The relaxation method can be used in multigridform. A multigrid method is illustrated for one turbulence model.


## INTRODUCTION

Most numerical methods nowadays for turbulent compressible Navier-Stokes equations are based on TVD-schemes and implicit time stepping. The time step allowed by these methods often can be very large. This observation indicates that it is possible to develop methods that solve directly the steady equations without the use of time stepping. In this paper we develop such a relaxation method. To ensure solvability by a relaxation method, the system of discrete equations has to be so-called positive. To obtain positiveness, the different parts of the equations have to be treated carefully: the convective part, the diffusive part and the source part. These three parts have to be split into positive and non-positive contributions. The positive contributions form the left hand side of the system; the non-positive contributions form the right hand side. The relaxation method acts on the left hand side of the equations. The right hand side is updated in a defect-correction cycle.

## THE DISCRETIZATION

We consider the set of steady compressible Navier-Stokes equations for turbulent flow, coupled to a k-$\epsilon$ turbulence model. In two dimensions, the equations take the form:

$$\frac{\partial F}{\partial x} + \frac{\partial G}{\partial y} - \frac{\partial F_v}{\partial x} - \frac{\partial G_v}{\partial y} = S,$$

where

$$F = \begin{pmatrix} \rho u \\ \rho u u + p + \frac{2}{3}\rho k \\ \rho u v \\ \rho u E + (p + \frac{2}{3}\rho k)u \\ \rho u k \\ \rho u \epsilon \end{pmatrix} \quad , \quad G = \begin{pmatrix} \rho v \\ \rho u v \\ \rho v v + p + \frac{2}{3}\rho k \\ \rho v E + (p + \frac{2}{3}\rho k)v \\ \rho v k \\ \rho v \epsilon \end{pmatrix} \quad ,$$

$$F_v = \begin{pmatrix} 0 \\ \tau_{xx} \\ \tau_{xy} \\ u\tau_{xx} + v\tau_{xy} + q_x \\ (\mu + \mu_t/\sigma_k)k_x \\ (\mu + \mu_t/\sigma_\epsilon)\epsilon_x \end{pmatrix} \quad , \quad G_v = \begin{pmatrix} 0 \\ \tau_{xy} \\ \tau_{yy} \\ u\tau_{xy} + v\tau_{yy} + q_y \\ (\mu + \mu_t/\sigma_k)k_y \\ (\mu + \mu_t/\sigma_\epsilon)\epsilon_y \end{pmatrix} \quad , \quad S = \begin{pmatrix} 0 \\ 0 \\ 0 \\ 0 \\ S_k \\ S_\epsilon \end{pmatrix} .$$

The components of the stress tensor $\tau$ and the heat flux vector $q$ are given by

$$\tau_{xx} = (\mu + \mu_t)\left[2u_x - \frac{2}{3}(u_x + v_y)\right] \quad , \quad \tau_{yy} = (\mu + \mu_t)\left[2v_y - \frac{2}{3}(u_x + v_y)\right],$$

$$\tau_{xy} = \tau_{yx} = (\mu + \mu_t)(u_y + v_x),$$

$$q_x = \left(\frac{\mu}{Pr} + \frac{\mu_t}{Pr_t}\right)h_x \quad , \quad q_y = \left(\frac{\mu}{Pr} + \frac{\mu_t}{Pr_t}\right)h_y.$$

The molecular viscosity is denoted by $\mu$ and the turbulent viscostity by $\mu_t$ . $Pr$ is the laminar Prandtl number (taken as .71 for air) and $Pr_t$ the turbulent Prandtl number (taken as .91 for air). The static enthalpy is denoted by $h$. $E$ stands for the total internal energy per unit mass:

$$E = \frac{1}{\gamma - 1}\frac{p}{\rho} + \frac{1}{2}u^2 + \frac{1}{2}v^2 + k.$$

The source terms of the turbulence model are

$$S_k = P_k - \rho\epsilon - \mathcal{D}, \quad S_\epsilon = C_{\epsilon_1}f_1 P_k\frac{\epsilon}{k} - C_{\epsilon_2}f_2\rho\frac{\epsilon^2}{k} + \mathcal{E},$$

where $C_{\epsilon_1}$, $C_{\epsilon_2}$, $C_\mu, \sigma_k$ and $\sigma_\epsilon$ are the standard $k$-$\epsilon$ model constants; $f_1$, $f_2$ and $f_\mu$ are the so-called wall proximity damping functions and $\mathcal{D}$ and $\mathcal{E}$ are the low Reynolds number terms. In the next sections, the discretization of the different parts (convection, diffusion and source) will be discussed and their role in the relaxation method will be detailed.


## Convective terms

The convective part is treated by a flux-difference splitting method. Here, the polynomial flux-difference splitting [1] is employed which is a variant of Roe-splitting. Using the polynomial character of the convective fluxes F and G with respect to the vector of primitive variables $W^T = \{\rho, u, v, p, k, \epsilon\}$, differences of flux vectors can be expanded as:

$$\Delta F = \begin{bmatrix} \bar{u} & \bar{\rho} & 0 & 0 & 0 & 0 \\ \bar{u}^2 + \frac{2}{3}\bar{k} & \overline{\rho u} + \bar{\rho}\bar{u} & 0 & 1 & \frac{2}{3}\bar{\rho} & 0 \\ \bar{u}\bar{v} & \bar{\rho}\bar{v} & \overline{\rho u} & 0 & 0 & 0 \\ \bar{q}\bar{u} + \frac{5}{3}\bar{k}\bar{u} & a_{42} & \overline{\rho u v} & \frac{\gamma}{\gamma-1}\bar{u} & \overline{\rho u} + \frac{2}{3}\bar{\rho}\bar{u} & 0 \\ \bar{k}\bar{u} & \bar{\rho}\,\bar{k} & 0 & 0 & \overline{\rho u} & 0 \\ \bar{\epsilon}\bar{u} & \bar{\rho}\,\bar{\epsilon} & 0 & 0 & 0 & \overline{\rho u} \end{bmatrix} \Delta W = A_1'\Delta W,$$

where $\quad \bar{q} = \frac{1}{2}\overline{u^2} + \frac{1}{2}\overline{v^2}, \quad a_{42} = \bar{\rho}\,\bar{q} + \overline{\rho u}\,\bar{u} + \frac{\gamma}{\gamma-1}\bar{p} + \bar{\rho}\bar{k} + \frac{2}{3}\overline{\rho k}.$

The bar denotes the algebraic mean value. $\Delta G$ is treated similarly. Most of the terms in the above expression are self-evident. Some ambiguity arises in treating the triad terms. The following choices are made. The term $\rho uu$ in the momentum equation is seen as the product of the mass flux $\rho u$ with the velocity component $u$. This leads to $\Delta \rho uu = \overline{\rho u}\Delta u + \bar{u}\Delta \rho u = \overline{\rho u}\Delta u + \bar{u}(\bar{u}\Delta\rho + \bar{\rho}\Delta u)$. The other possible grouping into $\rho$ and $u^2$ being a density and twice a kinetic energy has no physical meaning. The energy flux is

$$\rho u E + (p + \tfrac{2}{3}\rho k)u = \tfrac{1}{\gamma-1}pu + \tfrac{1}{2}\rho u u^2 + \tfrac{1}{2}\rho u v^2 + \rho u k + (p + \tfrac{2}{3}\rho k)u.$$

The term $\frac{1}{2}\rho uu^2$ is seen as the product of a mass flux $\rho u$ with a kinetic energy $\frac{1}{2}u^2$. Similarly the term $\rho uk$ is a product of a mass flux and the turbulence kinetic energy. The term $(p + \frac{2}{3}\rho k)u$ is the product of the effective pressure (molecular plus turbulent pressure) and the velocity $u$. This means that the groupings in the term $\frac{2}{3}\rho ku$ and the convective term $\rho uk$ are different.

The above choice of groupings (i.e. following the physical meaning of the terms) leads to the simplest expressions afterwards. The basic splitting is done with respect to the primitive variables $W$. A transformation to a splitting with respect to the conservative variables $U^T = \{\rho, \rho u, \rho v, \rho E, \rho k, \rho\epsilon\}$ is obtained from

$$\Delta U = \begin{bmatrix} 1 & 0 & 0 & 0 & 0 & 0 \\ \bar{u} & \bar{\rho} & 0 & 0 & 0 & 0 \\ \bar{v} & 0 & \bar{\rho} & 0 & 0 & 0 \\ \bar{q} + \bar{k} & \bar{\rho}\bar{u} & \bar{\rho}\bar{v} & \frac{1}{\gamma-1} & \bar{\rho} & 0 \\ \bar{k} & 0 & 0 & 0 & \bar{\rho} & 0 \\ \bar{\epsilon} & 0 & 0 & 0 & 0 & \bar{\rho} \end{bmatrix} \Delta W = T\Delta W.$$

The expression $\Delta F = A_1'\Delta W$ transforms into $\Delta F = A_1'T^{-1}\Delta U = A_1\Delta U$. Similarly $\Delta G = A_2\Delta U$. The flux-differences are split into a positive and a negative part according to the sign of the eigenvalues of the matrices $A_1$ and $A_2$ : $\Delta F = A_1^+\Delta U + A_1^-\Delta U$. Similarly for $\Delta G$.

We consider a vertex-centred finite volume discretization. Using first order upwind differencing, the convective flux balance has a positive form. This means that the matrix coefficient of the central node has positive eigenvalues and that the matrix coefficients of the neighbouring nodes have non-positive eigenvalues. The resulting system can be solved by any relaxation method. The second order correction to the fluxes is constructed by the flux-extrapolation technique involving a minmod limiter. This contribution has no definite character and is therefore placed in the right hand side. Full details on the splitting and the second order correction are given in [1].

## Diffusive terms

The treatment of the diffusive terms is illustrated on the momentum-x equation. On a surface of the control volume according to figure 1, the viscous flux terms combine into

$$n_x \tau_{xx} + n_y \tau_{xy} = (\mu + \mu_t)\left[n_x(\frac{4}{3}u_x - \frac{2}{3}v_y) + n_y(u_y + v_x)\right].$$

The derivatives are expressed in the local coordinate system $(\xi, \eta)$:

$$u_x = \frac{1}{J}\left[\ \Delta_\xi u \Delta_\eta y - \Delta_\eta u \Delta_\xi y\right] = g_{11}\Delta_\xi u + g_{12}\Delta_\eta u,$$

$$u_y = \frac{1}{J}\left[-\Delta_\xi u \Delta_\eta x + \Delta_\eta u \Delta_\xi x\right] = g_{21}\Delta_\xi u + g_{22}\Delta_\eta u,$$

with $J = \Delta_\xi x \Delta_\eta y - \Delta_\eta x \Delta_\xi y$.

For the surface in fig 1:
$$\Delta_\xi u = u_d - u_c, \quad \Delta_\eta u = \tfrac{1}{4}(u_a + u_b - u_e - u_f).$$

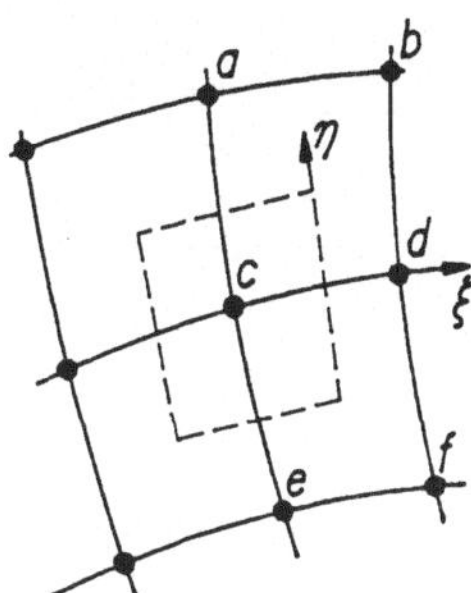


Fig 1. Control volume.

We obtain:

$$
\begin{aligned}
n_x \tau_{xx} + n_y \tau_{yy} &= (\mu + \mu_t)\left[(\frac{4}{3}n_x g_{11} + n_y g_{21})\Delta_\xi u + (-\frac{2}{3}n_x g_{21} + n_y g_{11})\Delta_\xi v\right] \\
&+ (\mu + \mu_t)\left[(\frac{4}{3}n_x g_{12} + n_y g_{22})\Delta_\eta u + (-\frac{2}{3}n_x g_{22} + n_y g_{12})\Delta_\eta v\right].
\end{aligned}
$$

For the other equations, a similar deduction can be made which results in a viscous contribution on the surface:

$$B'_\xi \Delta_\xi \tilde{W} + B'_\eta \Delta_\eta \tilde{W},$$

where $\tilde{W}^T = \{\rho, u, v, e, k, \epsilon\}$, with e the internal energy.

As the tangential derivatives $\Delta_\eta \tilde{W}$ do not contain the central node, they do not form a positive system and are placed in the right hand side. The normal contribution can be put in the left hand side since $B'_\xi$ has positive eigenvalues. To determine these eigenvalues, the matrix $B'_\xi$ is written as

$$
B'_\xi = \begin{bmatrix}
0 & 0 & 0 & 0 & 0 & 0 \\
0 & b_{22} & b_{23} & 0 & 0 & 0 \\
0 & b_{32} & b_{33} & 0 & 0 & 0 \\
0 & b_{42} & b_{43} & b_{44} & 0 & 0 \\
0 & 0 & 0 & 0 & b_{55} & 0 \\
0 & 0 & 0 & 0 & 0 & b_{66}
\end{bmatrix}.
$$

We obtain

$$
\begin{aligned}
\lambda_1 &= 0, \\
\lambda_4 &= b_{44} = (n_x g_{11} + n_y g_{21})\gamma(\mu/Pr + \mu_t/Pr_t), \\
\lambda_5 &= b_{55} = (n_x g_{11} + n_y g_{21})(\mu + \mu_t/\sigma_k), \\
\lambda_6 &= b_{66} = (n_x g_{11} + n_y g_{21})(\mu + \mu_t/\sigma_\epsilon).
\end{aligned}
$$

As $n_x = \Delta_\eta y / \Delta s = \alpha\, g_{11}$ and $n_y = -\Delta_\eta x / \Delta s = \alpha\, g_{21}$, where $\alpha > 0$ is a proportionality factor, we have

$$
\begin{aligned}
\lambda_4 &= \alpha(g_{11}^2 + g_{21}^2)\gamma(\mu/Pr + \mu_t/Pr_t) > 0, \\
\lambda_5 &= \alpha(g_{11}^2 + g_{21}^2)(\mu + \mu_t/\sigma_k) > 0, \\
\lambda_6 &= \alpha(g_{11}^2 + g_{21}^2)(\mu + \mu_t/\sigma_\epsilon) > 0.
\end{aligned}
$$

For the second and third eigenvalues, the submatrix $B_\xi^{''}$ must be considered:

$$
B_\xi^{''} = \begin{bmatrix} b_{22} & b_{23} \\ b_{32} & b_{33} \end{bmatrix} = (\mu + \mu_t) \begin{bmatrix} \frac{4}{3}n_x g_{11} + n_y g_{21} & -\frac{2}{3}n_x g_{21} + n_y g_{11} \\ n_x g_{21} - \frac{2}{3}n_y g_{11} & n_x g_{11} + \frac{4}{3}n_y g_{21} \end{bmatrix}.
$$

With the same proportionality factor the submatrix can be written as:

$$
B_\xi^{''} = \alpha\frac{(\mu + \mu_t)}{3} \begin{bmatrix} 4g_{11}^2 + 3g_{21}^2 & g_{11}g_{21} \\ g_{11}g_{21} & 3g_{11}^2 + 4g_{21}^2 \end{bmatrix}.
$$

The eigenvalues are:

$$
\begin{aligned}
\lambda_2 &= \frac{4}{3}(\mu + \mu_t)\alpha(g_{11}^2 + g_{21}^2) > 0, \\
\lambda_3 &= (\mu + \mu_t)\alpha(g_{11}^2 + g_{21}^2) > 0.
\end{aligned}
$$

The contribution of the differences in the $\xi$ direction in the diffusive flux balance forms a positive system. The resulting positive system in the left hand side is expressed in the variables $\{\rho, u, v, e, k, \epsilon\}$. The transformation to the conservative variables is done by

$$
\Delta U = \begin{bmatrix}
1 & 0 & 0 & 0 & 0 & 0 \\
\bar{u} & \bar{\rho} & 0 & 0 & 0 & 0 \\
\bar{v} & 0 & \bar{\rho} & 0 & 0 & 0 \\
\bar{q} + \bar{k} + \bar{e} & \bar{\rho}\bar{u} & \bar{\rho}\bar{v} & \bar{\rho} & \bar{\rho} & 0 \\
\bar{k} & 0 & 0 & 0 & \bar{\rho} & 0 \\
\bar{e} & 0 & 0 & 0 & 0 & \bar{\rho}
\end{bmatrix} \Delta \tilde{W} = \tilde{T}\Delta\tilde{W}.
$$

To transform we use $\Delta\tilde{W} = \tilde{T}^{-1}\Delta U$. The transformation does not change the positive character of the left hand side.


Source terms.


Whereas the construction of the convective and diffusive Jacobians is rather straightforward, this is not the case for the source terms. For the source terms, a proper linearization must be chosen. The Jacobian of the negative source term is then to be brought into the left-hand side to increase the diagonal dominance of the system of equations [2]. Dependent on the $k$-$\epsilon$-model, a different linearization is necessary. Until now we have

$$
\sum_k (-A_k^- + B_{\xi,k})\Delta s_k\, U_{ij} - \sum_k (-A_k^- + B_{\xi,k})\Delta s_k\, U_k = RHS + S\, Vol,
$$

where the sum extends over the surfaces of the control volume, $\Delta s_k$ denotes the length of the surface and $k$ refers to a surrounding node. The term RHS collects the second order contributions from the inviscid terms and the tangential contributions from the viscous terms. $Vol$ denotes the volume of the control volume. The source term still has to be treated.

A typical relaxation scheme used in the sequel is three Gauss-Seidel steps on the left hand side (inner iteration) between updates of the right-hand side (outer or defect-correction iteration).

Put into $\delta$-formulation, we obtain

$$\sum_k (-A_k^- + B_{\xi,k})\Delta s_k \; \delta U_{ij} + FLX = RHS + S \, Vol,$$

where $FLX$ is the flux balance based on first order inviscid fluxes and the normal part of the viscous fluxes. This term is partly on an old iteration level and partly on a new iteration level. The coefficients $A_k^-$ and $B_{\xi,k}$ are also partly on new and old iteration levels as these are updated during the inner iterations.

The source term $S$ is split into positive and negative terms. The negative term is put into the left hand side and takes part in the inner iteration:

$$S \;=\; S^+ + S^- + \frac{\partial S^-}{\partial U}\delta U_{ij}.$$

So we obtain:

$$\left\{\sum_k (-A_k^- + B_{\xi,k})\Delta s_k - \frac{\partial S^-}{\partial U}Vol\right\} \delta U_{ij} + FLX - S^- Vol \;=\; RHS + S^+ \, Vol.$$

Source terms of a k-$\epsilon$ model can be written as

$$S_k \;=\; P_k - \rho\epsilon - \mathcal{D}, \qquad S_\epsilon = [C_{\epsilon_1} f_1 P_k - C_{\epsilon_2} f_2 \rho\epsilon]\frac{1}{T} + \mathcal{E},$$

with

$$P_k \;=\; \left\{\mu_t\left[\frac{\partial u_i}{\partial x_j} + \frac{\partial u_j}{\partial x_i} - \frac{2}{3}\delta_{ij}\frac{\partial u_k}{\partial x_k}\right] - \frac{2}{3}\delta_{ij}\rho k\right\}\frac{\partial u_i}{\partial x_j}, \quad \mu_t \;=\; C_\mu f_\mu \rho k T.$$

In our study, 3 models are used: Launder-Sharma (LS) [3], Lam-Bremhorst (LB) with the modification of the $f_2$-function [4] and Yang-Shih (YS) [5]. The five basic constants are : $C_{\epsilon_1} = 1.44$, $C_{\epsilon_2} = 1.92$, $\sigma_k = 1$, $\sigma_\epsilon = 1.3$, $C_\mu = .09$. Table 1 gives the low Reynolds terms together with the boundary conditions for these models and table 2 gives the damping functions. Upstream of the leading edge, the wall distance coming in the definition of $R_y$ is taken as the distance from the upstream node to the leading edge. The negative source terms taken into consideration for linearization are the same for all models:

$$S_k^- \;=\; -[\rho\epsilon + \mathcal{D}], \qquad S_\epsilon^- \;=\; -C_{\epsilon_2} f_2 \rho\epsilon\frac{1}{T}.$$

We consider first the LS & LB model. Following Vandromme [2], we write these terms as:

$$S_k^- \;=\; -\left[\frac{C_\mu f_\mu}{\mu_t}\right](\rho k)^2 - \left[\frac{\mathcal{D}}{k}\right]k, \qquad S_\epsilon^- \;=\; -[C_{\epsilon_2} f_2]\frac{(\rho\epsilon)^2}{\rho k}.$$

Table 1: Low Reynolds terms

| k-$\epsilon$ | $\mathcal{T}$ | $\epsilon_w - B.C.$ | $\mathcal{D}$ | $\mathcal{E}$ |
|---|---|---|---|---|
| LS | $\frac{k}{\epsilon}$ | 0 | $2\mu\left(\frac{\partial\sqrt{k}}{\partial y}\right)^2$ | $2\mu\nu_t\left(\frac{\partial^2 u}{\partial y^2}\right)^2$ |
| LB | $\frac{k}{\epsilon}$ | $\frac{\partial\epsilon}{\partial y}=0$ | 0 | 0 |
| YS | $\frac{k}{\epsilon}+\sqrt{\frac{\nu}{\epsilon}}$ | $2\mu\left(\frac{\partial\sqrt{k}}{\partial y}\right)^2$ | 0 | $\mu\nu_t\left(\frac{\partial^2 u}{\partial y^2}\right)^2$ |

Table 2: Damping functions

| k-$\epsilon$ | $f_\mu$ | $f_1$ | $f_2$ |
|---|---|---|---|
| LS | $\exp\left[\frac{-3.4}{(1+R_T/50)^2}\right]$ | 1 | $1-.3\exp(-R_T^2)$ |
| LB | $[1-\exp(-0.0165 R_y)]^2\left(1+\frac{20.5}{R_T}\right)$ | $1+\left(\frac{.05}{f_\mu}\right)^3$ | $1-\exp\left(-R_T^2-10^{-10}\right)$ |
| YS | $\left[1-\exp\left(-a_1 R_y - a_3 R_y^3 - a_5 R_y^5\right)\right]^{\frac{1}{2}}$ | 1 | $1-.22\exp\left(-\frac{R_T^2}{36}\right)$ |

$$R_T = \frac{kT}{\nu} \qquad R_y = \frac{\sqrt{k}y}{\nu} \qquad a_1 = 1.5\text{x}10^{-4},\, a_3 = 5.0\text{x}10^{-7},\, a_5 = 1.0\text{x}10^{-10}$$

Considering the quantities in square brackets to be constant, a linearization which guarantees positiveness, is:

$$\frac{\partial(S_k^-;S_\epsilon^-)}{\partial(\rho k;\rho\epsilon)} = \begin{bmatrix} \frac{\partial S_k^-}{\partial\rho k} & \frac{\partial S_k^-}{\partial\rho\epsilon} \\ \frac{\partial S_\epsilon^-}{\partial\rho k} & \frac{\partial S_\epsilon^-}{\partial\rho\epsilon} \end{bmatrix} = \begin{bmatrix} -\frac{2C_\mu f_\mu}{\mu_t}(\rho k) - \frac{D}{k} & 0 \\ C_{\epsilon_2} f_2\left(\frac{\rho\epsilon}{\rho k}\right)^2 & -2C_{\epsilon_2} f_2\frac{\rho\epsilon}{\rho k} \end{bmatrix}.$$

For the YS model, the same combinations are kept constant, but due to the different expression of $\mathcal{T}$, this results in a more complex Jacobian:

$$\frac{\partial(S_k^-;S_\epsilon^-)}{\partial(\rho k;\rho\epsilon)} = \begin{bmatrix} -\frac{(\sqrt{\nu}+\sqrt{\nu+4kT})}{2kT} & 0 \\ C_{\epsilon_2} f_2\frac{1}{T^2} & -C_{\epsilon_2} f_2\left(2T-\frac{1}{2}\sqrt{\frac{\nu}{\epsilon}}\right)\frac{1}{T^2} \end{bmatrix}.$$

## TEST CASE

Transitional flow with a zero-pressure gradient was calculated over a flat plate with a freestream turbulence level of 3% (T3A test case from Savill [6]). A stretched grid of 385 x 97 points was used. The grid extends upstream of the plate, with the sharp leading edge at station 97. The first grid point in the direction normal to the plate lies at about $y^+ = yu_\tau/\nu = 1$, where $u_\tau$ is the friction velocity. Stretching was applied normal to the plate and in the flow direction near the leading edge. Uniform inlet profiles for total temperature, total pressure, $k$ and $\epsilon$ were specified. At inlet, Mach number was extrapolated from the flow field. The values of $k$ and $\epsilon$ at the inlet were calculated with the equations for $k$ and $\epsilon$ for uniform flow with velocity $U$:

$$U\frac{\partial k}{\partial x} = -\epsilon, \quad U\frac{\partial\epsilon}{\partial x} = -C_{\epsilon_2}\frac{\epsilon^2}{k},$$

where at the leading edge the following values were matched to be in accordance with the
experiments (for L = 1m) :

$$k_e = .03(\frac{3}{2}U_e^2), \quad \epsilon_e = .378m^2/s^3, \quad U_e = 5.4m/s.$$

The upper and right boundaries are outlet boundaries. There, pressure was imposed.
Velocity components, temperature and turbulent quantities were extrapolated. The part
of the lower boundary upstream of the leading edge was treated as a symmetry line. At
the plate, no-slip and adiabatic boundary conditions were imposed. Density and pressure
were obtained by characteristic combinations of the equations [1].

Three lexicographic Gauss-Seidel relaxations with underrelaxation factor .9 were used
as inner iteration. The first relaxation starts from the left bottom point and ends at
right upper point. The second relaxation has the reversed ordering. The third relaxation
has the same ordering as the first one. Figure 2 shows the obtained distribution of the
skin friction coefficient for the three models. The upper and lower lines correspond with
fully laminar and fully turbulent flow fields. Figure 2 also shows the distribution of the
turbulence kinetic energy during the transition. The profiles are at the position $Re_x =$
3850, 76000, 170700, 271250, 375650, 483600. Transition point and transition length are
not well predicted by all models, when compared to experimental results [6]. This shows
that the models still have to be much improved. The convergence results are similar for
all models and are discussed in the next section.

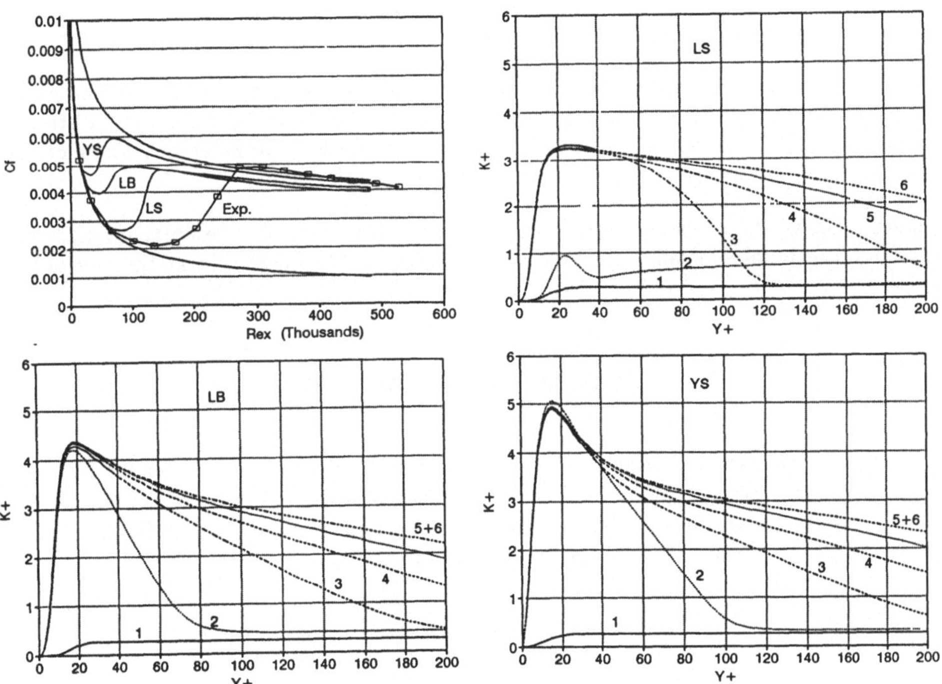

Fig. 2 $C_f$ and $k$ predicted by the models.

# MULTIGRID FORMULATION

A standard multigrid method using four grids (385 x 97; 193 x 49; 97 x 25; 49 x 13), W-cycle, full weighting as restriction for residuals and bilinear interpolation as prolongation, was employed. The multigrid acts on the left hand side of the set of equations. The right hand side is updated in a defect correction cycle. The procedure is the same as the one used for the Euler equations in [1], except for the ordering of the relaxation. Three Gauss-Seidel relaxations are used as prerelaxation and postrelaxation.

A relaxation on the current grid is taken as one local work unit. A residual evaluation plus the associated grid transfer is also taken as one local work unit. Local work units are counted as fractions of work units on the level of the finest grid proportional to the number of cells. The update of the right hand side of the system of equations in the defect correction is also taken as one work unit. With these rules, the cost of the cycle is found to be 14.0625 work units. In a single grid calculation, the defect correction cycle is counted as 4 work units.

The calculation starts from uniform flow. First a laminar solution is calculated up to a sufficient level of convergence. With this solution, initial values of $k$ and $\epsilon$ are calculated according to the boundary layer laws [6]:

$$k \; = \; k_i \left( \frac{u}{u_i} \right)^2 , \quad \epsilon \; = \; .3k\frac{\partial u}{\partial y} \quad with \quad \epsilon \geq \epsilon_i ,$$

where the subscript i refers to inlet conditions.

The turbulence equations are not solved close to solid boundary on the coarser grids. Nodes in the region $R_y < 60$ are excluded. This was found to be necessary to prevent divergence of the multigrid method. The reason is the very singular behaviour of the low-Reynolds number equations close to a solid boundary.

Figure 3 shows the convergence behaviour for the YS-model. The residual shown is the maximum residual over all equations and all nodes on the finest grid at the end of the cycle.

# ACKNOWLEDGEMENT

The research reported here was granted under contract 9.0001.91 by the Belgian National Science Foundation (N.F.W.O.) and under contract IUAP/17 as part of the Belgian National Programme on Interuniversity Poles of Attraction, initiated by the Belgian State, Prime Minister's Office, Science Policy Programming.

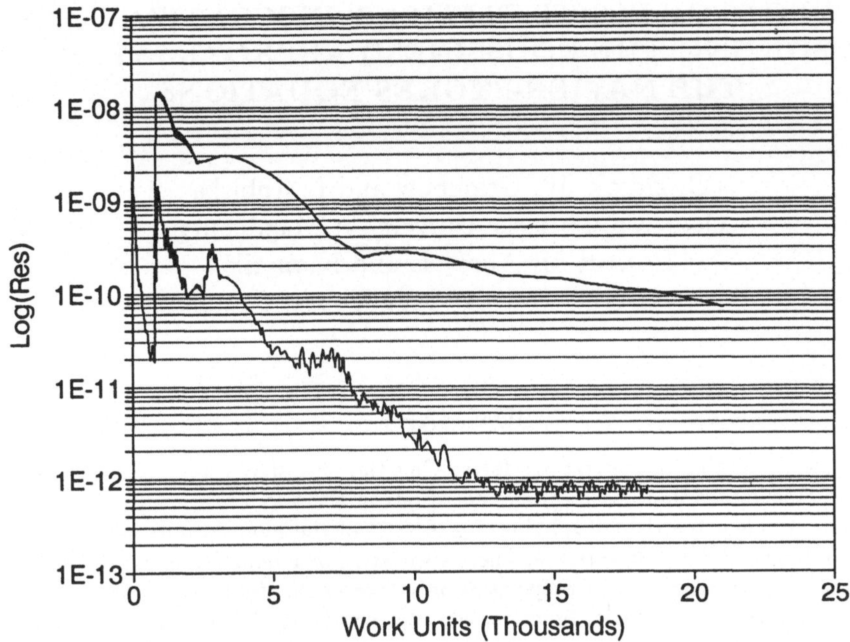


Fig. 3. Convergence history for single and multigrid calculations.

## REFERENCES


[1] Dick E., Multigrid solution of steady Euler equations based on polynomial flux-difference splitting, *Int. J. Num. Methods Heat Fluid Flow 1* (1991), 51-62.

[2] Vandromme D., Turbulence modeling for compressible flows and implementation in Navier-Stokes solvers, *VKI-LS 1991-02*.

[3] Patel V.C., Rodi W. and Scheuerer G., Turbulence Models for near-wall and low Reynolds number flows: a review, *AIAA Journal 23* (1984), 1308-1319.

[4] Sieger K., Schulz A., Crawford M.E. and Wittig S., Comparitive study of low-Reynolds number k-$\epsilon$ turbulence models for predicting heat transfer along turbine blades with transition, *Proceedings Int. Symp. Heat transfer in turbomachinery* (Athens, Aug. 1992).

[5] Yang Z. and Shih T.H., A k-$\epsilon$ calculation of transitional boundary layers, *ICOMP-92-08*.

[6] Savill A.M., A synthesis of T3 Test Case Predictions, in: Pironneau O. *et al. (eds.)*, *Numerical simulation of Unsteady flows and transition to turbulence* (Cambridge University Press, 1992), 404-442.

# ASPECTS OF FINITE ELEMENT DISCRETIZATIONS FOR SOLVING THE BOUSSINESQ APPROXIMATION OF THE NAVIER-STOKES EQUATIONS


O. Dorok, W. Grambow and L. Tobiska

Otto von Guericke Universität Magdeburg

Institut für Analysis und Numerik

Postfach 4120, D-39016 Magdeburg, Germany


## SUMMARY


We consider stable discretizations for solving the Boussinesq approximation of the stationary, incompressible Navier-Stokes equations in the twodimensional case. For the continuous problem the right hand side $f \in L^2(\Omega)^2$ of the momentum equation can be splitted in the form $f = \nabla \Phi + \mathbf{curl}\ \Psi$, where a variation of $\Phi$ does not change the velocity $u$. For the discrete problem this property is only true in the limit case, $h$ tends to zero, unless exact divergencefree trial functions for approximating the velocity field are used. The main objective of the paper is to discuss the influence of this phenomenon on the accuracy of the approximated velocity field $u_h$ when using only discrete divergencefree trial functions. For some benchmark problems the results of numerical calculations are also presented.


## INTRODUCTION

We consider stable numerical methods for solving the Boussinesq approximation of the stationary incompressible Navier-Stokes equations

$$
\begin{aligned}
-\nu \Delta u + u \cdot \nabla u + \nabla p &= \alpha f(T) \quad \text{in } \Omega, \\
\nabla \cdot u &= 0 \quad \text{in } \Omega, \\
-\lambda \Delta T + u \cdot \nabla T &= 0 \quad \text{in } \Omega,
\end{aligned}
\tag{1}
$$

$$
u_{|\Gamma} = 0, \qquad T_{|\Gamma_D} = T_D, \qquad \frac{\partial T}{\partial n}_{|\Gamma_N} = 0,
\tag{2}
$$

where $\Omega$ is a bounded twodimensional domain with Lipschitz continuous boundary $\Gamma = \Gamma_D \cup \Gamma_N$, $\Gamma_D \cap \Gamma_N = \emptyset$.

Starting with stable combinations for approximating the velocity and pressure field, respectively, we use additionally an upwind technique in order to handle the dominance of the convective terms in the case of higher Reynolds/Rayleigh numbers. Another approach for getting a stable approximation consists of adding appropriate terms of Galerkin-least-squares type to the standard Galerkin discretization and thus allowing the use of arbitrary pairs of finite element spaces for the discrete velocities and pressures, respectively. In any case the right hand side $f \in L^2(\Omega)^2$ of the continuous problem can be splitted into

$$
f = \nabla \Phi + \mathbf{curl}\ \Psi,
$$

where a variation of $\Phi$ does not change the solution $u$. In the discrete problem this property only holds for $h \to 0$ unless exact divergencefree trial functions for approximating the velocity field are used. The main objective of the paper is to study the consequences of this phenomena on the accuracy of the calculated velocity field both theoretically and numerically.

The plan of the paper is the following. First we formulate stable finite element methods for solving problems of the form (1), (2). Then, we derive error bounds for the velocity field in the special case of a no flow problem. The general analysis covers nonconforming and conforming finite element methods without and with stabilization techniques. In particular the dependency of the error constants on the Reynolds/Rayleigh number is explicitly given. In the following section special properties of the nonconforming P1-P0 discretization for solving the linear Stokes problem are discussed. Finally, some numerical experiments show that for higher Reynolds/Rayleigh numbers the influence of the discretization error becomes important.

## FINITE ELEMENT METHODS

For simplicity let us consider discretization methods for the Navier-Stokes equation

$$
\begin{aligned}
-\nu\Delta u + u \cdot \nabla u + \nabla p &= f & &\text{in } \Omega, \\
\nabla \cdot u &= 0 & &\text{in } \Omega, \\
u &= 0 & &\text{on } \Gamma,
\end{aligned}
\tag{3}
$$

where $\nu$ is the inverse of the Reynolds number and $f$ is a given body force. The generalization to problems of the form (1), (2) is straightforward. We are looking for solutions $u, p$ in the spaces

$$
V := H_0^1(\Omega)^2, \quad Q := L_0^2(\Omega) = \{ q \in L^2(\Omega) : \int_\Omega q \, dx = 0 \}
$$

and denote the space of divergencefree functions by

$$
W := \{ v \in V : \forall q \in Q \quad (q, \operatorname{div} v) = 0 \}
$$

with $(.,.)$ the inner product in $L^2(\Omega)$ and $L^2(\Omega)^2$, respectively. Then, the weak formulation of (3) reads

Find $[u, p] \in V \times Q$ such that for all $[v, q] \in V \times Q$

$$
\nu(\nabla u, \nabla v) + n(u, u, v) - (p, \nabla \cdot v) + (q, \nabla \cdot u) = < f, v >
\tag{4}
$$

where the nonlinear term is given by

$$
n(w, u, v) := \frac{1}{2}[(w \cdot \nabla u, v) - (w \cdot \nabla v, u)].
$$

Let us now consider finite element spaces $V_h \approx V$ and $Q_h \approx Q$ satisfying the discrete LBB-condition

$$
\exists \beta > 0 \quad \forall h, \forall q_h \in Q_h : \quad \sup_{v_h \in V_h} \frac{(q_h, \nabla \cdot v_h)_h}{|v_h|_{1,h}} \geq \beta \, \|q_h\|_0
\tag{5}
$$

and let

$$W_h := \{\, v_h \in V_h \; : \; \forall q_h \in Q_h \quad (q_h, \nabla \cdot v_h)_h = 0 \,\}$$

denote the space of discrete-divergencefree functions. Note that we consider here both conforming and nonconforming finite element approximations. The index $h$, for example in $(.,.)_h$ and $|.|_{1,h}$, will be used to indicate that the corresponding inner products, seminorms and norms are calculated as the sum of integrals over all elements. In particular in case of a nonconforming finite element method the compatibility condition

- $< q, [v_h]_E >_E := \int_E q\,[v_h]_E\,ds = 0 \quad \forall v_h \in V_h, \; q \in P_{k-1}, \quad k \geq 1$

is supposed to be satisfied, which guaranties that $|.|_{1,h}$ is a norm on $V_h$. Here $P_{k-1}$ denotes the set of polynomials of degree smaller or equal to $k-1$ and $[v_h]_E$ the jump of $v_h$ crossing an edge $E$ of the triangulation. Now the discrete problem is given by

Find $[u_h, p_h] \in V_h \times Q_h$ such that for all $[v_h, q_h] \in V_h \times Q_h$

$$\nu(\nabla u_h, \nabla v_h)_h + n_h(u_h, u_h, v_h) - (p_h, \nabla \cdot v_h)_h + (q_h, \nabla \cdot u_h)_h =< f, v_h >, \qquad (6)$$

where $n_h$ is a discretization of the convective part $n$. In general $n_h : V_h^3 \to \mathbb{R}$ is not supposed to be a trilinear form. Thus, it is possible to include special upwind techniques by our investigations. Concerning $n_h$ we will assume that

- $\forall u_h \in V_h \; : \; (v_h, w_h) \to n_h(u_h, v_h, w_h)$ is a continuous bilinear form on $V_h^2$,

- $\exists L > 0 \quad \forall u_{1h}, u_{2h}, v_h, w_h \in V_h$

$$|n_h(u_{1h}, v_h, w_h) - n_h(u_{2h}, v_h, w_h)| \leq L|u_{1h} - u_{2h}|_{1,h}\,|v_h|_{1,h}\,|w_h|_{1,h},$$

- $\forall u_h \in W_h, \, v_h \in V_h \qquad n_h(u_h, v_h, v_h) \geq 0.$

Note that the finite element method developed in [2] satisfy these assumptions.

Finally, we consider the Galerkin-least-square stabilization of the Navier-Stokes equation for a conforming finite element discretization $V_h \subset V$, $Q_h \subset Q$ which need not satisfy the discrete LBB-condition in general.

Find $[u_h, p_h] \in V_h \times Q_h$ such that for all $[v_h, q_h] \in V_h \times Q_h$

$$\left.\begin{aligned}
\nu(\nabla u_h, \nabla v_h) + n(u_h, u_h, v_h) - (p_h, \nabla \cdot v_h) \\
+ (q_h, \nabla \cdot u_h) + \tau(\nabla \cdot u_h, \nabla \cdot v_h) \\
+ \delta \sum_T h_T^2 (-f - \nu\Delta u_h + u_h \cdot \nabla u_h + \nabla p_h, u_h \cdot \nabla v_h + \nabla q_h)_T
\end{aligned}\right\} =< f, v_h > \qquad (7)$$

For choosing the design parameter $\tau$ and $\delta$ we refer to [8], [3].

## ANALYSIS OF THE NO FLOW PROBLEM

Unfortunately, the uniqueness of the solution of the problems (4), (6) and (7), respectively, can only be guaranted for small data which implies $\nu > \nu_0 > 0$. On the other hand the standard local error analysis requires a detail knowledge on the corresponding linearized operators, in particular on the behaviour of the norm of the inverse operators with respect to $\nu$, which is not yet available [5]. Therefore, we focus here on the special case of a no flow problem, i.e. the velocity field of the solution of the continuous problem (4) is identically zero. This situation happens if the following necessary and sufficient condition on the right hand side $f$ is fulfilled.

**LEMMA 1** *The continuous problem (4) admits a solution with vanishing velocity field if and only if the right hand side of $f$ belongs to the set*

$$W^0 = \{\, g \in V^* \ : \ \forall v \in W \quad < g, v >= 0 \,\} \,. \tag{8}$$

*Moreover, if $f \in W^0$ then there is a function $\Phi \in L^2(\Omega)$ such that*

$$\forall v \in V : \quad < f, v >= -(\Phi, \nabla \cdot v) \tag{9}$$

*and for each $\nu > 0$ the unique solution of (4) is given by $[u, p] = [0, \Phi_0]$, where*

$$\Phi_0 = \Phi - \frac{1}{|\Omega|} \int_\Omega \Phi(x)\, dx \,. \tag{10}$$

**Proof:** Setting in (4) $u$ and $q$ equal to zero, we get

$$\forall v \in V \quad < f, v >= -(p, \nabla \cdot v)$$

implying that $f \in W^0$. Now let $f$ belong to $W^0$. Then, setting $q = 0$ in (4) we obtain for each $v \in W$

$$\nu(\nabla u, \nabla v) + n(u, u, v) = 0, \tag{11}$$

such that $\nu|u|_1^2 = 0$ for each possible solution $u \in W$. From [4] (Chapter I, Lemma 2.1) we know that for each $f \in W^0$ there is a $\Phi \in L^2(\Omega)$ such that (9) holds. The pressure field $p \in Q$ satisfies

$$-(p, \nabla \cdot v) = -(\Phi, \nabla \cdot v) \quad \forall v \in V,$$

thus, by means of the continuous version of the Babuška-Brezzi condition $p = \Phi_0$. $\qquad \square$

In the following let $C$ denote a constant which is independent on the mesh size $h$ and the Reynolds number $\mathrm{Re} = 1/\nu$.

**THEOREM 1** *Let the assumptions given above be fulfilled, let $Q_h$ consists of piecewise polynomials of degree $k - 1$ and let the right hand side of (4) be $f = \nabla\Phi$ with sufficiently regular $\Phi$. Then, all solutions of the discrete (stabilized) problem (6) satisfies the estimate*

$$|u - u_h|_{1,h} \leq C \, \mathrm{Re}\, h^k \, |\Phi|_k. \tag{12}$$

*Moreover, if for a conforming method, i.e. $V_h \subset V$, the function $\Phi_0$ given by (10) belongs to $Q_h$ or $W_h \subset W$, then we have $u = u_h = 0$.*

**Proof:** From (6) we get by setting $v_h = u_h$ and elementwise integrating by parts

$$\begin{aligned}
\nu|u_h|_{1,h}^2 \ &\leq \ \sum_T \left\{ \sum_{E \subset \partial T} < \Phi, u_h \cdot n >_E -(\Phi, \nabla \cdot u_h)_T \right\} \\
&\leq \ \sum_E < \Phi, [u_h \cdot n_E]_E >_E - \sum_T (\Phi, \nabla \cdot u_h)_T
\end{aligned} \tag{13}$$

where $n_E$ denotes a fixed direction of the normal on $E$ and $[\varphi \cdot n_E]_E$ is the jump of $\varphi \cdot n_E$ on $E$ which does not depend on the choosen normal direction.

Let us consider first a conforming finite element method, in which the first sum vanishes and the second sum can be represented in the form

$$-\sum_T (\Phi, \nabla \cdot u_h)_T = -(\Phi - q_h, \nabla \cdot u_h)$$

for all $q_h \in Q_h$. Now, if $\Phi$ belongs to $Q_h$ or if $W_h \subset W$ we conclude $u_h = 0$. In all other cases we use the estimate

$$|(\Phi - q_h, \nabla \cdot u_h)| \leq C \, \|\Phi - q_h\|_0 \, |u_h|_{1,h}$$

and obtain

$$|u - u_h|_{1,h} \leq C \operatorname{Re} \inf_{q_h \in Q_h} \|\Phi - q_h\|_0.$$

In case of a nonconforming finite element method we have also to consider the first sum in (13)

$$\sum_E < \Phi, [u_h \cdot n_E]_E >_E = \sum_E < \Phi - q, [u_h \cdot n_E]_E >_E \quad \forall q \in P_{k-1}.$$

Choosing $q$ as the $L^2(E)$-projection of $\Phi$ onto $P_{k-1}|_E$ we get

$$| < \Phi - q, [u_h \cdot n_E]_E >_E | \leq C \, h^k \, |u_h|_{1,T} |\Phi|_{k,T} \quad E \subset \partial T.$$

Summarizing all estimates we finally obtain

$$|u - u_h|_{1,h} \leq C \operatorname{Re} h^k \, |\Phi|_k,$$

consequently Theorem 1 is verified. $\qquad\square$

Now we consider the conforming Galerkin-least-squares finite element method (7) with piecewise polynomials of degree $l$ and $k-1$ for approximating the velocity and the pressure fields, respectively.

**THEOREM 2** *Let the right hand side of (4) be $f = \nabla \Phi$ with sufficiently regular $\Phi$. Choosing the design parameter $\tau = 1$ and $0 < \nu\delta < \delta_0$, where $\delta_0$ is given by the inverse inequality*

$$\delta_0 h_T^2 \|\Delta u_h\|_{0,T}^2 \leq |u_h|_{1,T}^2 \quad \forall T, \forall u_h \in V_h.$$

*Then, all solutions of the discrete problem (7) satisfy the the error estimates*

$$|u - u_h|_1 \;\leq\; C \sqrt{\operatorname{Re}} \, h^k \, |\Phi|_k, \tag{14}$$

$$\|\nabla \cdot u_h\|_0 \;\leq\; C \, h^k \, |\Phi|_k. \tag{15}$$

**Proof:** See [1]. $\qquad\square$


## AN EXAMPLE WITH AN IRROTATIONAL BODY FORCE

In this section we study a special case of the problem (6) in more detail. The problem under consideration is characterized by

- Stokes Flow on the unit square $\Omega = (0,1)^2$ with the irrotational body force $f = \nabla\Phi$, $\Phi \in H^4(\Omega)$ and

- a nonconforming P1-P0 finite element method on a uniform mesh.

$\Omega$ is devided into $n^2$ cells $\Omega_{i,j} = \{(x,y) \; : \; ih \le x \le (i+1)h; \; jh \le y \le (j+1)h\}$; $i,j = 0,1,\ldots,n-1$ with the edge length $h = 1/n$. Then, each cell $\Omega_{i,j}$ is decomposed into four triangles $\Omega_{i,j,l}$; $l = 1,2,3,4$ by its diagonals. $V_h$ and $Q_h$ consists of piecewise linear and piecewise constant functions. In the velocity-pressure formulation the values of the velocity components at the midpoints of the edges and the values of the pressure at the barycentres of the triangles, respectively, are choosen as degrees of freedom.

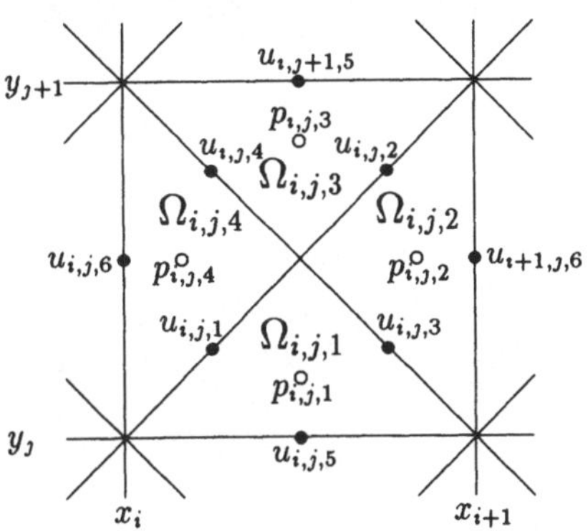


Figure 1: *Cell* $\Omega_{i,j}$

It is possible to find a local basis in the space of discrete divergencefree functions $W_h$, which is spanned by the set of functions:

- $\psi_{i,j} \in W_h$; $i,j = 1,2,\ldots,n-1$ (connected with each vertex of a cell).
  $\psi_{i,j}$ at the midpoints of the edges, having the vertex $P_{i,j} = (x_i, y_j)$ as one endpoint, is equal to a scaled normal vector to this edge directed in a counterclockwise sense around $P_{i,j}$. The length of this vector is equal one over the edge length. At all other nodes $\psi_{i,j}$ is equal to zero.

- $\delta_{i,j} \in W_h$; $i,j = 0,1,\ldots,n-1$ (connected with each midpoint of a cell).
  The definition of $\delta_{i,j}$ corresponds that of $\psi_{i,j}$ replacing
  the vertex $P_{i,j}$ by $P_{i+1/2,j+1/2} = (x_i + h/2, y_j + h/2)$.

- $\beta_{i,j,k} \in W_h$; $i,j = 0,1,\ldots,n-1; k = 1,\ldots,4$
  $\beta_{i,j,k}$, $k = 1,\ldots,4$ at the corresponding inner midpoint of a cell is equal to the tangential to this edge multiplied by $1/\sqrt{2}$ and directed to the vertices of the cell. $\beta_{i,j,k}$ vanishes at all other nodes.

- $\alpha_{i,j,5}, \alpha_{j,i,6} \in W_h$; $i = 0,1,\ldots,n-1; j = 1,\ldots,n-1$
  $\alpha_{i,j,5}, \alpha_{j,i,6}$ at the corresponding inner midpoints of a boundary edge of a cell is equal to the tangential in the direction of the axsis of coordinates. $\alpha_{i,j,k} = 0$ at all other nodes.

Using the discrete divergencefree basis $u_h$ can be represented in the form

$$u_h = \sum_{i,j=1}^{n-1} c_{i,j}\psi_{i,j} + \sum_{i,j=0}^{n-1}\left(d_{i,j}\delta_{i,j} + \sum_{k=1}^{4} b_{i,j,k}\beta_{i,j,k}\right) + \sum_{i=0}^{n-1}\sum_{j=1}^{n-1}(a_{i,j,5}\alpha_{i,j,5} + a_{j,i,6}\alpha_{j,i,6}) \qquad (16)$$

and have to satisfy the equation

$$\nu(\nabla u_h, \nabla v_h)_h = (f, v_h) \qquad \forall v_h \in W_h\,.$$

Because of the special mesh it is possible to eliminate the unknowns $\beta_{i,j,k}$ and $d_{i,j}$ from the algebraic system of equations by a static condensation locally on the cells. Approximating $f = \nabla\Phi$ by the first terms of a Taylor series at the points $P_{i,j}$, $P_{i+1/2,j} = (ih + h/2, jh)$ and $P_{i,j+1/2} = (ih, jh + h/2)$, respectively, we obtain the algebraic system of equations

$$\nu\left[\frac{20}{h^2}c_{i,j} - \frac{4}{h^2}(c_{i,j-1}+c_{i+1,j}+c_{i,j+1}+c_{i-1,j}) - \frac{1}{h^2}(c_{i+1,j+1}+c_{i+1,j-1}+c_{i-1,j+1}+c_{i-1,j-1})+\right.$$
$$\left.\frac{3}{2h}(-a_{i+1,j-1,6}-a_{i+1,j,6}+a_{i-1,j-1,6}+a_{i-1,j,6}+a_{i-1,j+1,5}+a_{i,j+1,5}-a_{i-1,j-1,5}-a_{i,j-1,5})\right]$$
$$= \frac{h^4}{720}[\Phi_{xyyy} - \Phi_{xxxy}]\,|_{P_{i,j}} + O(h^6) \qquad (17)$$

$$\nu\left[6a_{i,j,5} + \frac{1}{2}(a_{i,j+1,5}-2a_{i,j,5}+a_{i,j-1,5}) + \frac{3}{2h}(-c_{i+1,j+1}+c_{i+1,j-1}-c_{i,j+1}+c_{i,j-1})\right]$$
$$= \frac{h^2}{3}\Phi_x + \frac{h^4}{480}[5\Phi_{xxx} - \Phi_{xyy}]\,|_{P_{i+\frac{1}{2},j}} + O(h^6)$$

$$\nu\left[6a_{i,j,6} + \frac{1}{2}(a_{i+1,j,6}-2a_{i,j,6}+a_{i-1,j,6}) + \frac{1}{h}(c_{i+1,j+1}-c_{i-1,j+1}+c_{i+1,j}-c_{i-1,j})\right]$$
$$= \frac{h^2}{3}\Phi_y + \frac{h^4}{480}[5\Phi_{yyy} - \Phi_{xxy}]\,|_{P_{i,j+\frac{1}{2}}} + O(h^6)\,.$$

For solving the system (17) approximately, we assume asymptotic expansions of the form

$$c_{i,j} \;\asymp\; Re\sum_{l=0}^{\infty} c_{i,j}^l h^{l+2}; \qquad i,j = 0, 1, \ldots, n$$

$$a_{i,j,5} \;\asymp\; Re\sum_{l=0}^{\infty} a_{i,j,5}^l h^{l+2}; \qquad i = 0, 1, \ldots, n-1;\ j = 0, 1, \ldots, n$$

$$a_{i,j,6} \;\asymp\; Re\sum_{l=0}^{\infty} a_{i,j,6}^l h^{l+2}; \qquad i = 0, 1, \ldots, n;\ j = 0, 1, \ldots, n-1,$$

where the coefficients $c_{i,j}^l$ and $a_{i,j,k}^l$ can be recovered by smooth functions $C^l(x,y)$ and $A_k^l(x,y)$ in the following way:

$$c_{i,j}^l = C^l(P_{i,j}) \quad a_{i,j,5}^l = A_5^l(P_{i+1/2,j}) \quad a_{i,j,6}^l = A_6^l(P_{i,j+1/2})\,.$$

By comparision of coefficients we get from (17) partial differential equations for the functions $C^l$, $A_k^l$, for example for $l = 0$ and $l = 1$ we have

$$\Delta\Delta C^0 \;=\; -\frac{1}{36}[\Phi_{xxxy} - \Phi_{xyyy}]; \qquad \text{in } \Omega; \qquad C^0|_\Gamma = 0; \qquad \frac{\partial C^0}{\partial\vec{n}}\Big|_\Gamma = \frac{\partial\Phi}{\partial\vec{t}}\Big|_\Gamma$$

$$A_5^0 \;=\; \frac{1}{18}\Phi_x + C_y^0; \qquad A_6^0 = \frac{1}{18}\Phi_y - C_x^0; \qquad (18)$$

$$C^1 \;=\; 0; \qquad A_k^1 = 0\,.$$

The coefficients $c^l_{i,j}$, $d^l_{i,j}$, $b^l_{i,j,k}$ and $a^l_{i,j,k}$ can be calculated for $l = 0,1$ by means of the solution $C^0$ of the fourth order problem. In the primitive variables we get for the first terms of the velocity $u_h$

$$u_h \; \asymp \; Re \, h^2 \left( \mathrm{curl}\, C^0 + \frac{5}{144} F_+ \begin{bmatrix} 1 \\ 1 \end{bmatrix} + \right) \cdots \quad \text{at } P_{i+1/4,j+1/4} \; P_{i+3/4,j+3/4}$$

$$u_h \; \asymp \; Re \, h^2 \left( \mathrm{curl}\, C^0 + \frac{5}{144} F_- \begin{bmatrix} 1 \\ -1 \end{bmatrix} + \right) \cdots \quad \text{at } P_{i+3/4,j+1/4} \; P_{i+1/4,j+3/4}$$

$$u_h \; \asymp \; Re \, h^2 \left( \mathrm{curl}\, C^0 + \frac{\Phi_x}{18} \begin{bmatrix} 1 \\ 0 \end{bmatrix} \right) + \cdots \quad \text{at } P_{i+1/2,j}$$

$$u_h \; \asymp \; Re \, h^2 \left( \mathrm{curl}\, C^0 + \frac{\Phi_y}{18} \begin{bmatrix} 0 \\ 1 \end{bmatrix} \right) + \cdots \quad \text{at } P_{i,j+1/2},$$

where the abbreviation $F_\pm = \Phi_x \pm \Phi_y$ was used.

This result corresponds to the expected discretization error of the Stokes problem. In the special case that $\Phi_{xxxy} = \Phi_{xyyy}$ and $\frac{\partial \Phi}{\partial t}|_\Gamma = 0$ the solution $C^0$ becomes identically zero, consequently, the first order terms of the expansion of $u_h \cdot n$ at the mesh points vanish. For this special situation the flux through the edges is of higher order equal to zero (see the examples in the next section).

## NUMERICAL EXPERIMENTS

For discretizing problem (1), (2) we use the nonconforming Crouzeix-Raviart element satisfying the discrete LBB-condition (5) and combine it with an upstream technique analyzed for the isothermal case in [6], [7]. The algebraic system of equations is solved by a nonlinear multigrid method described in [2].
In all test problems $\Omega$ will be the unit square. An interpolation of the nonconforming P1-approximation $u_h$ onto a P1-conforming approximation $\bar{u}_h$ defined by

$$\bar{u}_h(P_i) = \frac{1}{N} \sum_{j=1}^{N} u_h|_{T_j}(P_i), \quad P_i \text{ is a vertex of the triangle } T_j, \tag{19}$$

is used to represent the flow fields. The streamfunction is calculated in the following way: First we fix a value of the streamfunction $\Psi$ on a boundary vertex $P_i$ of the mesh by zero. Then the value of the streamfunction in a neighbour vertex $P_j$ is computed by

$$\Psi(P_j) = \int_{P_i}^{P_j} u_h \cdot n \, ds + \Psi(P_i). \tag{20}$$

This procedure is done successively for all vertices of the mesh.

**Example 1:** Stokes flow with $f = \nabla\Phi$, $\Phi = 2xy(1-x)(1-y)$.

The exact solution is $(u,p) = (0,\Phi - \frac{1}{18})$. The Figures (2) and (3) show the vector plots of the velocity field for $Re = 10$ and the meshsize $h = \frac{1}{2}$ and $h = \frac{1}{16}$. In Figure (4) and Figure (5) we see the vector plots for $Re = 100$. Comparing the vector plots of the velocity fields the following behaviour can be noticed:

- For a fixed Reynolds number $Re$ the velocity approximation $u_h$ tends to zero if the meshsize tends to zero (compare Figure (2) and Figure (3)).

- For higher Reynolds number $Re$ we need a finer meshsize in order to get the same accuracy of the velocity approximation $u_h$ (compare the Figures (3),(4) and (5)).

Both observations are in agreement with the theoretical estimate

$$|u - u_h|_{1,h} = O(Re\, h) \tag{21}$$

given in Theorem 1.

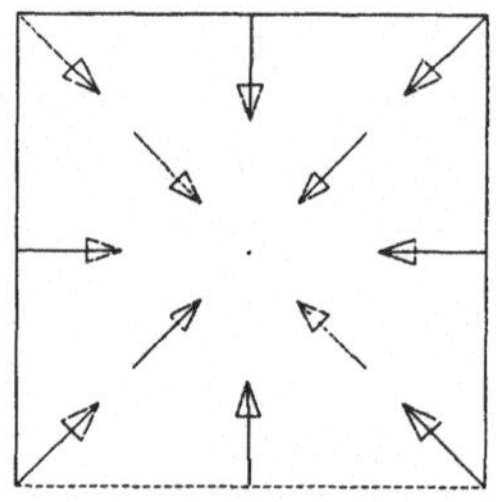

Figure 2: Vector plot of the velocity field, $Re = 10$, $2\times 2$ cells

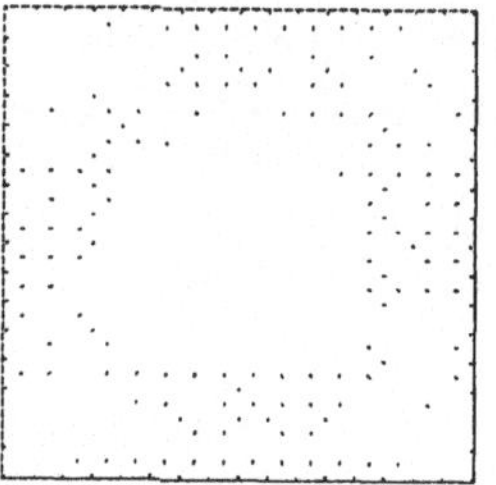

Figure 3: Vector plot of the velocity field, $Re = 10$, $16\times16$ cells

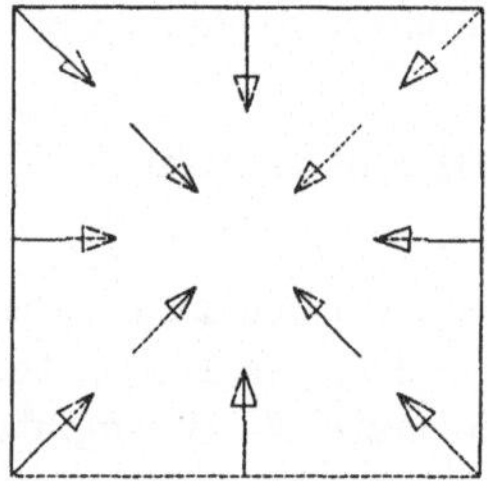

Figure 4: Vector plot of the velocity field, $Re = 100$, $2\times2$ cells

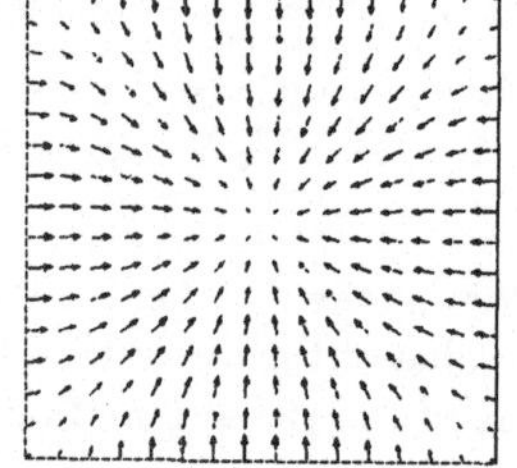

Figure 5: Vector plot of the velocity field, $Re = 100$, $16\times16$ cells

**Example 2:** Navier-Stokes flow with $f_1 := curl\,\Phi$ and $f_2 := f_1 + 100\nabla\Phi$,
$\Phi = 2xy(1 - x)(1 - y)$.

Let $(u,p)$ be a solution for the right hand side $f = f_1$, then $(u,p + 100\Phi - \frac{50}{9})$ is a solution of (3) with the right hand side $f = f_2$. The Figures (6) - (9) show the vector plots of the velocity fields and the streamlines in the two cases of the right hand side $f$. We remark that $\|\nabla\Phi\|_{0,2,\Omega}^2 = \|curl\,\Phi\|_{0,2,\Omega}^2 = \frac{4}{45}$. Thus the gradient of $\Phi$ dominates the right hande side $f$ for $f = f_2$. We have the following situation:

- The corresponding discrete velocity fields are completely different, as indicated in Figure (6) and (7).

- The flow field of the discrete solution $u_h$ in the case $f = f_1$ (Fig. 6) is "acceptable". The flow field in the case $f = f_2$ (Fig. 7) is wrong.

- The plotted streamlines in Figure (8) and (9) are in good agreement.

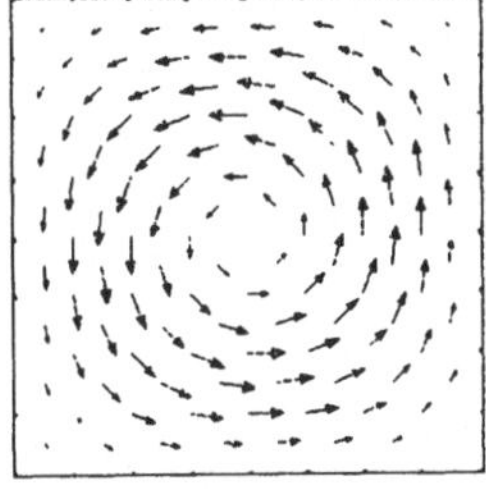

Figure 6: Vector plot of the velocity field for $f_1$, $Re = 10$, 8×8 cells

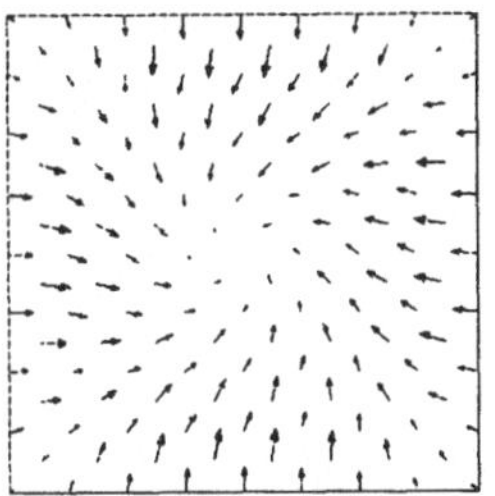

Figure 7: Vector plot of the velocity field for $f_2$, $Re = 10$, 8×8 cells

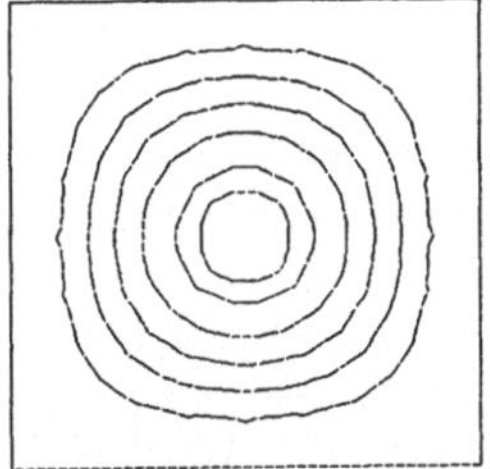

Figure 8: Streamlines for $f_1$, $Re = 10$, 8×8 cells

Figure 9: Streamlines for $f_2$, $Re = 10$, 8×8 cells

Thus we have seen that only looking at the streamlines does not guarantee the quality of the calculated discrete velocity $u_h$. The reason for this behaviour is due to the fact that a wrong tangential component of $u_h$ does not influence the streamfunction of the velocity as it can be seen from (20).

**Example 3:** Navier-Stokes flow with $f = f_2 + \nabla\Theta$ with $\Theta = 10x + 10y$.

Again the additional term $\nabla\Theta$ does not change the exact velocity field u. However, comparing with example 2 we see that:

- The term $\nabla\Theta$ strongly influence the discrete velocity field <u>and</u> the plotted stream-lines (see Figure (10) and (11)).

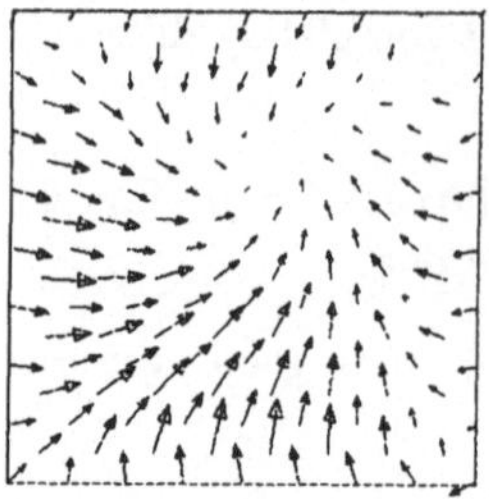

Figure 10: Vector plot of the velocity field, $Re = 10$, 8×8 cells

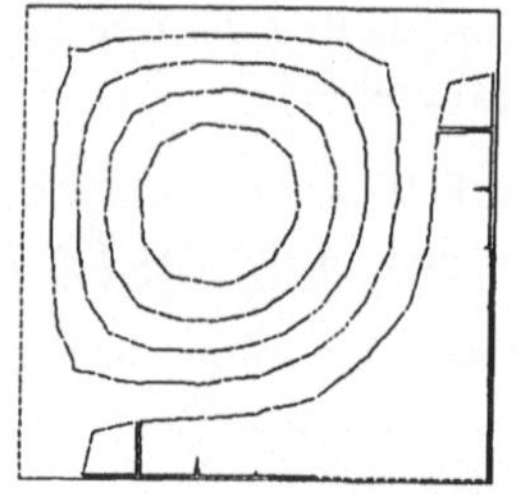

Figure 11: Streamlines, $Re = 10$, 8×8 cells

The different behaviour of the two gradients $\nabla\Phi$ and $\nabla\Theta$ are caused by the pr operty $0 = \Phi|_\Gamma \neq \Theta|_\Gamma$. That follows by the results of the previous s ection.

**Example 4:** Nonisothermal Navier-Stokes flow

Now we come back to the general problem (1), (2) with

$$\Gamma_D = \{(x,y) \in \Gamma : x = 0 \quad \text{or} \quad x = 1\}. \tag{22}$$

Let $T_H$ and $T_C$, $T_H > T_C$, be the prescribed constant temperature values on the left and the right wall of the cavity. By a scaling argument we can assume that

$$\nu = \frac{Pr}{Ra^{\frac{1}{2}}}, \qquad \lambda = \frac{1}{Ra^{\frac{1}{2}}}, \qquad \alpha f(T) = Pr\, T \begin{pmatrix} 0 \\ 1 \end{pmatrix}, \tag{23}$$

where $Pr$, $Ra$ denote the Prandtl and Rayleigh number, respectively. From the physical point of view only the difference $T_H - T_C$ is responsible for the behaviour of the velocity field by fixed Prandtl and Rayleigh number and not the absolute values of $T_H$ and $T_C$. So we expect the same picture for the plotted streamlines for example in the cases $T_H = 0.5$, $T_C = -0.5$ and $T_H = 1.0$, $T_C = 0.0$. The streamlines are represented for nearly the same values of the streamfunction $\Psi$ in Figure (12) and (13). What is the reason for the different Figures ?

- Introducing the transformation $T = \tilde{T} + 0.5$ we can switch from the first formulation of the problem to the second. By a careful consideration we see that the two problems for calculating (u,p,T) and $(\tilde{u}, \tilde{p}, \tilde{T})$ distinguish only in the right hand side by an additive gradient $\nabla\Phi = \begin{pmatrix} 0 \\ 0.5Pr \end{pmatrix}$.

- From the examples above we have seen that taking off the "correct" gradient from the right hand side can considerably improve the accuracy of the calculated velocity field.

Because an symmetric solution of the problem is expected we think that the results represented in Figure (12) are more accurate.

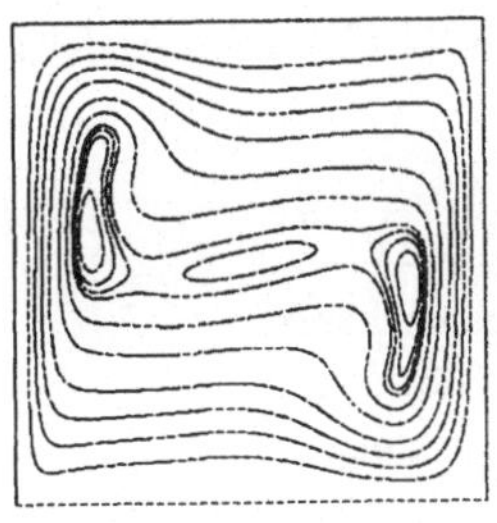

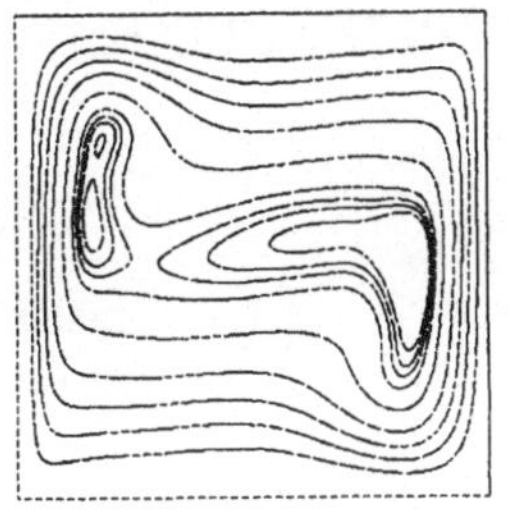

Figure 12: Streamlines, $Ra=10^6$, Pr=0.71, $T_H=0.5$, $T_C=-0.5$, 128×128 cells

Figure 13: Streamlines, $Ra=10^6$, Pr=0.71, $T_H=1.0$, $T_C=0.0$, 128×128 cells

# References


[1] O. Dorok, W. Grambow, and L. Tobiska. Aspects of Finite Element Discretizations for Solving the Boussinesq Approximation of the Navier-Stokes Equations. *Preprint Otto-von-Guericke-Universität Magdeburg, Math 5/94*, Februar 1994.

[2] O. Dorok, F. Schieweck, and L. Tobiska. A Multigrid Method for Solving the Boussinesq Approximation of the Navier-Stokes Equations. *Preprint TU Magdeburg*, Math 18/93, October 1993.

[3] L.P. Franca. Incompressible flows based upon stabilized methods. UCD/CCM 4, Center for Computational Mathematics, University of Colorado at Denver, December 1993.

[4] V. Girault and P.-A. Raviart. *Finite Element Methods for Navier-Stokes equations.* Springer-Verlag, Berlin-Heidelberg-New York, 1986.

[5] C. Johnson, R. Rannacher, and M. Boman. Numerics and hydrodynamic stability: Towards error control in CFD. Technical Report Preprint 93-12(SFB 359), IWR Heidelberg, März 1993.

[6] F. Schieweck and L. Tobiska. A nonconforming finite element method of upstream type applied to stationary Navier-Stokes equations. *M²AN*, 23:627–647, 1989.

[7] A. Thiele and L. Tobiska. A weighted upwind finite element method for solving the stationary Navier-Stokes equations. *WZ TU Magdeburg*, 33:13–20, 1989.

[8] L. Tobiska and R. Verfürth. Analysis of a Streamline Diffusion Finite Element Method for the Stokes and Navier-Stokes Equations. *Preprint TU Magdeburg , Math 1/92*, Januar 1992.

# STUDY OF EXTENDED FLOW SEPARATION ON PARALLEL MACHINES


D. Drikakis and F. Durst

Lehrstuhl für Strömungsmechanik

Universität Erlangen–Nürnberg

Cauerstr. 4, D–91058 Erlangen

Germany


## SUMMARY


A two dimensional Navier-Stokes code has been parallelized and applied for the simulation of supersonic flow over a flat plate at large angles of incidence. The objective of this work is to present results and experience from parallel computations done on this flow field, which characterized by large separation regions. The numerical algorithm used is an implicit flux vector splitting method with high order upwind extrapolation schemes. Parallelization was obtained by grid partitioning and parallel computations were performed on different grid sizes for three parallel machines. Results for the *parallel* and *numerical* efficiency, as well as, for the computing time are shown.


## INTRODUCTION

The viscous supersonic flow over a flat plate is one of the most discussed flows in aerodynamics. In the past, a large number of papers has dealt with the separated flow over a plate (e.g. [1–3]). In studying this flow, these works were limited to small angles of incidence and most were involved with analytical methods in the context of the *triple deck theory* [1]. Due to the significant progress of numerical analysis during the last two decades, CFD methods have reached a mature stage where simulation of physical and engineering problems can be achieved with great accuracy. Furthermore, the rapid progress of high performance computers, especially parallel computers, has offered the capability of obtaining faster solutions to large scale computations, in which fine grids are used.

Recently the authors studied numerically [4] the supersonic flow over a flat plate at large angles of incidence for which large separation regions appear. This study was performed by two parallel Navier-Stokes codes [5] based on high order Flux Vector Splitting (FVS) [6] and Riemann solver methods [7]. These methods were used for space discretization while an unfactored implicit algorithm was used for the time integration. In this work results for the supersonic flow were presented for flow angles up to 20° and for Reynolds $Re$ numbers up to $10^5$.

The present paper deals with results concerning the parallel computing used in this flow simulation. The parallelization is analysed in terms of the *numerical, parallel,* and *total* efficiencies and results from parallel calculations on three parallel machines are shown.

## NUMERICAL ALGORITHM

The supersonic flow over the flat plate has been simulated by solving the Navier-Stokes equations in conservative form and using general curvilinear coordinate system:

$$(JQ)_t + E_\xi + G_\zeta = \frac{1}{Re}(R_\xi + S_\zeta) \qquad (1)$$

where $Q = J(\rho, \rho u, \rho w, e)^T$ is the unknown solution vector and $E, G, R$, and $S$ are the inviscid and viscous flux vectors, respectively. The equations are solved in dimensionless form by introducing the Reynolds ($Re$), and the Mach ($M_\infty$) numbers. Furthermore, $J = x_\xi z_\zeta - x_\zeta z_\xi$ is the Jacobian of the transformation $\xi = \xi(x,z)$ and $\zeta = \zeta(x,z)$ from Cartesian coordinates (x,z) to generalized coordinates ($\xi, \zeta$).

Different numerical methods have been used for the discretization of the inviscid fluxes $E$ and $G$. For the supersonic flow over the plate comparisons between a modified Steger-Warming FVS method [6], the van Leer FVS and a Riemann solver [7] can be found in Reference [4]. The motivation of these comparisons was to examine the numerical uncertainties introduced in the flow structure by different space discretization schemes. The conclusion was that for the finest grid all the methods resulted in the same predictions. In the present study the modified Steger-Warming FVS scheme was used. Hence, the numerical uncertainties related to the space discretization were minimized. The time integration was obtained by an unfactored implicit method. A Newton form of the Navier-Stokes equations was employed by constructing a sequence of approximations between two time steps, while the inversion of the system of equations was obtained by a Gauss-Seidel relaxation scheme.

## SUPERSONIC FLOW WITH EXTENDED SEPARATION

A complete study of the supersonic flow over the plate at large angles of incidence can be found in Reference [4]. In this section a brief description of the flow is given while parallelization issues and results are discussed in the next sections. The calculations were performed for a Mach number $M_\infty = 2$ and for Reynolds numbers in the regime $Re=10^4$–$10^5$. The angle of incidence ($a$) was varied from $5^\circ$ to $20^\circ$. The complete flow structure can be described as follows:

According to gas dynamics theory, a shock wave and expansion waves are formed at the leading edge of the plate on the lower and the upper side, respectively. Then, the boundary layer starts to develop on the upper and lower sides of the plate. For flow angles $a > 5^\circ$ separation of the flow occurs on the upper side of the plate and at a certain distance from the leading edge. The separation region extends up to the trailing edge of the plate. On the upper side of the trailing edge a shock wave forms and interacts with the separated boundary layer. On the lower side of the trailing edge the flow accelerates through the expansion waves. In figure (1) the pressure distribution on the lower side of the plate is shown for angles $a = 10^o$, $20^o$ and for a Reynolds number $Re = 10^5$. This figure shows a higher pressure jump at the trailing edge for a larger angle of incidence. A shear layer develops at the wake region and pressure recovery occurs at a certain distance away from the trailing edge.

At $a = 5^o$ a small separation bubble at the trailing edge of the plate appears. The numerical solution captures the separation bubble at one grid point before the trailing edge. The appearance of separation at the trailing edge for $a = 5^o$ has been noted in the past by Daniels [2], who observed it using analytical methods in the context of the "triple deck theory". As the angle of incidence is increased the separation point moves upstream. In figure (2) the beginning of separation as a function of the angle of incidence and the Reynolds number is shown. It is seen that for Reynolds numbers less than $10^5$ separation occurs only when the

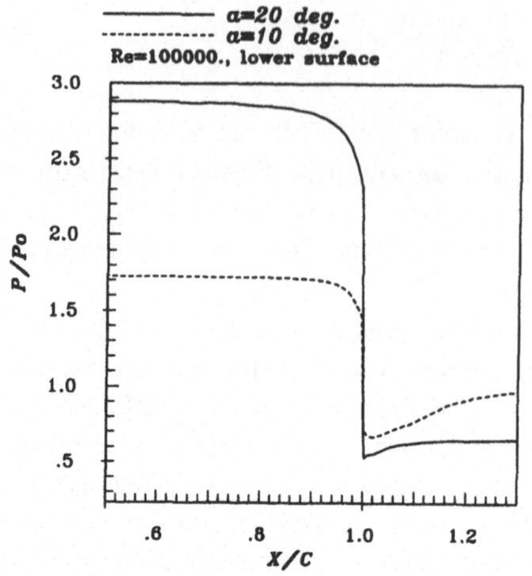


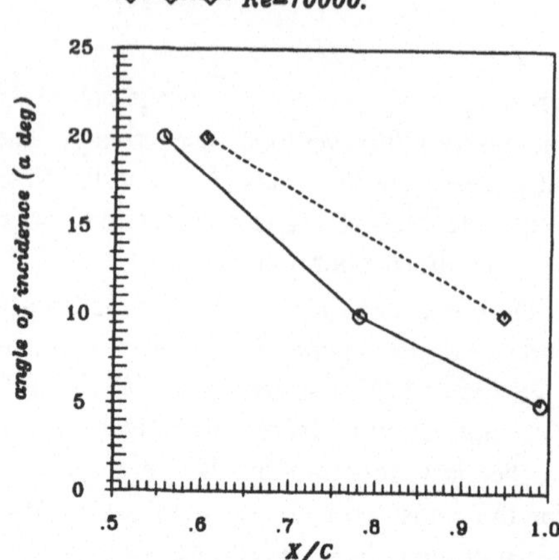


Fig. 1 Pressure distribution at the lower side of the trailing edge for $Re = 10^5$.

Fig. 2 Separation position as function of the angle of incidence and the Reynolds number.

angle of incidence is: $a > 5°$. For a higher Reynolds number ($Re = 10^5$) the adverse pressure gradient is large enough to lead into separation.

## PARALLELIZATION ISSUES

The parallelization of the Navier-Stokes algorithm is based on the grid partitioning technique. The computational domain is subdivided into non-overlapping subdomains and each subdomain is assigned to one processor. The CFD code has been divided into a numerical and a parallel code. Furthermore, the parallel code has been divided into communication and high level routines. The communication routines are used for the distribution of geometric data to the processors and collection of flow data from the *slave* processors to the *master*. The communication has been separated into *local* and *global* communication. The *local* communication is used for exchanging boundary data between the neighbouring subdomains (i.e. the processors). Each time a processor updates a variable which is needed by a neighbour processor, it is copied to the neighbouring processor's memory. The number of control volumes which are stored in the neighbouring processor depends on the type and the order of accuracy of the discretization method. For the present solver an upwind scheme third-order of accuracy is used. This upwind scheme defines the variables at the cell face of each CV using extrapolation of neighbouring values:

$$Q^l_{i+\frac{1}{2}} = aQ_i - bQ_{i-1} + cQ_{i-2} + dQ_{i+1} \tag{2}$$

$$Q^r_{i+\frac{1}{2}} = aQ_{i+1} - bQ_{i+2} + cQ_{i+3} + dQ_i \ . \tag{3}$$

The superscripts $l,r$ denote the left and right sides of the cell face while $a,b,c,d$ are coefficients defining the order of the extrapolation. The left or right side is chosen according to the sign of the eigenvalue. From the above relations it is evident that two columns and two lines of the bounding control volumes (CVs) needs to be exchanged on each processor in the x- and z-direction, respectively. The subdomains do not overlap i.e. each processor calculates only variable values which are not calculated by other processors. This requires, in the case of MIMD computers with distributed memory, an overlap of storage, with each processor storing data from one or more CV layers belonging to neighbouring subdomains along its boundary. The iterative unfactored solution requires exchange of the values of $Q$ after each outer iteration and the exchange of the variation of solution $\Delta Q$ after each inner iteration. In the case where an unfactored implicit solution is used, the inner iterations are the sub-iteration states of the Gauss-Seidel relaxation. For the flow under investigation local communication is also needed for the CVs in the wake region, because the subdomains on the upper and lower sides of the wake are neighbouring. The global communication is for collecting the residuals from the *slave* processors and transferring them to the *master*. In the case of global communication, only a certain number of processors is involved in communication at any time between the beginning and the end of information gathering or scattering. Global communication is a limiting factor for massive parallelization, unless communication and computation are allowed to take place simultaneously. For the algorithm described in this study, global communication is performed after each outer iteration. Global communication propagates in one direction similar to a wavefront, and then in the other direction, sequentially from processor to processor.

The computational grid is generated by an elliptic method solving two Poissons' equations for the coordinates (x,z). In the numerical simulation three computational grids with $120\times40$, $240\times80$, and $480\times160$ points were used, respectively. For the parallelization of the Navier-Stokes solver a decomposition algorithm has been developed. Acceleration of the numerical convergence was achieved by using a parallel version of the mesh-sequencing technique [8]. In the mesh-sequencing technique the solution of the equations is initially calculated on a sequence of coarser grids. The solution on the coarser meshes is used as an initial guess for the solution on the fine mesh. Acceleration of the convergence is achieved because a better initial condition than that of the uniform flow field condition is defined on the fine mesh by previously solving the equations on the coarser meshes.

The total efficiency ($E_n^{tot}$) is used for the performance measurement of the parallel computations:

$$E_n^{tot} = \frac{T_1}{nT_n} \tag{4}$$

where $T_1$ and $T_n$ are the computation times using one and $n$ processors, respectively. In the ideal case the speed-up of a parallel solver is $n$, which corresponds to an efficiency of 100%. In reality the efficiency is less than 100%. The total efficiency can be expressed as a product of three factors: *parallel*($E_n^{par}$), *numerical* ($E_n^{num}$), and *load balancing* efficiency $E_n^{lb}$. The product of these factors gives the total efficiency $E_n^{tot}$:

$$E_n^{tot} = E_n^{par} E_n^{num} E_n^{lb}. \tag{5}$$

The parallel efficiency represents the time loss in a parallel computation. This is due to communication lag between processors during which computation cannot take place. The numerical efficiency represents the increase in the number of iterations necessary to fulfil the convergence criterion. This increase is due to the changes in the algorithm which are required for its parallelization. The load balancing efficiency represents the time some processors stay idle due to the different problem size per processor (i.e. number of grid points). In the present work the subdomains are defined by the same number of grid points and therefore the *load balancing* efficiency is equal to 100%. The numerical efficiency is defined as the ratio of the total number of floating point operations per CV in the serial algorithm to the total number of operations in the parallel algorithm on $n$ processors, required to reach the same convergence criterion. It does not depend on the performance characteristics of the computer. In order to parallelize a numerical solution procedure, it may be necessary to modify the serial algorithm. This usually leads to an increase in the number of both inner and outer iterations which are required to obtain a solution of prescribed accuracy compared to the calculation on one processor. The numerical efficiency is easier to measure if a constant number of inner iterations is considered. Then it is defined by the ratio of the number of outer iterations. A more exact value can be obtained by measuring the total and parallel efficiencies and calculating $E_n^{num}$ from Eq. (5). The total efficiency is easily determined by measuring the computing time necessary to reach a converged solution. The parallel efficiency can be measured by using a fixed number of outer iterations on one and $n$ processors. In that case $E_n^{num}=1$, and therefore, the total efficiency is equal to the parallel efficiency.

## RESULTS OF PARALLEL COMPUTATIONS

The present calculations were performed on three parallel machine architectures. The first is the Meiko Computing Surface with 64 T800 transputers, each with a clock rate of 25 MHz and 4MB of memory. The four transputer links are connected to routing chips which can be programmed to establish the desired configuration. Each transputer can be connected to at most four physical neighbours and one transputer is connected to the host. The configurations used were the ring and the surface of the cylinder (thorus). The communication possibilities between the processors are: (i) four hardwired links (channels) with very short set-up time but communication only with their four nearest neighbours; (ii) transports, a soft link that can be established at run time to any processor but with an order of magnitude longer set-up time. A development toolset, CSTools, makes possible the implementation of parallel applications and supports the communication process on transputers without taking care of the physical architecture constraints. Due to faster communication the channel's communication is used, which allows only connections with four direct neighbours. The second computer is the Parsytec machine which uses the T805 transputer, having a clock rate of 30 MHz. Here four communication possibilities exist: (i) *dumb links*, similar but somewhat slower than the previously Meiko channels, (ii) *message ports* with more software support and flexibility but still slower; (iii) input/output of the *Helios* operating system, the most comfortable but the slowest option; and (v) the *Parix* operation system, the fastest communication possibility. The third parallel computer is a KSR–1 from Kendall Square Research. It consists of proprietary processors with 40 MFlops peak performance. Each processor has 32 MBytes of memory. It is a so-called virtual shared memory machine. It was programmed with the TCGMSG message-passing library from Argonne National Laboratories.

Table 1: Performance characteristics of computers used.

| Parallel machine | $t^{st}$ ($\mu$s) | $R_{tr}$ (MB/s) | $1/\tau$ (MFlops) | $t^{st}/\tau$ |
|---|---|---|---|---|
| Meiko CS | 22 | 1.4 | 0.45 | 10 |
| Parsytec MC3 | 70 | 1.2 | 0.35 | 24.5 |
| KSR1 | 110 | 7.3 | 5 | 550 |

Calculations were performed for three grid sizes ($120\times40$, $240\times80$, $480\times160$) and different number of processors on the three parallel platforms. In figures (3a), (3b), and (3c) the total efficiency as function of the number of processors is shown for the Meiko CS, Parsytec MC3, and KSR1, respectively. The efficiency reduces as the number of processors increase and increases for larger grid sizes. This behaviour is due to the following factor: when the number of CVs in each direction is increased by a factor of two, the calculation time of each processor increases by a factor of four, but the number of boundary CVs, and, therefore the communication time, increases by a factor of two (ignoring set-up time). Thus, the calculation time varies linearly with the number of CVs while the communication time varies as the square root of the number of CVs. Therefore the ratio of communication to calculation time is reduced as the grid is refined, leading to an increase in efficiency.

The major parameters characterizing the parallel machines and influencing the performance of a parallel algorithm are: (i) the set-up time required to enable message passing, $t^{st}$; (ii) the time needed to perform one floating-point operation, $\tau$, and (iii) the rate at which data is transferred between processors, $R_{tr}$. Measurements of performance characteristics of the three parallel machines are shown in Table (1). From the figures (3a,b,c) and the table (1) it is seen the highest efficiencies are always achieved for the smaller $t^{st}/\tau$ ratio. The set-up time is the crucial parameter influencing communication when the amount of transferred data is low. The difference would diminish if the grid was further refined. In table (2) the computing time is shown for the three parallel machines.

The number of iterations required for convergence of the Navier-Stokes solver is not the same for calculations performed on one and $n$ processors, respectively. This is due to the relaxation procedure at the subdomains' boundaries. In the parallel version of the Gauss-Seidel line-relaxation algorithm the values $\Delta Q$ at the subdomain boundaries are taken from the previous relaxation steps. The effect of the grid partitioning to the numerical efficiency is shown in table (3) where the number of iterations and the *parallel* efficiency are shown using different grid sizes and number of processors on the Parsytec MC3. The numerical efficiency reduces when the number of subdomains increases. This effect seems to become stronger for coarser grids and larger number of processors.

Table 2: Computing time on the parallel machines.

| Machine/Grid | $240 \times 80$ | $480 \times 160$ |
|---|---|---|
| Parsytec (40 procs.) | 1.60 h | 11.20 h |
| Meiko (40 procs.) | 1.50 h | 10.0 h |
| KSR1 (8 procs.) | 0.82 h | 5.26 h |

Table 3: Number of iterations and *parallel* efficiency on the Parsytec MC3.

| Grid | No. of procs. | iterations | $E_n^{par}$ % |
|---|---|---|---|
| 480 × 160 | 10 | 5,000 | 99 |
|  | 20 | 5,085 | 98 |
|  | 40 | 5,164 | 96 |
| 240 × 80 | 10 | 3,400 | 98 |
|  | 20 | 3,435 | 97 |
|  | 40 | 3,507 | 95 |
| 120 × 40 | 10 | 1,093 | 97 |
|  | 20 | 1,135 | 95 |
|  | 40 | 1,177 | 90 |

## CONCLUSIONS

A Navier-Stokes solver has been parallelized and used for simulating the supersonic flow over a flat plate. For this flow extended separation occurs for angles of incidence $a \geq 10^o$. Results from parallel computations performed on three parallel platforms were shown. The performance was studied in terms of the total efficiency factor while the *parallel* ($E_n^{par}$), and *numerical* ($E_n^{num}$) efficiencies were also obtained. High efficiencies are achieved by the present Navier-Stokes solver. For the performance of a parallel system, the crucial parameter is the ratio of the communication to the calculation time ($t^{st}/\tau$). Consequently, transputer systems show better total efficiency than the KSR1 machine because they require less set up time for the communication.

## ACKNOWLEDGEMENTS

The authors would like to thank the Bavarian Ministry of Education and the Bavarian Science Foundation for their financial support.

## REFERENCES

[1] Stewartson, K.: "On the flow near the trailing edge of a flat plate", Proc. Roy. Soc. Lond., (1968), A 306, pp. 275–289.

[2] Daniels, P. G.: "Numerical and asymptotic solutions for the supersonic flow near the trailing edge of a flat plate at incidence", J. Fluid Mech., (1974), 63, pp. 641–656.

[3] Riley, N., Stewartson, K.: "Trailing edge flows", J. Fluid Mech., (1969), 39, pp. 193–207.

[4] Drikakis, D., Durst, F.: "A numerical study of the viscous supersonic flow past a flat plate at large angles of incidence", Phys. Fluids A, (1994), in print.

[5] Drikakis, D., Schreck, E.: "Development of parallel implicit Navier-Stokes solvers on MIMD multi-processor systems", AIAA Paper 93–0062, 31st Aerospace Sciences Meeting and Exhibit, January 11–14, Reno, NV (1993).

[6] Drikakis, D., Tsangaris, S.: "On the solution of Navier-Stokes equations using improved flux vector splitting methods", Applied Mathematical Modeling, (1993), 17, pp. 282–297.

[7] Eberle, A.: "Characteristic Flux Averaging Approach to the Solution of Euler's Equations", VKI Lecture Series, Comp. Fluid Dynamics, 1987–04, (1987).

[8] Drikakis, D., Schreck, E.: "Parallel Multi-Level Calculations for Viscous Compressible Flows", ASME Conference, Washington DC, June, 1993, FED Vol. 156, CFD Algorithms and Applications for Parallel Processors, Editors O. Baysal, and V. Saxena, (1993).

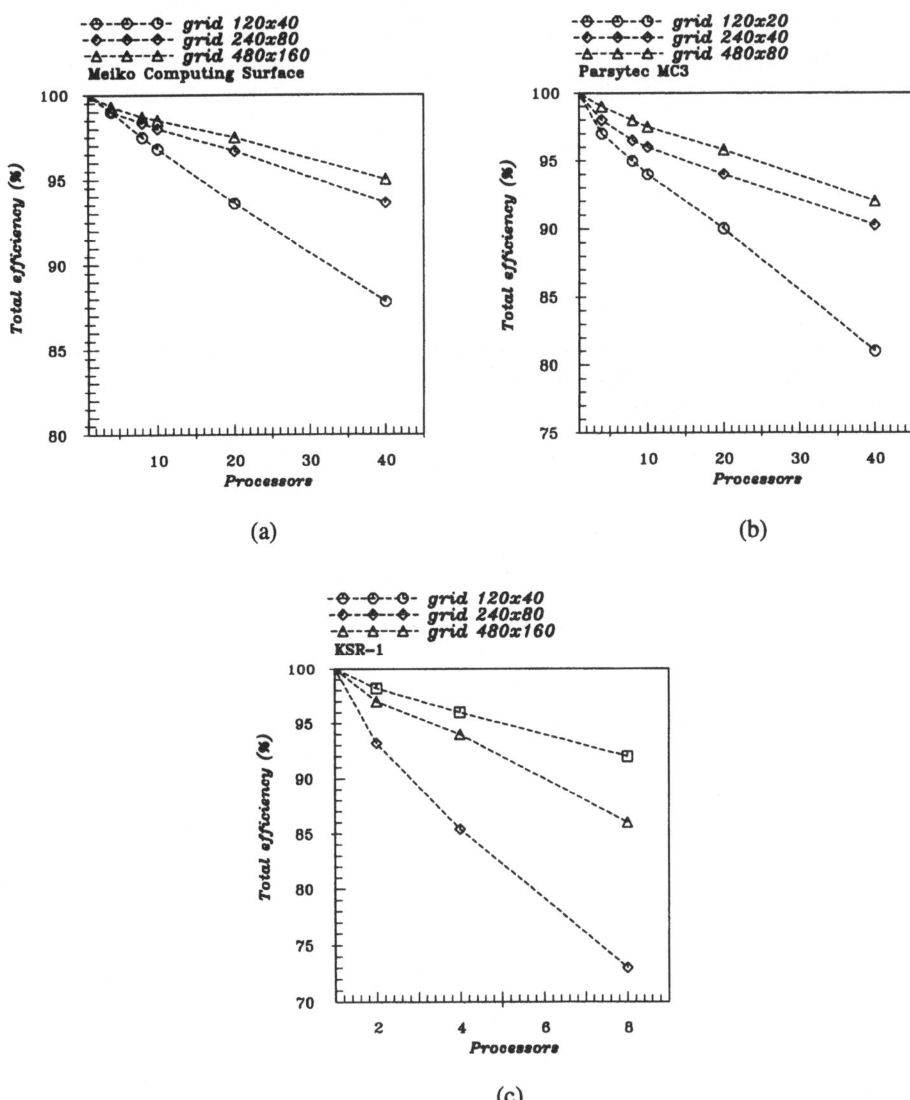



Fig. 3 Total efficiency of the parallel computations on the Meiko CS, Parsytec MC3, and KSR1, for diffrent grid sizes.

# OPERATOR SPLITTING METHOD FOR COMPRESSIBLE EULER AND NAVIER–STOKES EQUATIONS


M. Feistauer

Charles University Prague, Faculty of Mathematics and Physics
Sokolovská 83, 186 00 Praha 8, Czech Republic

P. Knobloch

Otto von Guericke University of Magdeburg, Department of Mathematics
PF 4120, D–39 016 Magdeburg, Germany



## SUMMARY

The paper is concerned with a method for numerical solution of compressible transonic and hypersonic viscous flow with high Reynolds numbers. The method is based on the finite volume Osher–Solomon scheme applied on a nonuniform grid and used for the discretization of hyperbolic convective terms. The solution of the complete viscous compressible system is carried out via operator inviscid-viscous splitting. Some numerical results are presented.


## 1. FORMULATION OF THE PROBLEM

Let us consider viscous compressible flow in a bounded domain $\Omega \subset \mathbb{R}^2$ and time interval $(0, T)$. The governing system consisting of the continuity equation, Navier–Stokes equations and energy equation can be written in the form

$$\frac{\partial w}{\partial t} + \frac{\partial f(w)}{\partial x} + \frac{\partial g(w)}{\partial y} = \frac{\partial R(w, \nabla w)}{\partial x} + \frac{\partial S(w, \nabla w)}{\partial y} \quad \text{in } Q_T = \Omega \times (0, T), \qquad (1)$$

where

$$w = (\rho, \rho u, \rho v, e)^{\mathrm{T}} \qquad (2)$$

$$f(w) = (\rho u, \rho u^2 + p, \rho u v, (e + p) u)^{\mathrm{T}}, \qquad (3)$$

$$g(w) = (\rho v, \rho u v, \rho v^2 + p, (e + p) v)^{\mathrm{T}}, \qquad (4)$$

$$p = (\kappa - 1)(e - \rho(u^2 + v^2)/2), \quad e = \rho(c_v \theta + (u^2 + v^2)/2), \qquad (5)$$

$$R(w, \nabla w) = (0, \tau_{xx}, \tau_{xy}, \tau_{xx} u + \tau_{xy} v + k \, \partial\theta/\partial x)^{\mathrm{T}}, \qquad (6)$$

$$S(w, \nabla w) = (0, \tau_{xy}, \tau_{yy}, \tau_{xy} u + \tau_{yy} v + k \, \partial\theta/\partial y)^{\mathrm{T}}, \qquad (7)$$

$$\tau_{xx} = 2\mu \, \partial u/\partial x + \lambda(\partial u/\partial x + \partial v/\partial y), \qquad (8)$$

$$\tau_{yy} = 2\mu \, \partial v/\partial y + \lambda(\partial u/\partial x + \partial v/\partial y), \qquad (9)$$

$$\tau_{xy} = \mu(\partial v/\partial x + \partial u/\partial y). \qquad (10)$$

We use the standard notation: $x$, $y$ – Cartesian coordinates, $u$, $v$ – the velocity components in the directions $x$, $y$, $\rho$ – density, $p$ – pressure, $\theta$ – absolute temperature, $e$ – total energy, $\tau_{xx}$, $\tau_{xy}$, $\tau_{yy}$ – the components of the kinematic part of stress tensor; we assume that $\kappa > 1$, $c_v$, $k$, $\mu > 0$ are constants and $\lambda = -2\mu/3$. The vector functions $f$, $g$ represent inviscid Euler fluxes, $R$ and $S$ are viscous terms.

System (1) is equipped by the *initial condition*

$$w(x, y, 0) = w_0(x, y), \quad (x, y) \in \Omega, \tag{11}$$

and the *boundary conditions*: a) On the *inlet* we prescribe $\rho$, $u$, $v$, $\theta$. b) On *impermeable walls* the no-slip conditions $u = v = 0$ and the adiabatic condition $\partial\theta/\partial n = 0$ are considered. c) On the rest of the boundary we use the conditions

$$\tau_{xx} n_x + \tau_{xy} n_y = 0 = \tau_{xy} n_x + \tau_{yy} n_y, \quad \partial\theta/\partial n = 0. \tag{12}$$

The solution of the problem is carried out by means of the inviscid-viscous operator splitting. This means that system (1) is split into two systems

$$\frac{1}{2}\frac{\partial w}{\partial t} + \frac{\partial f(w)}{\partial x} + \frac{\partial g(w)}{\partial y} = 0 \quad \text{(Euler equations)}, \tag{13}$$

$$\frac{1}{2}\frac{\partial w}{\partial t} + \frac{\partial R(w, \nabla w)}{\partial x} + \frac{\partial S(w, \nabla w)}{\partial y} = 0 \quad \text{(pure diffusion system)} \tag{14}$$

which are considered and discretized separately.

## 2. DISCRETIZATION OF INVISCID SYSTEM (13)

Let us substitute $t := t/2$ in (13) and set

$$\begin{aligned}
&\mathbb{A}(w) = D\dot{f}(w)/Dw, \quad \mathbb{B}(w) = Dg(w)/Dw, \\
&\mathcal{P}(w, n) = n_x f(w) + n_y g(w), \quad \mathbb{P}(w, n) = D\mathcal{P}/Dw = n_x\mathbb{A}(w) + n_y\mathbb{B}(w), \\
&n = (n_x, n_y) \in \mathbb{R}^2.
\end{aligned} \tag{15}$$

System (13) is *hyperbolic*, which means that the eigenvalues $\lambda_1, \ldots, \lambda_4$ of $\mathbb{P}$ are real and there exists a nonsingular matrix $\mathbb{T}$ such that

$$\mathbb{P} = \mathbb{T}\mathbb{D}\mathbb{T}^{-1}, \quad \mathbb{D} = \mathrm{diag}(\lambda_1, \ldots, \lambda_4). \tag{16}$$

Moreover, system (13) is *rotationally symmetric*: if we put

$$\mathbb{Q} = \begin{pmatrix} 1 & 0 & 0 & 0 \\ 0 & n_x & n_y & 0 \\ 0 & -n_y & n_x & 0 \\ 0 & 0 & 0 & 1 \end{pmatrix}, \quad n = (n_x, n_y), \ |n| = 1, \tag{17}$$

then we have

$$\mathcal{P}(w, n) = \mathbb{Q}^{-1} f(\mathbb{Q}w), \quad \mathbb{P}(w, n) = \mathbb{Q}^{-1}\mathbb{A}(\mathbb{Q}w)\mathbb{Q}. \tag{18}$$

System (13) is discretized by the cell-centred *finite volume method*. We approximate $\Omega$ by a polygonal domain $\Omega_h$ and construct a partition $\mathcal{T}_h = \{T_i\}$, $i \in \mathcal{I}$ (= an index set), of $\Omega_h$ consisting of a finite number of polygons called finite volumes. Here we use quadrilateral meshes. We can write $\partial T_i = \bigcup_{j \in s(i)} \partial T_{ij}$, where $\partial T_{ij}$ are sides of $T_i \in \mathcal{T}_h$ and $s(i)$ is an index set. For neighbouring volumes $T_i, T_j \in \mathcal{T}_h$ we write $\partial T_{ij} = T_i \cap T_j$. We use the following notation: $|T_i|$ = the measure of $T_i$, $\ell_{ij}$ = the length of $\partial T_{ij}$, $n_{ij} = (n_{xij}, n_{yij})$ = unit outer normal to $\partial T_i$ on $\partial T_{ij}$. Further, we consider a partition $0 = t_0 < t_1 < \ldots$ of the time interval $(0, T)$ and put $\tau_k = t_{k+1} - t_k$.

The finite volume discretization is based on the integration of (13) over the set $T_i \times (t_k, t_{k+1})$ and the use of Green's theorem. Approximating $w$ by a piecewise constant vector function with values $w_i^k \approx w(\cdot, t_k)|T_i$, we obtain the relations

$$w_i^{k+1} = w_i^k - \frac{\tau_k}{|T_i|} \sum_{j \in s(i)} H(w_i^k, w_j^k, n_{ij}) \ell_{ij}, \quad i \in \mathcal{I}, \; k = 0, 1, \dots. \tag{19}$$

The term $H(w_i^k, w_j^k, n_{ij})$ is called numerical flux. It approximates the flux $\mathcal{P}(w, n)$ through the side $\partial T_{ij}$ in the direction $n = n_{ij}$.

We express the numerical flux $H$ in the form

$$H(w_1, w_2, n) = \mathbb{Q}^{-1} f_R(\mathbb{Q}w_1, \mathbb{Q}w_2), \tag{20}$$

where $f_R$ is the *approximate Riemann solver* for the system with one space dimension obtained by the transformation of (13) determined by the matrix $\mathbb{Q}$. The transformation of $w$ yields $q = \mathbb{Q}w = (\rho, \rho\tilde{u}, \rho\tilde{v}, e)$, $\tilde{u} = u\,n_x + v\,n_y$, $\tilde{v} = -u\,n_y + v\,n_x$. We symbolically write

$$f_R(q_1, q_2) = f(q_2) - \int_{q_1}^{q_2} \mathbb{A}^+(q)\,\mathrm{d}q. \tag{21}$$

The matrix $\mathbb{A}^+$ is defined with the aid of (16), where we set $n = (1, 0)$, i. e. $\mathbb{A} = \mathbb{T}\mathbb{D}\mathbb{T}^{-1}$, $\mathbb{D} = \mathrm{diag}(\lambda_1, \dots, \lambda_4)$, $\lambda_1, \dots, \lambda_4 =$ the eigenvalues of $\mathbb{A}$. Then we write

$$\mathbb{A}^+ = \mathbb{T}\mathbb{D}^+\mathbb{T}^{-1}, \quad \mathbb{D}^+ = \mathrm{diag}(\lambda_1^+, \dots, \lambda_4^+), \quad \lambda^+ = \max(0, \lambda). \tag{22}$$

(For details, see [2, Par. 7.3, 7.2.110, 7.2.114].) The integral in (21) is either approximated with the use of a suitable numerical quadrature or is evaluated along a suitable path. Here we use the Osher–Solomon scheme ([7, 8, 4]). This scheme is based on the possibility to connect two states $q_1$, $q_2$ in a unique way by the piecewise smooth curve $q(\xi)$ in the state space and to evaluate the integral in (21) along this curve. The curve $q$ consists of four smooth parts $q^{(k)}$, $k = 1, \dots, 4$, which are tangential to the eigenvectors of the matrix $\mathbb{A}(q)$. Then, with the use of the Riemann invariants, it is possible to express the approximate Riemann solver as linear combinations of values of the flux $f$ at uniquely determined and analytically expressed points as we can see from the following table:

Table 1: Osher–Solomon approximate Riemann solver $f_R(q_1, q_2)$

| | $\tilde{u}_2 \geq -c_2$ | | $\tilde{u}_2 < -c_2$ | |
| | $\tilde{u}_1 \leq c_1$ | $\tilde{u}_1 > c_1$ | $\tilde{u}_1 \leq c_1$ | $\tilde{u}_1 > c_1$ |
|---|---|---|---|---|
| $c_A \leq \tilde{u}_A$ | $f(q_1^S)$ | $f(q_1)$ | $f(q_2) - f(q_2^S) + f(q_1^S)$ | $f(q_1) - f(q_2^S) + f(q_2)$ |
| $0 < \tilde{u}_A < c_A$ | $f(q_A)$ | $f(q_1) - f(q_1^S) + f(q_A)$ | $f(q_2) - f(q_2^S) + f(q_A)$ | $f(q_1) - f(q_2^S) + f(q_2) - f(q_1^S) + f(q_A)$ |
| $-c_B \leq \tilde{u}_A \leq 0$ | $f(q_B)$ | $f(q_1) - f(q_1^S) + f(q_B)$ | $f(q_2) - f(q_2^S) + f(q_B)$ | $f(q_1) - f(q_2^S) + f(q_2) - f(q_1^S) + f(q_B)$ |
| $\tilde{u}_A < -c_B$ | $f(q_2^S)$ | $f(q_1) - f(q_1^S) + f(s_2^S)$ | $f(q_2)$ | $f(q_1) + f(q_2) - f(q_1^S)$ |

The states $q_A$, $q_B$, $q_1^S$, $q_2^S$ are determined as follows. We put

$$c = (\gamma p/\rho)^{1/2}, \quad s = p/\rho^\gamma, \quad \alpha = (s_2/s_1)^{1/2\gamma},$$
$$z_1 = (\gamma - 1)\tilde{u}_1/2 + c_1, \quad z_2 = (\gamma - 1)\tilde{u}_2/2 - c_2. \tag{23}$$

Then

$$c_A = (z_1 - z_2)/(1 + \alpha), \quad c_B = \alpha\, c_A, \quad \rho_A = (c_A/c_1)^{2/(\gamma-1)}\rho_1, \tag{24}$$
$$\rho_B = \rho_A/\alpha^2, \quad \tilde{u}_A = 2(z_1 - c_A)/(\gamma - 1), \quad \tilde{u}_B = \tilde{u}_A,$$
$$\tilde{v}_A = \tilde{v}_1, \quad \tilde{v}_B = \tilde{v}_2;$$
$$c_1^S = 2z_1/(\gamma + 1), \quad \rho_1^S = (c_1^S/c_1)^{2/(\gamma-1)}\rho_1, \quad \tilde{u}_1^S = c_1^S, \quad \tilde{v}_1^S = \tilde{v}_1; \tag{25}$$
$$c_2^S = -2z_2/(\gamma + 1), \quad \rho_2^S = (c_2^S/c_2)^{2/(\gamma-1)}\rho_2, \quad \tilde{u}_2^S = -c_2^S, \quad \tilde{v}_2^S = \tilde{v}_2. \tag{26}$$

The mentioned construction is possible under the condition

$$c_1 + c_2 + (\gamma - 1)(\tilde{u}_1 - \tilde{u}_2)/2 > \max\{0, (\tilde{v}_1 - \tilde{v}_2)/2\}. \tag{27}$$

The numerical flux $H$ obtained in the described way is consistent and conservative.

For the edge $\partial T_{iB} \subset \partial\Omega_h \cap \partial T_i$ a modified approach is used. We determine the transformed boundary state $q_B = \mathbb{Q}\, w_B$ in the following way: a) We prescribe $\ell$ components of $q_B$, where $\ell$ is the number of negative eigenvalues of $\mathbb{A}(q_i)$. ($q_i$ is the transformed state on $T_i$.) b) The rest of components of $q_B$ is determined with the use of a curve in the state space consisting of subcurves which are tangential to the eigenvectors of $\mathbb{A}(q)$ corresponding to negative eigenvalues. Then (21) implies that $f_R(q_i, q_B) = f(q_B)$. It is suitable to distinguish several cases:

1. *Subsonic inlet*: We prescribe $\rho_B$, $\tilde{u}_B$, $\tilde{v}_B$ and compute $p_B$ from the formulae

$$p_B = \rho_I\, c_I^2/\gamma, \quad \rho_I = (c_I^2 \rho_i/\gamma p_i)^{1/(\gamma-1)}\rho_i, \quad c_I = c_i + \frac{\gamma - 1}{2}(\tilde{u}_i - \tilde{u}_B). \tag{28}$$

2. *Supersonic inlet*: We prescribe $\rho_B$, $\tilde{u}_B$, $\tilde{v}_B$, $c_B$.

3. *Subsonic outlet*: We prescribe $p_B$ and use the formulae

$$\rho_B = \rho_i(p_B/p_i)^{1/\gamma}, \quad \tilde{u}_B = \tilde{u}_i + \frac{2}{\gamma - 1}\left(c_i - \sqrt{\gamma p_B/\rho_B}\right), \quad \tilde{v}_B = \tilde{v}_i. \tag{29}$$

4. *Supersonic outlet*: $q_B = q_i$.

5. *Impermeable wall*:

$$\tilde{u}_B = 0, \quad c_B = c_i + (\gamma - 1)\tilde{u}_i/2, \quad \rho_B = (c_B^2 \rho_i/\gamma\, p_i)^{1/(\gamma-1)}\rho_i, \quad p_B = \rho_B\, c_B^2/\gamma. \tag{30}$$

In this case we have $f(q_B) = (0, p_B, 0, 0)^{\mathrm{T}}$.

## 3. DISCRETIZATION OF VISCOUS TERMS AND COMPLETE SCHEME

Equation (14) is discretized again by the finite volume method which yields the scheme

$$w_i^{k+1} = w_i^k + \frac{\tau_k}{|T_i|}\sum_{j \in s(i)}\left(R_{ij}^k n_{xij} + S_{ij}^k n_{yij}\right)\ell_{ij}, \tag{31}$$

where

$$R_{ij}^k \approx R(w, \nabla w)\,|\,\partial T_{ij} \times \{t_k\}, \tag{32}$$
$$S_{ij}^k \approx S(w, \nabla w)\,|\,\partial T_{ij} \times \{t_k\}.$$

The derivatives $\partial w/\partial x$, $\partial w/\partial y$ are approximated with the aid of the constant values $w_i^k$ on $T_i$, $i \in \mathcal{I}$, at $t = t_k$. For an internal vertex $A \in \Omega_h$ we use the formulae

$$\frac{\partial w}{\partial x}(A, t_k) \approx \frac{1}{|V|} \int_V \frac{\partial w(\cdot, t_k)}{\partial x}\, dxdy = \frac{1}{|V|} \int_{\partial V} w(\cdot, t_k)\, n_x\, dS \approx \tag{33}$$

$$\approx \frac{1}{|V|} \sum_{j=1}^{n_A} w_{i(j)}^k \left(A_y^{(j+1)} - A_y^{(j)}\right)$$

(similar formulae are used for the approximation of $\partial w/\partial y$). For the notation see Figure 1: $A^{(n_A+1)} = A^{(1)}$, $V = $ polygon with vertices $A^{(1)}, \ldots, A^{(n_A)}$, $|V| = \mathrm{meas}(V)$, $i(j) = $ index of the finite volume $T_{i(j)}$ containing the segment $A^{(j)} A^{(j+1)}$. In [4] it was shown that for $w \in C^2(\Omega_h)$ the discretization error in (33) is $O(h)$ and, if $V$ is a regular $2k$-polygon and $w \in C^3(\Omega_h)$, then the error is $O(h^2)$, where $h = \max\{\mathrm{diam}\, T_i;\ i \in \mathcal{I}\}$.

If $\partial T_{ij}$ is an edge connecting two internal vertices $A$, $B \in \Omega_h$, then we use (33) and write

$$(\nabla w)_{ij}^k \approx (\nabla w(A, t_k) + \nabla w(B, t_k))/2, \quad w_{ij}^k \approx (w_i^k + w_j^k)/2 + (\nabla w)_{ij}^k \left(C_{ij} - (C_i + C_j)/2\right), \tag{34}$$
$$R_{ij}^k = R(w_{ij}^k, (\nabla w)_{ij}^k), \quad S_{ij}^k = S(w_{ij}^k, (\nabla w)_{ij}^k),$$

where $C_i$, $C_j$ and $C_{ij}$ denote the centres of $T_i$, $T_j$ and $\partial T_{ij}$, respectively.

The description of the calculation of viscous fluxes through edges having at least one endpoint on $\partial \Omega_h$ is rather complex and, in view of various types of boundary conditions, it is necessary to distinguish several particular cases. For the detailed analysis we refer to [4].

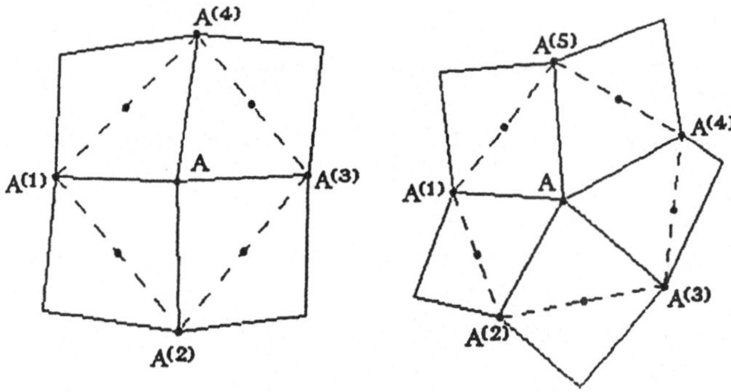



Figure 1

The inviscid scheme (19) is combined with the viscous scheme (31) via fractional steps. The resulting method has the following form:

$$w_i^{k+1/2} = w_i^k - \frac{\tau_k}{|T_i|} \sum_{j \in s(i)} H_{ij}(w_i^k, w_j^k, n_{ij})\, \ell_{ij} \tag{35}$$

& inviscid boundary conditions,

$$w_i^{k+1} = w_i^{k+1/2} + \frac{\tau_k}{|T_i|} \sum_{j \in s(i)} \left(R_{ij}^{k+1/2} n_{xij} + S_{ij}^{k+1/2} n_{yij}\right) \ell_{ij}$$

& complete boundary conditions for viscous flow.

Starting from the initial conditions

$$w_i^0 = \frac{1}{|T_i|} \int_{T_i} w(x,y,0)\,dx dy, \quad T_i \in \mathcal{T}_h, \tag{36}$$

we compute successively $w_i^k$, $i \in \mathcal{I}$, $k = 0,1,\dots$ . Usually we are interested in a steady state solution and try to obtain it by the time stabilization for $t_k \to \infty$. It is possible to consider (35) as a finite dimensional dynamical system.

Since the method (35) is completely explicit, it is necessary to apply a stability condition. Using linearization and analogy with a scalar problem, we arrive at the stability conditions

$$\frac{\tau_k}{|T_i|} \max_{j \in s(i)} \ell_{ij} \max_{m=1,\dots,4} \left\{ \lambda_m(\mathbb{Q}(n_{ij})\,w_j^k) \right\} \leq \mathrm{CFL} \approx 0.85, \tag{37}$$

$$4\frac{\tau_k}{|T_i|} \max(\mu, k) \leq \mathrm{CFL}, \quad T_i \in \mathcal{T}_h.$$

Here $\lambda_m(q)$ denote the eigenvalues of $\mathbb{A}(q)$.

# 4. RESULTS

The method was tested by the flow through the GAMM channel (10 % circular arc in the channel of width 1 m) for air, i.e. $\gamma = 1.4$, $\mu = 1.72 \cdot 10^{-5}\,\mathrm{kg\,m^{-1}\,s^{-1}}$, $k = 2.4 \cdot 10^{-2}\,\mathrm{kg\,m\,s^{-3}\,K^{-1}}$. Figures 2 – 6 show the results of the computation of inviscid flow with inlet Mach number $M_{\mathrm{inlet}} = 0.69$. In Figures 7 – 9 we see the results obtained for viscous flow.

Interesting results were obtained for viscous flow with $M_{\mathrm{inlet}} = 0.67$ and $\mu = 1.7 \cdot 10^{-3}\,\mathrm{kg\,m^{-1}\,s^{-1}}$. It means that the Reynolds number is $10^2$ times lower as before. In this case we do not get the convergence to steady state as above. The results are shown in Figures 10 – 13. Finally, Figures 14 – 17 show inviscid and viscous flow results for $\mu = 1.72 \cdot 10^{-5}\,\mathrm{kg\,m^{-1}\,s^{-1}}$ and $M_{\mathrm{inlet}} = 1.6$. The mesh for viscous computation was obtained from the inviscid case by a refinement at the walls by the law of geometric progression. For inviscid as well as viscous flow the convergence to steady state was achieved.

# 5. CONCLUSION

The comparison of the results with other works ([1], [3], [5]) indicate that the Osher–Solomon inviscid solver used as the part of the method for the solution of viscous flow leads to a sufficiently robust scheme. Further improvements can be achieved with the use of higher order methods (MUSCL, Runge–Kutta), a suitable adaptive strategy (to obtain a precise resolution of discontinuities and boundary layers) and multigrid (to increase the convergence to steady state). Possible modification can be obtained with the use of triangular grids or dual meshes to triangular grids, which would give the possibility of the finite element approximation of viscous terms. Let us note that another finite volume approach leads to the so-called cell-vertex schemes. See, e. g., [6].

# REFERENCES


[1] BRISTEAU, M. O., GLOWINSKI R., PERIAUX J., VIVIAND H. (eds.): "Numerical Simulation of Compressible Navier–Stokes Flows", Notes on Numerical Fluid Mechanics, Volume 18, Vieweg, Braunschweig, 1987.

[2] FEISTAUER, M.: "Mathematical Methods in Fluid Dynamics", Pitman Monographs and Surveys in Pure and Applied Mathematics Series 67, Longman Scientific & Technical, Harlow, 1993.

[3] FELCMAN, J.: "Finite volume solution of the inviscid compressible fluid flow", ZAMM, $\underline{72}$ (1993), pp. 513–516.

[4] KNOBLOCH, P.: "Numerical Modelling of Viscous Compressible Flow", Thesis, Fac. of Math. and Physics, Charles University, Prague 1993 (in Czech).

[5] KOREN, B.: "Upwind schemes for the Navier–Stokes equations", in: Notes on Numerical Fluid Mechanics, Volume $\underline{24}$, (Ballman, J., Jeltsch, R. – eds.), Vieweg, Braunschweig–Wiesbaden, 1989, pp. 300–309.

[6] MACKENZIE, J. A., MORTON, K. W.: Finite volume solutions of convection-diffusion test problems. Math. Comp., $\underline{60}$ (1992), pp. 187–230.

[7] OSHER, S., SOLOMON, F.: "Upwind difference schemes for hyperbolic systems of conservation laws", Math. Comp., $\underline{38}$ (1982), pp. 339–377.

[8] SPEKREIJSE, S. P.: "Multigrid solution of the steady Euler equations", CWI Tracts, Volume $\underline{46}$, Amstedam, 1988.

## Acknowledgements

This work was supported by the grant No. 201/93/2177 of the Czech Grant Agency.

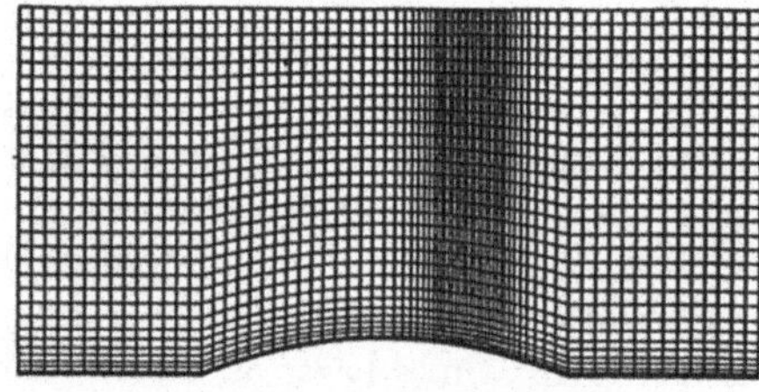

Figure 2: Mesh ($M_{\mathrm{inlet}} = 0.69$, inviscid flow)

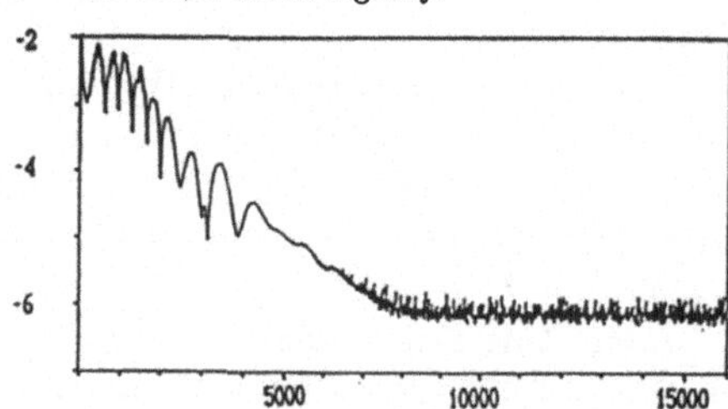

Figure 3: Convergence history (inviscid flow)

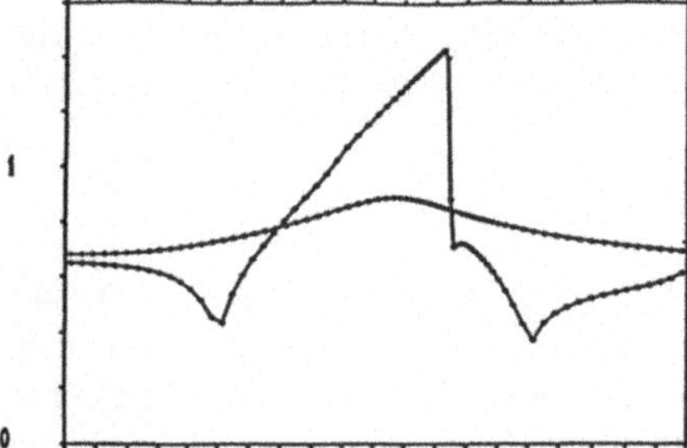

Figure 4: Mach number on the walls (inviscid flow)

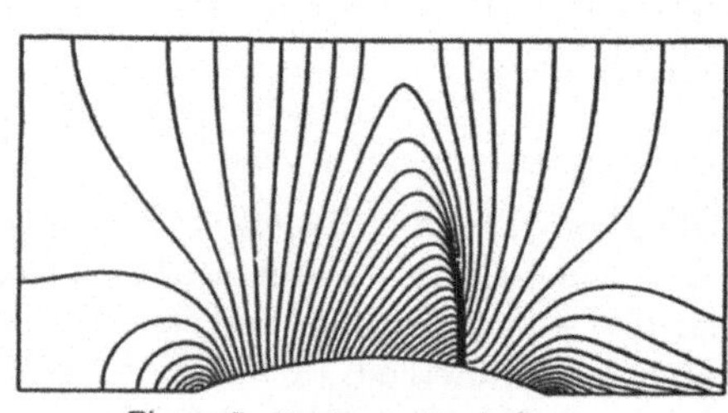

Figure 5: Mach number isolines (inviscid flow)

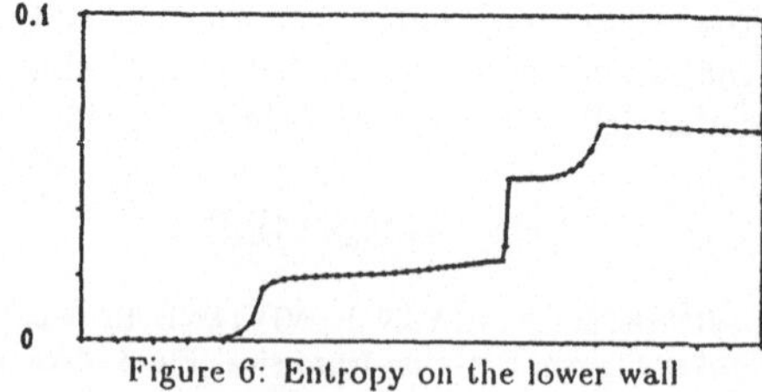

Figure 6: Entropy on the lower wall (inviscid flow)

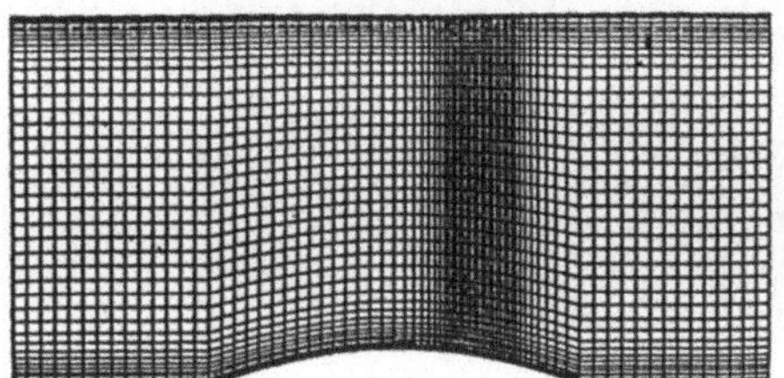

Figure 7: Mesh ($M_{\text{inlet}} = 0.69$, viscous flow)

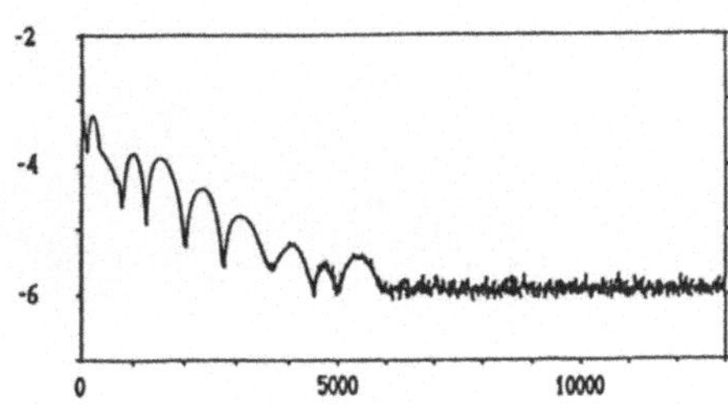

Figure 8: Convergence history (viscous flow)

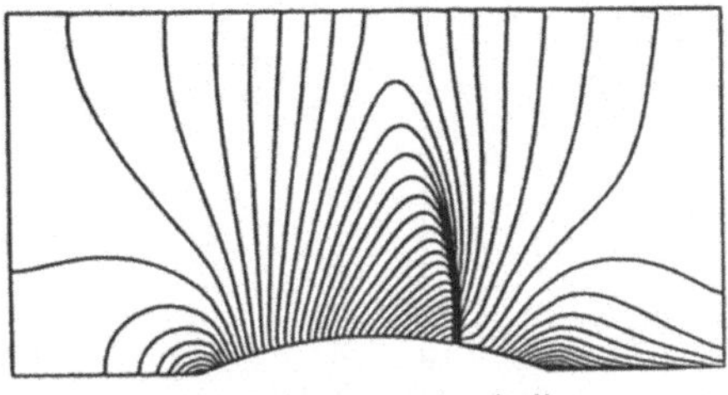

Figure 9: Mach number isolines (viscous flow)

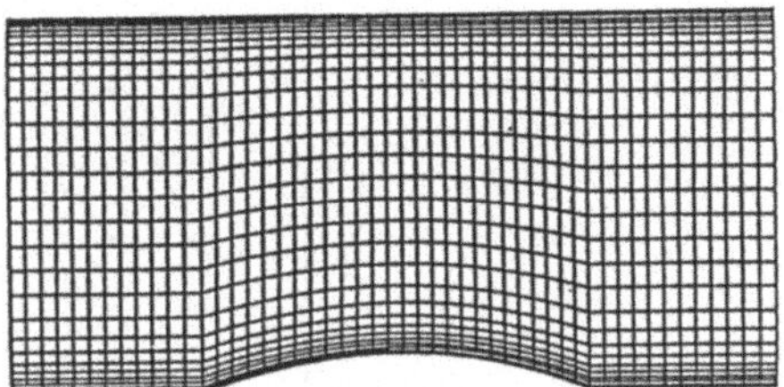

Figure 10: Mesh ($M_{\text{inlet}} = 0.67$, viscous flow)

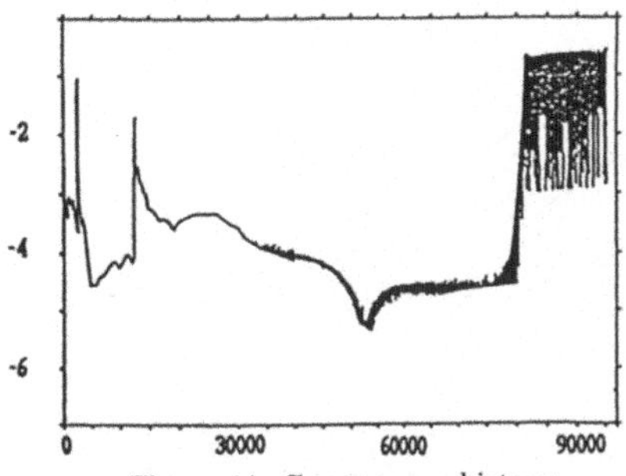

Figure 11: Convergence history (viscous flow)

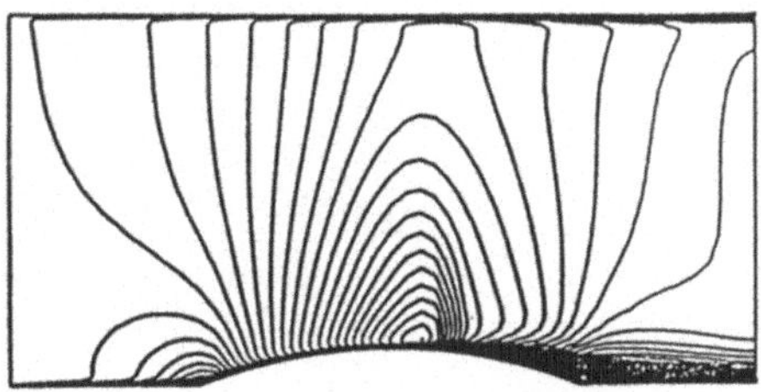

Figure 12: Mach number isolines (62 000 time steps)

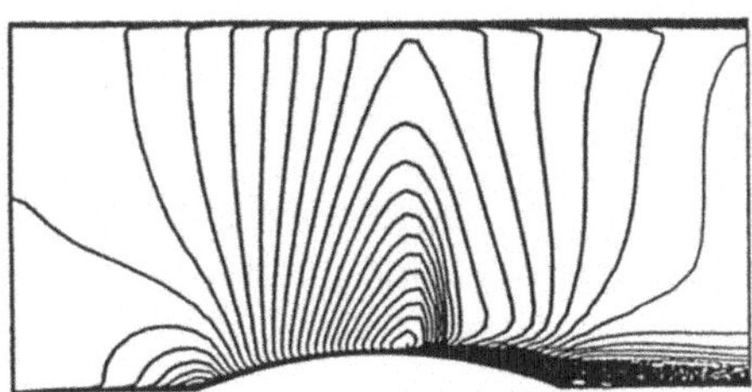

Figure 13: Mach number isolines (95 000 time steps)

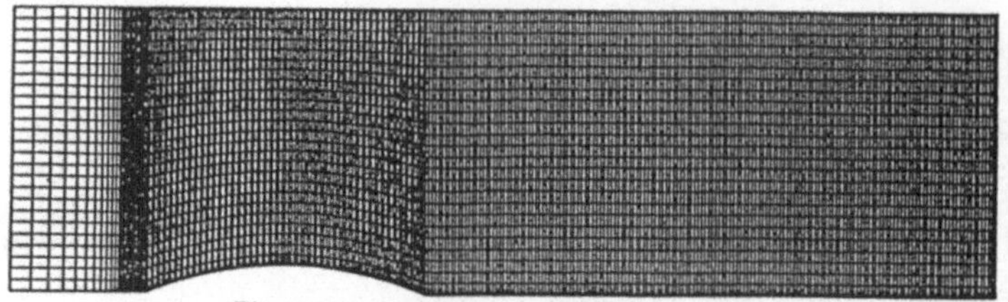

Figure 14: Mesh ($M_{\text{inlet}} = 1.6$,
inviscid flow)

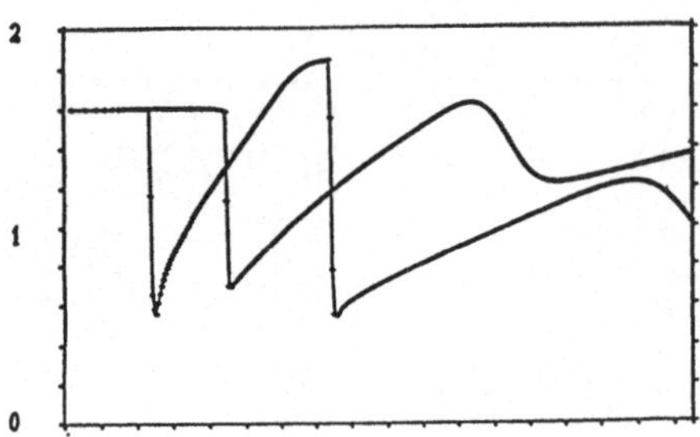

Figure 15  Mach number on the walls
(inviscid flow)

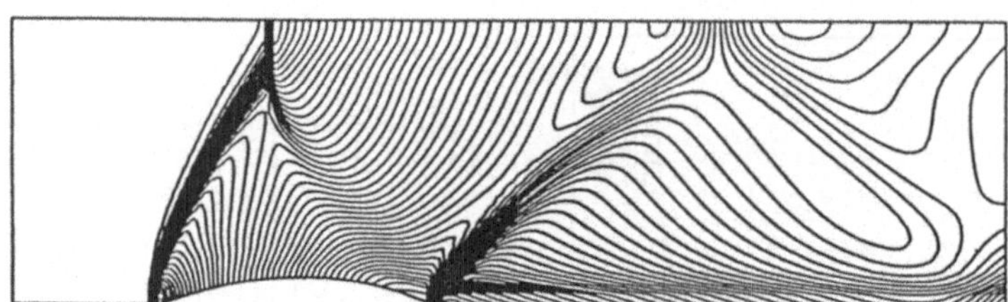

Figure 16: Mach number isolines
(inviscid flow)

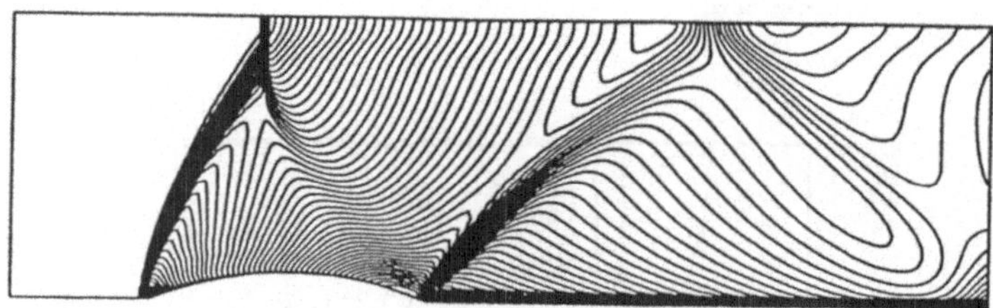

Figure 17: Mach number isolines
(viscous flow)

# AN EFFICIENT NUMERICAL ALGORITHM FOR VELOCITY VORTICITY 3D UNSTEADY NAVIER-STOKES EQUATIONS: APPLICATION TO THE STUDY OF A SEPARATED FLOW AROUND A FINITE RECTANGULAR PLATE


**Jean Fontaine** and **Ta Phuoc Loc**

LIMSI - CNRS

B.P 133

91403 ORSAY cedex - FRANCE



## SUMMARY

The 3D external viscous flow around a plate is predicted using the velocity-vorticity formulation of the Navier-Stokes equations. The problem is spatially discretized on a staggered grid using second order finite differences. We do not consider any hypothesis on the spatial nature of the flows. A second order Euler backward scheme is employed for the time marching. A decoupled algorithm is presented for the computation of the six components. One component of the velocity vector is calculted through the computation of one 3D elliptic equation. The others are explicitly determined. Using an influence matrix technique, the method ensures the free-divergence constraint of the fields and the relation $\omega = \nabla \times \mathbf{u}$. Flows at Re=100 are analysed and compared with the experiments.


## INTRODUCTION

While primitive variables are widely employed when solving three-dimensional flows, we present in this paper a new numerical method based on the velocity-vorticity formulation. In a general review [1], Gatski *et al.* present with many references different methods for solving the Navier-Stokes equations based on the $(V,\omega)$ variables. Despite some drawbacks of the velocity-vorticity approach (six differential equations have to be solved *a priori* ), one should mention the following advantages :

1. For external flows, it seems that the definition of boundary conditions at infinity is simpler because the pressure term is avoided.
2. In the case of vortex dominated flows, one may mention that this formulation is closer to the physical reality, particularly for the flows around an obstacle which is generally the main source of the vorticity.

3. For incompressible flows, the form of the equations are not changed because of the non-inertial effects only enter into the solution of the problem through the discretized initial and boundary conditions [2].

In the proposed algorithm, an original numerical approach is introduced in order to determinate the velocity field by means of the computation of *only one 3D relation* for one component. *The other ones are explicitly calculated.* But, considering an obstacle in the flow, the determination of the velocity can not be uncoupled with the vorticity field because of the coupled boundary condition in the $(V,\omega)$ formulation. Thus, an influence matrix technique is introduced to allow a decoupled computation of the Navier-Stokes equations. *With this algorithm, we are able to ensure the free-divergence constraint of the fields and the definition of the vorticity ($\omega = \nabla \times \mathbf{u}$) at each time step.*

## 2. GOVERNING EQUATIONS

If we consider an incompressible Newtonian fluid, we can write the non-dimensional Navier-Stokes equations using the primitive variables (problem $\mathcal{P}$) :

$$\frac{\partial \mathbf{u}}{\partial t} + (\mathbf{u}.\nabla)\mathbf{u} = -\nabla p + \frac{1}{Re}\nabla^2\mathbf{u} \qquad \text{in } \Omega \qquad (2.1)$$

$$(\mathcal{P}) \qquad \nabla.\mathbf{u} = 0 \qquad \text{in } \Omega \qquad (2.2)$$

$$\mathbf{u} = \mathbf{f} \qquad \text{on } \Gamma \qquad (2.3)$$

$$\mathbf{u}(x,t=0) = \mathbf{u_0}(x) \qquad x \in \Omega \qquad (2.4)$$

where $\Omega$ is a connected open of $\mathbb{R}^3$ and $\Gamma$ its boundary which is supposed regular. The Reynolds number Re is defined as $Re=UL/v$, where U is the velocity at the infinity, L is a representative length of an object and $v$ the kinematic viscosity. $\mathbf{f}$ and $\mathbf{u_0}$ are respectively the boundary and the initial conditions.

Using the relation $\mathbf{u} \times (\nabla \times \mathbf{u}) = \frac{1}{2}\nabla(\mathbf{u}.\mathbf{u}) - (\mathbf{u}.\nabla)\mathbf{u}$ and applying the curl operator, a transport equation for the vorticity is deduced :

$$\frac{\partial \omega}{\partial t} + \nabla \times (\omega \times \mathbf{u}) = \frac{1}{Re}\nabla^2\omega \qquad (2.5)$$

where $\omega$ is the vorticity vector defined by $\omega = \nabla \times \mathbf{u}$.

Now, the pressure term is eliminated. The problem of the definition of good boundary conditions for the pressure is avoided. This equation seems to be closer to the physical reality than the primitive formulation for vortex dominated flows.

The velocity field is defined with :

$$\begin{cases} \nabla . \mathbf{u} = 0 \\ \nabla \times \mathbf{u} = \omega \, . \end{cases} \tag{2.6}$$

Applying the curl operator to the second equation, a three-dimensional elliptic equation is obtained, using the divergence constraint relation for the velocity:

$$\nabla^2 \mathbf{u} = -\nabla \times \omega \, . \tag{2.7}$$

This equation is numerically more attractive than the first equation of (2.6).

A $(\mathcal{P}')$ problem derived from the primitive formulation $(\mathcal{P})$, can be defined using the velocity and the vorticity fields :

$$(\mathcal{P}') \quad \begin{array}{lll} \dfrac{\partial \omega}{\partial t} + \nabla \times (\omega \times \mathbf{u}) = \dfrac{1}{\mathrm{Re}} \nabla^2 \omega & \text{in } \Omega & (2.5) \\[2mm] \nabla^2 \mathbf{u} = -\nabla \times \omega & \text{in } \Omega & (2.7) \\[2mm] \nabla . \omega_0 = 0, \; \omega = \omega_0, \; \mathbf{u} = \mathbf{u}_0 \text{ and } \omega_0 = \nabla \times \mathbf{u}_0 & \text{in } \Omega & (2.8) \\[2mm] \nabla . \omega = 0 \text{ and } \nabla . \mathbf{u} = 0 & \text{on } \Gamma & (2.9) \\[2mm] \mathbf{u} = \mathbf{u}_b & \text{on } \Gamma & (2.10) \\[2mm] \omega . \mathbf{n} = (\nabla \times \mathbf{u}) . \mathbf{n} \, . & \text{on } \Gamma & (2.11) \end{array}$$

The equivalence between the $(\mathcal{P})$ and $(\mathcal{P}')$ problems, and the definition of the initial and boundary conditions required for $(\mathcal{P}')$, have been demonstrated using strong formulation by Daube *et al.* [3]. These authors assume that :

- $\Omega$ is a connected open domain of $\mathbb{R}^3$ and $\Gamma$ the boundary of $\Omega$ is regular.

- $\omega \in H^1(\Omega)^3$ and $\mathbf{u} \in H^2(\Omega)^3$.

It has been chosen to study the flow past a simple airfoil : a thin plate. The description of this obstacle is quiet easy using finite differences on a cartesian grid. In the other hand, this case presents some mathematical difficulties. The differential geometry is complex. The plate corresponds to a cut of the domain with a boundary defined on the two sides. In this kind of geometry (corners, edges), the existence of a solution of the

Navier-Stokes equations written with the primitive variables (v,p) has been proved by Dauge [4]. Solutions can be find in the classical spaces ($v \in H_0^1$ and $p \in L^2$).

But, in the context of strong formulations, the equivalence between both formulations for this problem can not be ensured because $\omega \notin H^1(\Omega)^3$ on the edges and the corners of the plate. A corresponding variational approach *with weaker conditions on $\omega$*, has been formulated by Ruas [5] and can be used in this problem.

# 3. DISCRETIZED SCHEME

## 3.1 Time integration

The vorticity and the velocity field are solved in a decoupled way. An Euler-Backward scheme is employed to describe the transport of the vorticity in order to limit the stability condition. The time discretization for problem $(\mathcal{P}')$ is :

$$\frac{3\omega^{n+1} - 4\omega^n + \omega^{n-1}}{2\Delta t} + \nabla \times (\omega^{n+1} \times \mathbf{u}^*) = \frac{1}{\text{Re}} \nabla^2 \omega^{n+1}. \tag{3.1}$$

The components are computed in a decoupled way. We have, for example, for $\omega_z$ :

$$\omega_z^{n+1} + \frac{2\Delta t}{3}\left[\frac{\partial(u_x^*\omega_z^{n+1})}{\partial x} + \frac{\partial(u_y^*\omega_z^{n+1})}{\partial y}\right] - \frac{2\Delta t}{3\text{Re}}\left[\frac{\partial^2\omega_z^{n+1}}{\partial x^2} + \frac{\partial^2\omega_z^{n+1}}{\partial y^2} + \frac{\partial^2\omega_z^{n+1}}{\partial z^2}\right]$$

$$= \frac{4\omega_z^n - \omega_z^{n-1}}{3} + \frac{2\Delta t}{3}\left[\frac{\partial(u_z^*\omega_x^*)}{\partial x} + \frac{\partial(u_z^*\omega_y^*)}{\partial y}\right]. \tag{3.2}$$

The diffusion and the convection are implicitly computed. The stretching of the two other components of the vorticity, as the velocity $\mathbf{u}$ , are explicitly considered with a second order time extrapolation.

## 3.2 Spatial discretization

The variables are located on a staggered grid. The cell is represented on figure (1). Using this mesh, the definition of the vorticity ($\omega = \nabla \times \mathbf{u}$) can be verified on the edges of the cells and the free-divergence constraint of the fields can be ensured.

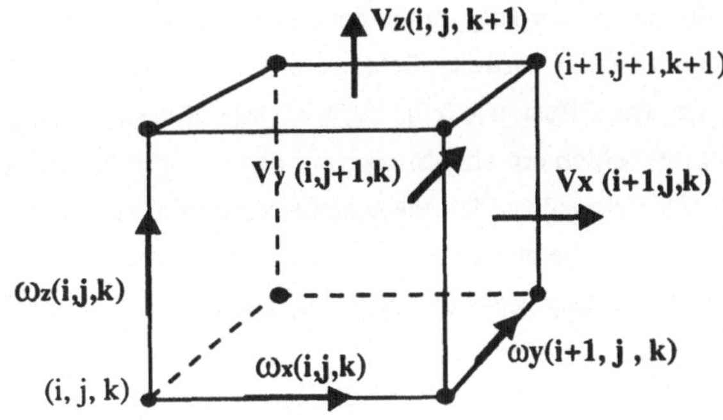

Fig.1 : the staggered cell for the velocity and the vorticity vectors

The discretization of the transport operators is centred with a second order approximation. Therefore, the scheme does not contain numerical viscosity.

### 3.3 Discretization of the boundary conditions

With this mesh, the problem of the boundary conditions at the infinity and on the plate is reduced to define the perpendicular component of the velocity and the two tangential components of the vorticity for each external face of the discretized domain (figure 2).

$$\text{for the velocity :} \qquad \mathbf{u} = \mathbf{u}_\Gamma^{per} \qquad\qquad \text{on } \Gamma \qquad (3.3)$$

$$\text{for the vorticity :} \qquad \omega = \omega_\Gamma^{tan} \qquad\qquad \text{on } \Gamma. \qquad (3.4)$$

In our study, the flow is supposed irrotationnal on the upstream boundaries and the viscous dissipation is supposed negligible on the downstream boundaries.

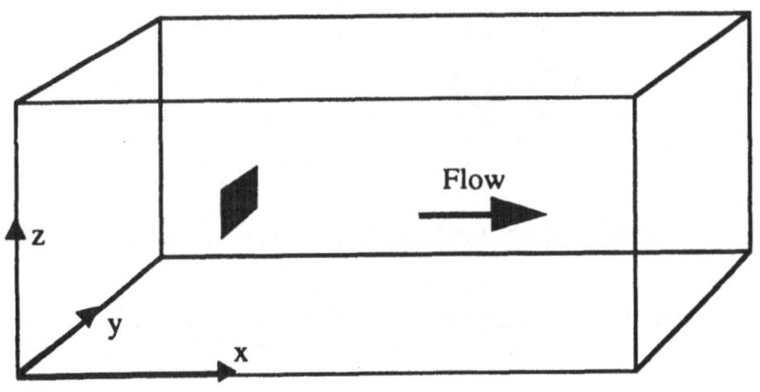


Fig.2 : the flow past the plate, representation of the discretized domain.

83

The discretization of the vorticity on the plate is not easy. In order to ensure the free divergence constraint of the vorticity field, complex definitions of the vorticity are required on the plate. The diffusion and the diffusive part of the stretching of the vorticity are the only processes which are able to take out of the plate the parietal vortcity. To maintain the vorticity field solenoidal, the scheme requires Dirichlet conditions for the two faces and the edges of the plate. For example, the transport per diffusion process of the z-component of the vortcity needs the following boundary conditions figure (3):

Diffusion for $\omega_z$ east :
$$\omega_{Z1}^* = \frac{1}{3h_x}\left(9u_y^*(i_{plate}, j, k) - u_y^*(i_{plate} + 1, j, k)\right)$$

Diffusion for $\omega_z$ west:
$$\omega_{Z2}^* = \frac{1}{3h_x}\left(9u_y^*(i_{plate} - 1, j, k) - u_y^*(i_{plate} - 2, j, k)\right) .$$

* denotes a second order time extrapolation; $h_x$ and $h_y$ are the space steps according to the x- and y-directions, $i_{plate}$ is the abcissa of the plate and $j_{edge}$ refers the position of one the vertical edge according the y-direction.

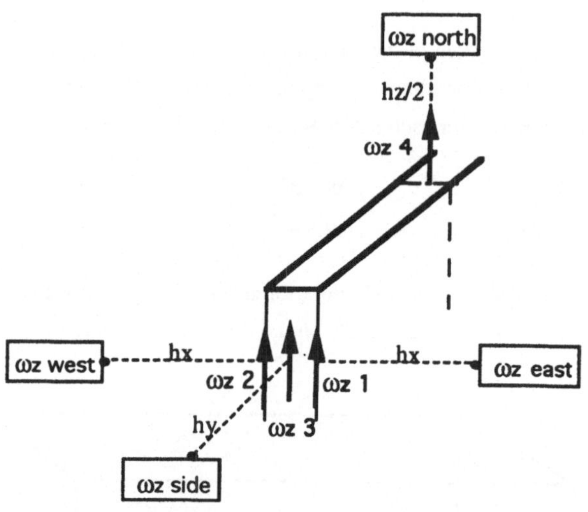


Fig. 3 : discretization of the boundaries on the plate for the vorticity

On the vertical edges, the following discretization has been chosen :

$$\omega_z \text{ side} : \omega_{Z3}^* = \frac{-1}{3h_y}\left(9u_x^*(i_{plate}, j_{edge} - 1, k) - u_x^*(i_{plate}, j_{edge} - 2, k)\right)$$
$$-\left(\frac{u_y^*(i_{plate}, j_{edge}, k) - u_y^*(i_{plate} - 1, j_{edge}, k)}{h_x}\right) .$$

On the horizontal edges, $\omega_z$ is not "directly" defined on a staggered grid. A particular definition of this boundary vorticity is required :

$$\omega_z \text{ north} : \omega_{Z4}^* = \frac{-1}{8h_z} \begin{pmatrix} 3u_y^*(i_{plate}, j, k_{edge} - 1) + 6u_y^*(i_{plate}, j, k_{edge}) \\ -u_y^*(i_{plate}, j, k_{edge} + 1) - 3u_y^*(i_{plate} - 1, j, k_{edge} - 1) \\ -6u_y^*(i_{plate} - 1, j, k_{edge}) + u_y^*(i_{plate} - 1, j, k_{edge} + 1) \end{pmatrix}.$$

## 4. THE NUMERICS

### 4.1 Solution for the transport of the Vorticiy

Two components of the vorticity, $\omega_y$ and $\omega_z$, are rapidly solved with a red-black Gauss-Seidel algorithm for Re=100. To obtain a residual equal to $10^{-6}$, five or six iterations are typically required. The governing operators for the two components can be described with seven coefficients only. The last component $\omega_x$ can be solved in the same way. Thus, the vorticity field is completely deteminated before the computation of the velocity. But, in order to reduce the CPU time, it is also possible with an other hypothesis on the nature of the vorticity at the infinity, to determinate *directly* the last component after the computation of the velocity field using the definition of the vorticity.

The main advantage for a calculation in a decoupled way is the relative low size memory required. The time step used is equal to $10^{-2}$.

### 4.2 Determination of the Velocity field

Up to date, eq.(2.7) is solved by means of the computation of three 3D elliptic equations. Thibeau [6] introduced an original decoupled method in order to solve this field. He proved that the velocity field can be solved through the computation of one 3D equation for one component while the other ones can be explicitly calculated. The 3D elliptic equation (for the first component for example) is :

$$\nabla^2 u_x = -[\nabla \times \omega]_x .\qquad(4.1)$$

This equation is solved with a multigrid method. The second component, $u_z$, can be solved quasi-explicitly using the definition of the vorticity because the first component of the velocity and the vorticity are known :

$$\omega_y = \frac{\partial u_x}{\partial z} - \frac{\partial u_z}{\partial x} \, . \qquad (4.2)$$

It is just required, before this direct method, to determinate $u_z$ in the planes $(x=h_x/2)$ and $(x=x_{max}-h_x/2)$. In these planes, using a staggered grid, the z-component of the velocity is not a boundary condition With the free-divergence constraint of the velocity and the definition of the definition of the vorticity, the following 2D elliptic equation is deduced :

$$\frac{\partial^2 u_z}{\partial y^2} + \frac{\partial^2 u_z}{\partial z^2} = \frac{\partial \omega_x}{\partial y} - \frac{\partial^2 u_x}{\partial z \partial x} \qquad (4.3)$$

and it is resolved with a multigrid method too. The last component, $u_y$, is directly calculated using the free-divergence constraint on the velocity field :

$$\frac{\partial u_x}{\partial x} + \frac{\partial u_y}{\partial y} + \frac{\partial u_z}{\partial z} = 0 \, . \qquad (4.4)$$

In this way, the determination of the velocity needs *only* the computation of a 3D-elliptic equation for one component. But, as the calculation of this field is decoupled with the computation of the vorticity field, it is not possible to verify the relation $\omega = \nabla \times \mathbf{u}$ *a posteriori* because of the presence of the object in the fluid domain. The velocity and the vorticity are coupled through condition (2.11). Thus, an influence matrix technique [7] is introduced to ensure this property. In our case, *the memory size required for its storage is very low. It does not depend on the Reynolds number, the time, the time step or the external conditions at the infinity.*

A residual equal to $10^{-6}$ for the computation of the first component of the velocity vector is typically obtained with 15 V-cycles. It requires 85% of the global CPU time.

The results presented are in the case of the flow past a square or rectangular plate with perpendicular incidence. Datas are reported for a square plate discretized with 17x17 points and a rectangular plate discretized with 33x17 points. 81x65x65 or 81x81x65 points are repectively used for the discretization of the external domain. The cartesian grid is regular. Calculations have been performed on the VP200 of CIRCE.

# 5. RESULTS

As it is oberved in the experiments [8], the recirculation zone past a plate with perpendicular incidence is complex. The nature of the wake depends on the aspect ratio of the plate and the Reynolds number. If the aspect ratio is not equal to one, it is observed a few non-dimensional seconds after an impulse start of the object, a symmetrical splitting of the recirculation zone. Figure 4 shows a representation of the y-component of the vorticity, which is parallel to the spanwise of the rectangular plate (aspect ratio : 2). One can notice on the typical horseshoe pattern the beginning of the vortex wake's splitting. Figure 5 shows a qualitative comparison between the experiments from the LMF of Poitiers (aspect ratio : 5; Re = 1000) and our results (aspect ratio : 2; Re = 100). Finally, the time evolution of the vortex sheet behind a square plate is represented on Figure 6. We can notice the preservation of the basic symmetries and the accuracy of the downstream numerical boundary conditions.

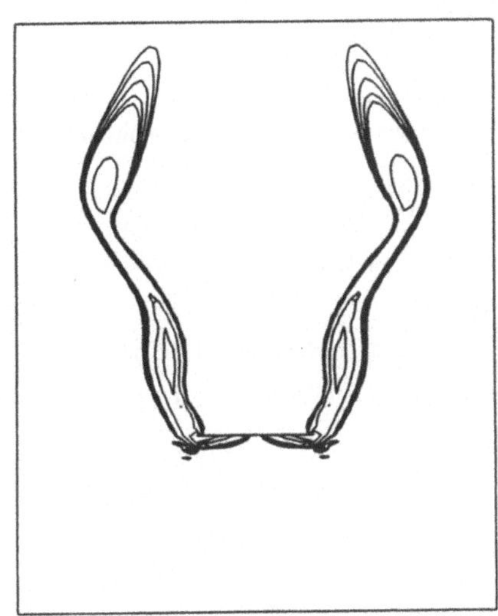

Fig 4 : Splitting of the vorticity wake

(y- component)

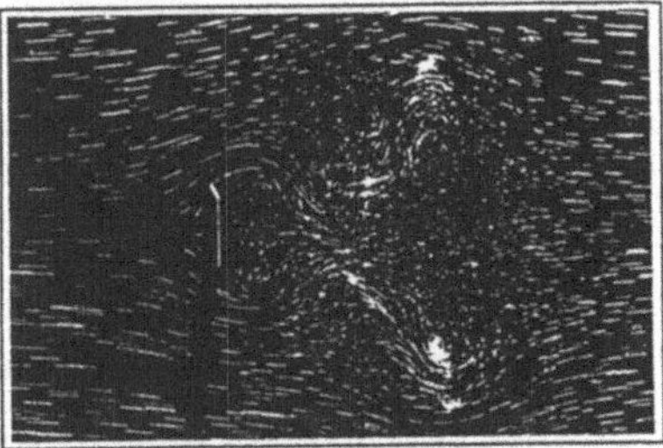

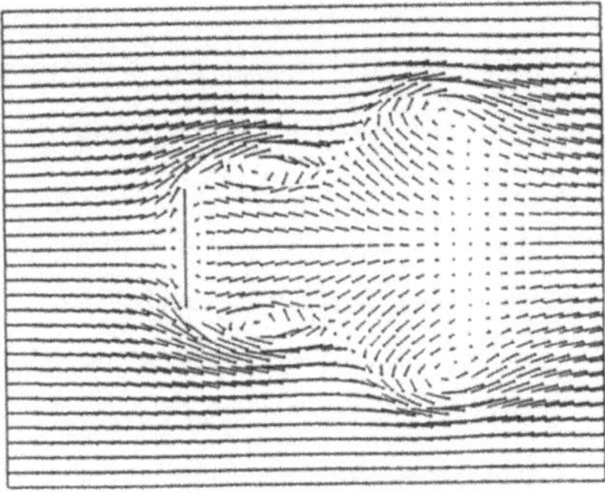

Fig 5 : Qualitative comparison

between the experiments

and the computational results

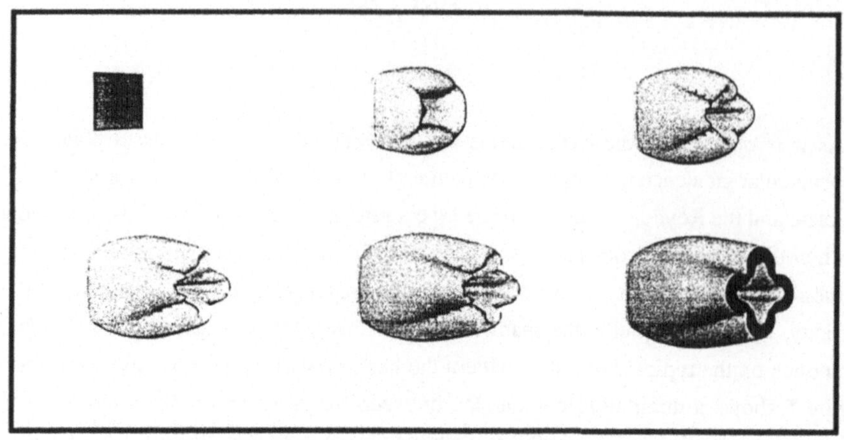

Fig 6 : Evolution of a vortex sheet $\|\omega\|$ past a square plate, flow with perpendicular incidence, Re =100, T = 0 to 5.

## REFERENCES

[1] SPEZIALE C.G., On the advantages of the velocity vorticity formulation of the equations of fluid dynamics. *J. Comput. Phys.* **73**, 476 (1987).

[2] GATSKI T.B., GROSCH C.E. & ROSE M.E. A numerical study of the two dimensional Navier-Stokes equations in vorticity velocity computations variables *J. Comput. Phys.* **48**, N°1 1 (1982).

[3] DAUBE O., GUERMOND J.L. & SELLIER A. Sur la formulation vitesse-tourbillon des équations de Navier-Stokes en écoulement incompressible. *C.R.Acad.Sci .Paris* **313**, série II, 377-382 (1991).

[4] DAUGE M. Stationary Stokes and Navier-Stokes systems on two- or three-dimensional domains with corners. Part I : linearized equations *SIAM J. Math. Anal.* **20**, N°1 74-97 (1989).

[5] RUAS V. Une formulation vitesse-tourbillon es équations de Navier-Stokes incompressibles tridimensionnelles. *C.R.Acad.Sci .Paris* **313**, série I, 639-644 (1991).

[6] THIBEAU S. Ecoulement visqueux dans une turbomachine. *Thèse de Doctorat de l'Uni. Paris 6.*

[7] DAUBE O. Resolution of the 2D Navier-Stokes equations in velocity vorticity form by means of an influence matrix technique. *J. Comput. Phys.* **103**, N°2 402-414 (1992).

[8] POLIDORI G. TEXIER A., COUTENCEAU M., PORCAR R., PRENEL J.P Visulisations d'écoulements tridimensionnels par tomographies multiples simultanées. *In Proc.* $5^{th}$ *Colloque National Visulisations et traitement d'images en mécanique des fluides, Poitiers* .

# INCOMPRESSIBLE FLOWS BASED UPON STABILIZED METHODS


Leopoldo P. Franca

Department of Mathematics, University of Colorado at Denver

P.O.Box 173364, Campus Box 170, Denver, CO 80217-3364, U.S.A.


## SUMMARY


We give a brief overview of our efforts in developing stabilized finite element methods for incompressible flows. In particular, we review some suggestions for stability parameters and comment once more on the curious relationships between some stable Galerkin and stabilized finite element methods.


## INTRODUCTION

Stabilized finite element methods are formed by adding perturbation terms to the standard Galerkin method, that are functions of the Euler-Lagrange equations evaluated elementwise, so that consistency is preserved. The perturbation terms are constructed to enhance stability of the original Galerkin formulation, allowing convergence of a variety of simple finite element interpolations in various applications.

For advective dominated equations it is well accepted to 'modify' the Galerkin method by adding 'artificial diffusion', whereas for elliptic problems with constraints, such as in the Stokes problem, approximation within the Galerkin method can be done employing the mixed method theory [2,4,6]. However, these two behaviors are present within the Navier-Stokes equations (i.e., locally advective dominated flows far from solid boundaries and locally diffusive dominated flows in stagnation regions) and the aforementioned numerical approaches seem conflicting one to the other: one admits addition of a perturbation term, the other suggests pursuing the Galerkin method framework.

*Vis-à-vis* these convictions, it could be considered a *sacrilège* to add a perturbation term to the Galerkin method to approximate the Stokes problem, as proposed in [19,20]. The rather intriguing result is that simple equal-low-order interpolations, as well as a wide range of combinations of simple piecewise polynomials for velocity and pressure are rendered convergent [14,16], an impossibility for the Galerkin method according to the mixed method theory. Nevertheless, this nonstandard methodology is actually related to the Galerkin method with a velocity space enriched by bubble functions [1], provided bubbles are eliminated at the element level and computations are performed in the reduced space without the bubbles [24]. In other words, static condensation of the MINI-element recovers the stabilized method [20] employing equal-order linear elements, a result that can be generalized for high order interpolations

[3]. Thus either the Galerkin method with bubbles and static condensation, or more directly stabilized methods can be used to deal with the Stokes problem.

Equally, it could be considered a *sacrilège* to employ the Galerkin method to approximate advective dominated equations [7,25]. However, this methodology falls within the artificial diffusion framework when we approximate the Galerkin method with the advected variable enriched by bubble functions and then eliminate the bubbles to compute in the reduced space [5]. Again, a stabilized method surfaces after condensation of the bubbles. Therefore either approach can be used to deal with advective dominated equations.

Therefore, our apparent dilemma to approximate the Navier-Stokes equations can be resolved by either employing a Galerkin method enriched with bubble functions (and using static condensation), or by using stabilized finite element methods. Both methods, separately, will deal with the two major numerical intricacies in these equations, namely advective-dominated flows and careful choices of velocity and pressure spaces. Both approaches are related, and therefore either could be used to solve our problem of interest: the Navier-Stokes equations.

Yet, neither one of these methods addresses the following question, phrased differently for the bubble and the artificial diffusion point-of-views, respectively: a) what are the shape of the bubbles?; or, b) how much perturbation should we add?

We know that there are 'virtual bubbles' [3] that can reproduce the asymptotic orders of the perturbation terms in the cases of locally advective dominated flows ($O(h)$) and locally diffusive dominated flows ($O(h^2)$) so that good accuracy properties are preserved. Still, this does not answer our question above.

Herein we review a few recent suggestions for the parameters of stabilized methods (the task seems considerably more complex for the corresponding appropriate bubble shape functions). We give designs including high order interpolations, that are amenable for $p$-adaptivity strategies. We review our suggestions of stability parameters for the model problems: Stokes equations and advective-diffusive scalar model, and combine them for the Navier-Stokes equations.


## MODEL PROBLEMS


## THE STOKES PROBLEM

Let us first consider the Stokes problem defined on an open bounded domain $\Omega \subset \mathbb{R}^N, N = 2, 3$ with a polygonal or polyhedral boundary $\Gamma$ and given by

$$-2\mu\nabla \cdot \boldsymbol{\varepsilon}(\mathbf{u}) + \nabla p = \mathbf{f} \qquad \text{in } \Omega, \tag{1}$$

$$\nabla \cdot \mathbf{u} = 0 \qquad \text{in } \Omega, \tag{2}$$

$$\mathbf{u} = \mathbf{0} \qquad \text{on } \Gamma \tag{3}$$

where $\mathbf{u}$ is the velocity, $p$ is the pressure, $\mu$ is the viscosity, $\boldsymbol{\varepsilon}(\mathbf{u})$ is the symmetric part of the velocity gradient and $\mathbf{f}$ is the body force. The homogeneous boundary condition (3) suffices for our discussion, and can be simply generalized for more complex situations such as non-homogenous Dirichlet combined with Neumann boundary conditions.

We approximate velocity and pressure by the standard finite element spaces given by

$$\mathbf{V}_h = \{\mathbf{v} \in H_0^1(\Omega)^N \,|\, \mathbf{v}_{|K} \in R_k(K)^N, \ K \in \mathcal{C}_h\}, \tag{4}$$

$$P_h = \{p \in C^0(\Omega) \cap L_0^2(\Omega) \,|\, p_{|K} \in R_l(K), \ K \in \mathcal{C}_h\}, \tag{5}$$

where $H_0^1(\Omega)$ is the Sobolev space of functions with square-integrable value and derivatives in $\Omega$ with zero value on the boundary $\Gamma$, the $H^1$-norm will be denoted by $\|\cdot\|_1$, $L_0^2(\Omega)$ is the space of functions with zero average and square-integrable values in $\Omega$, the $L^2$-norm will be denoted by $\|\cdot\|_0$, $C^0(\Omega)$ is the set of continuous functions in $\Omega$, $\mathcal{C}_h$ is a partition of $\overline{\Omega}$ into elements $K$ consisting of triangles (tetrahedrons in $I\!\!R^3$) or convex quadrilaterals (hexahedrons) performed in the usual way (i.e., no overlapping is allowed between any two elements of the partition; the union of all domains $K$ reproduces $\overline{\Omega}$, etc.). Quasiuniformity is *not* assumed. We also employ the following notation:

$$R_m(K) = \begin{cases} P_m(K) & \text{if } K \text{ is a triangle or tetrahedron,} \\ Q_m(K) & \text{if } K \text{ is a quadrilateral or hexahedron} \end{cases}$$

where for each integer $m \geq 0$, $P_m$ and $Q_m$ have the usual meaning.

A stabilized finite element method for this model can be written as: Find $\mathbf{u}_h \in \mathbf{V}_h$ and $p_h \in P_h$ such that

$$B(\mathbf{u}_h, p_h; \mathbf{v}, q) = F(\mathbf{v}, q) \qquad (\mathbf{v}, q) \in \mathbf{V}_h \times P_h \tag{6}$$

with

$$\begin{aligned} B(\mathbf{u}, p; \mathbf{v}, q) = (2\mu\,\boldsymbol{\varepsilon}(\mathbf{u}), \boldsymbol{\varepsilon}(\mathbf{v})) - (\nabla \cdot \mathbf{v}, p) + (\nabla \cdot \mathbf{u}, q) \\ + \sum_{K \in \mathcal{C}_h} \left(\nabla p - 2\mu \nabla \cdot \boldsymbol{\varepsilon}(\mathbf{u}), \tau(\nabla q - 2\mu \nabla \cdot \boldsymbol{\varepsilon}(\mathbf{v}))\right)_K \end{aligned} \tag{7}$$

and

$$F(\mathbf{v}, q) = (\mathbf{f}, \mathbf{v}) + \sum_{K \in \mathcal{C}_h} \left(\mathbf{f}, \tau(\nabla q - 2\mu \nabla \cdot \boldsymbol{\varepsilon}(\mathbf{v}))\right)_K \tag{8}$$

where $(\cdot, \cdot)$ denotes the $L^2(\Omega)$-inner product (between scalar, vector or second order tensor valued functions - context should be clear which is being used), and $(\cdot, \cdot)_K$ denotes the $L^2(K)$-inner product.

Following [20], the stability parameter can be written as

$$\tau = \frac{\alpha h_K^2}{2\mu} \tag{9}$$

where $h_K$ is the element diameter and $\alpha$ is a non-dimensional stability constant recommended by numerical experience to be $O(1)$ for linear velocities and equal to $C_k/2$

for high order interpolations. $C_k$, in turn, is the inverse estimate constant defined to be the largest constant such that

$$C_k \sum_{K \in C_h} h_K^2 \|\nabla \cdot \varepsilon(\mathbf{v})\|_{0,K}^2 \le \|\varepsilon(\mathbf{v})\|_0^2 \,, \quad \forall \mathbf{v} \in \mathbf{V}_h \,. \tag{10}$$

Such estimate holds for finite element polynomials defined on shape regular elements [9] (with no restrictions on size of elements) and computations of such constants can be found in [18].

Alternatively, following [15] one could define $\tau$ free from element diameters $h_K$ and inverse estimate constants $C_k$ by

$$\tau = \frac{1}{4\lambda_K \mu} \tag{11}$$

where

$$\lambda_K = \max_{0 \ne \mathbf{v} \in (R_k(K)/R)^N} \frac{\|\nabla \cdot \varepsilon(\mathbf{v})\|_{0,K}^2}{\|\varepsilon(\mathbf{v})\|_{0,K}^2}, \quad K \in C_h \,. \tag{12}$$

*Remarks*

1. The stabilized method (6)-(8) was proposed in [10] as a modification of the method introduced in [13,19,20]. From the stability analysis of this method any positive parameter $\tau$ renders this method stable [10,16]. However, from numerical experience [12], we know that better accuracy is obtained for a choice near the design given in (11)-(12).

2. The design given by (11)-(12) excludes linear velocity interpolations, in which case one should employ (9) with $\alpha = O(1)$. However, (11)-(12) can be employed for pressures that are linearly interpolated or higher, i.e., $l \ge 1$, provided that velocities are interpolated with piecewise polynomials of degree $k \ge 2$.

3. The parameter $\lambda_K$ is computed as the largest eigenvalue of the following generalized eigenvalue problem defined for each $K$: Find $\mathbf{w}_h \in (R_k(K)/I\!R)^N$ and $\lambda_K$ such that

$$(\nabla \cdot \varepsilon(\mathbf{w}_h), \nabla \cdot \varepsilon(\mathbf{v}))_K - \lambda_K(\nabla \mathbf{w}_h, \nabla \mathbf{v})_K = 0 \quad \forall \mathbf{v} \in (R_k(K)/I\!R)^N. \tag{13}$$

In practice to simulate the quotient space $(R_k(K)/I\!R)^N$, we fix the degrees of freedom of a node to zero and solve for the remaining ones. The largest eigenvalue is computed using the power method. Once we have the computed value from the first element, this is used as the guess for the computation on the next and so on, until the whole mesh is covered. In the case of a non-linear problem, such as the Navier-Stokes equations, this computation is done once and for all, after the coordinates and type of interpolations are input, before entering any non-linear algorithm (Newton-like) to solve the nonlinear set of algebraic equations. Determining the largest eigenvalue on each element is *not* a costly procedure, as one may think at first. The generalized eigenvalue problem is to be solved only for the largest eigenvalue, and each problem is formulated at the element level, therefore involving relatively small matrices.

4. For this model the relevant inverse estimate given by (10) provides us with the link to the definition of $\lambda_K$, (12), through (13), i.e.:

$$\lambda_K^{-1} = C_k h_K^2 \qquad K \in \mathcal{C}_h. \tag{14}$$

Therefore by using $\lambda_K$, we automatically carry the inverse estimate constant and the element diameter to our design of the stability parameter, and the design (11)-(12) is free from having to arbitrate parameters.


## THE ADVECTIVE-DIFFUSIVE PROBLEM

Let us now consider the advective-diffusive problem of finding $u$ such that

$$\mathbf{a} \cdot \nabla u - \kappa \Delta u = f \qquad \text{in } \Omega, \tag{15}$$

$$u = 0 \qquad \text{on } \Gamma, \tag{16}$$

where $\mathbf{a}(\mathbf{x})$ is the given velocity field with $\nabla \cdot \mathbf{a} = 0$, $\kappa$ is the diffusivity and $f(\mathbf{x})$ is a source function, and all the notations and assumptions from the previous section for $\Omega, \Gamma, \mathcal{C}_h$, etc., remain in force throughout.

The scalar field $u$ is approximated in the following standard finite element space:

$$V_h = \{v \in H_0^1(\Omega) \,|\, v_{|K} \in R_k(K), \ K \in \mathcal{C}_h\}, \tag{17}$$

The stabilized finite element method we wish to consider is the Galerkin-least-squares method studied for this model in [21]: find $u_h \in V_h$ such that

$$B(u_h, v) = F(v) \qquad v \in V_h \tag{18}$$

with

$$B(u, v) = (\mathbf{a} \cdot \nabla u, v) + (\kappa \nabla u, \nabla v) + \sum_{K \in \mathcal{C}_h} (\mathbf{a} \cdot \nabla u - \kappa \Delta u, \tau(\mathbf{a} \cdot \nabla v - \kappa \Delta v))_K \tag{19}$$

and

$$F(v) = (f, v) + \sum_{K \in \mathcal{C}_h} (f, \tau(\mathbf{a} \cdot \nabla v - \kappa \Delta v))_K. \tag{20}$$

Similarly to the last section we will consider two alternatives for the design of $\tau$. The first proposed in [12] is given by:

$$\tau(\mathbf{x}, \mathrm{Pe}_K(\mathbf{x})) = \frac{h_K}{2|\mathbf{a}(\mathbf{x})|_p} \xi(\mathrm{Pe}_K(\mathbf{x})) \tag{21}$$

$$\mathrm{Pe}_K(\mathbf{x}) = \frac{m_k |\mathbf{a}(\mathbf{x})|_p h_K}{2\kappa(\mathbf{x})} \tag{22}$$

$$\xi(\mathrm{Pe}_K(\mathbf{x})) = \begin{cases} \mathrm{Pe}_K(\mathbf{x}) & , 0 \le \mathrm{Pe}_K(\mathbf{x}) < 1 \\ 1 & , \mathrm{Pe}_K(\mathbf{x}) \ge 1 . \end{cases} \tag{23}$$

The convergence analysis of GLS with the design of $\tau$ in (27)-(31) can be done similarly as in [12]. The main difference here is that inverse estimates are not needed. Instead, by definition of $\lambda_K$ from eq.(30) we have

$$\lambda_K \|\nabla v\|_{0,K}^2 \geq \|\Delta v\|_{0,K}^2, \qquad K \in \mathcal{C}_h \qquad \forall v \in V_h \tag{34}$$

and this is used to establish stability as follows. First note that from (27)-(29), for $\mathrm{Pe}_K \geq 1$:

$$\tau = \frac{2}{\sqrt{\lambda_K}\,|\mathbf{a}(\mathbf{x})|_p}\frac{|\mathbf{a}(\mathbf{x})|_p}{4\sqrt{\lambda_K}\,\kappa}\frac{1}{\mathrm{Pe}_K} \leq \frac{1}{2\lambda_K\kappa(\mathbf{x})} \ . \tag{35}$$

Since by definition the equality sign of (35) holds for $\mathrm{Pe}_K < 1$ therefore it follows that the bound (35) is valid for all $\mathrm{Pe}_K \geq 0$.

Now, by definition (eq.(19)), and using that $\nabla \cdot \mathbf{a} = 0$ and that $\kappa$ is constant:

$$\begin{aligned}
B(v,v) =&(\mathbf{a} \cdot \nabla v, v) + \kappa\|\nabla v\|_0^2 + \|\tau^{1/2}\,\mathbf{a} \cdot \nabla v\|_0^2 - 2\sum_{K \in \mathcal{C}_h}(\mathbf{a} \cdot \nabla v, \tau\,\kappa\Delta v)_K \\
&+ \sum_{K \in \mathcal{C}_h}\|\tau^{1/2}\,\kappa\Delta v\|_{0,K}^2 \\
\geq &\kappa\|\nabla v\|_0^2 + \frac{1}{2}\|\tau^{1/2}\,\mathbf{a} \cdot \nabla v\|_0^2 - \sum_{K \in \mathcal{C}_h}\|\tau^{1/2}\,\kappa\Delta v\|_{0,K}^2 \ .
\end{aligned} \tag{36}$$

Note that from (34)-(35) the last term in (36) can be estimated as follows

$$\begin{aligned}
\sum_{K \in \mathcal{C}_h}\|\tau^{1/2}\,\kappa\Delta v\|_{0,K}^2 &= \sum_{K \in \mathcal{C}_h}\|(\tau\,\kappa)^{1/2}\kappa^{1/2}\Delta v\|_{0,K}^2 \\
&\leq \sum_{K \in \mathcal{C}_h}\frac{\kappa}{2\lambda_K}\|\Delta v\|_{0,K}^2 \qquad \text{(by (35))} \\
&\leq \frac{\kappa}{2}\|\nabla v\|_0^2 \ . \qquad \text{(by (34))}
\end{aligned}$$

Therefore combining this estimate with (36) implies

$$B(v,v) \geq \frac{1}{2}(\kappa\|\nabla v\|_0^2 + \|\tau^{1/2}\,\mathbf{a} \cdot \nabla v\|_0^2) \tag{37}$$

which is the (numerical) stability result for this method. Combining with the intrinsic consistency of stabilized methods, then the following convergence of $u_h$ solution of (18)-(20) to $u$ solution of (15)-(16) follows in the the same norm of the stability result above (37):

$$\begin{aligned}
\kappa\|\nabla(u_h - u)\|_0^2 &+ \|\tau^{1/2}\,\mathbf{a} \cdot \nabla(u_h - u)\|_0^2 \\
&\leq C\sum_{K \in \mathcal{C}_h}h_K^{2k}|u|_{k+1,K}^2\Big(\mathrm{H}(\mathrm{Pe}_K - 1)h_K\sup_{\mathbf{x} \in K}|\mathbf{a}|_p + \mathrm{H}(1 - \mathrm{Pe}_K)\kappa\Big)
\end{aligned} \tag{38}$$

$$|\mathbf{a}(\mathbf{x})|_p = \begin{cases} \left(\sum_{i=1}^{N}|a_i(\mathbf{x})|^p\right)^{1/p} & , \quad 1 \le p < \infty \\ \max_{i=1,N}|a_i(\mathbf{x})| & , \quad p = \infty \end{cases} \tag{24}$$

$$m_k = \min\left\{\frac{1}{3},\, 2\widetilde{C}_k\right\} \tag{25}$$

$$\widetilde{C}_k \sum_{K \in \mathcal{C}_h} h_K^2 \|\Delta v\|_{0,K}^2 \le \|\nabla v\|_0^2 \qquad v \in V_h . \tag{26}$$

The alternative design for $k \ge 2$ proposed in [15] is given by:

$$\tau = \frac{2}{\sqrt{\lambda_K}|\mathbf{a}(\mathbf{x})|_p}\xi(\mathrm{Pe}_K(\mathbf{x})) \tag{27}$$

$$\mathrm{Pe}_K(\mathbf{x}) = \frac{|\mathbf{a}(\mathbf{x})|_p}{4\sqrt{\lambda_K}\,\kappa(\mathbf{x})} \tag{28}$$

$$\xi(\mathrm{Pe}_K(\mathbf{x})) = \begin{cases} \mathrm{Pe}_K(\mathbf{x}) & , 0 \le \mathrm{Pe}_K(\mathbf{x}) < 1 \\ 1 & , \mathrm{Pe}_K(\mathbf{x}) \ge 1 \end{cases} \tag{29}$$

$$\lambda_K = \max_{0 \ne v \in R_k(K)/\mathbb{R}} \frac{\|\Delta v\|_{0,K}^2}{\|\nabla v\|_{0,K}^2} \qquad , K \in \mathcal{C}_h \tag{30}$$

$$|\mathbf{a}(\mathbf{x})|_p = \begin{cases} \left(\sum_{i=1}^{N}|a_i(\mathbf{x})|^p\right)^{1/p} & , \quad 1 \le p < \infty \\ \max_{i=1,N}|a_i(\mathbf{x})| & , \quad p = \infty . \end{cases} \tag{31}$$

*Remarks*

1. The design (27)-(31) excludes the linear interpolation case, $k = 1$. It is only valid for high order interpolations. In the case of linear interpolations, we recommend (21)-(25) with $m_1 = 1/3$.

2. The parameter $\lambda_K$ is calculated by computing the largest eigenvalue of the following generalized eigenvalue problem defined for each $K$: Find $w_h \in R_k(K)/\mathbb{R}$ and $\lambda$ such that

$$(\Delta w_h, \Delta v) - \lambda(\nabla w_h, \nabla v) = 0 \qquad \forall v \in R_k(K)/\mathbb{R} . \tag{32}$$

As before, the largest eigenvalue is computed using the power method.

3. By definition of $\lambda_K$ in eq.(30), it follows from (26) and (32) that

$$\lambda_K^{-1} = \widetilde{C}_k h_K^2 . \tag{33}$$

This is the link between the $\lambda_K$ parameter with the inverse estimate constant $\widetilde{C}_k$ and the mesh parameter $h_K$, similar to what we had for the Stokes problem.

where $H(\cdot)$ is the Heaviside function given by

$$H(x - y) = \begin{cases} 0, & x < y; \\ 1, & x > y. \end{cases} \tag{39}$$

To establish (38), besides (37) we need an interpolation estimate for this particular design of $\tau$ that can be obtained as Lemma 3.2 of [12] using the relation between $\lambda_K$ with $C_k$ and $h_K$ given by equation (33). This is the only instance that we need an inverse estimate in the analysis: to obtain the rates of convergence. Otherwise, the entire analysis goes through without ever needing (26) or (33).

## THE INCOMPRESSIBLE NAVIER-STOKES EQUATIONS

Let us consider the steady state incompressible Navier-Stokes given by:

$$(\nabla \mathbf{u})\mathbf{u} - 2\nu \nabla \cdot \boldsymbol{\varepsilon}(\mathbf{u}) + \nabla p = \mathbf{f} \qquad \text{in } \Omega, \tag{40}$$

$$\nabla \cdot \mathbf{u} = 0 \qquad \text{in } \Omega, \tag{41}$$

$$\mathbf{u} = 0 \qquad \text{on } \Gamma, \tag{42}$$

where $\mathbf{u}$ is the velocity, $p$ is the pressure, $\nu$ is the viscosity, $\boldsymbol{\varepsilon}(\mathbf{u})$ is the symmetric part of the velocity gradient and $\mathbf{f}$ is the body force.

Let us consider a partition $C_h$ of the domain $\Omega$ and the standard finite element spaces for velocity and pressure (4)-(5) as dicussed in the Stokes problem.

The stabilized finite element method as suggested in [11] can be written as: Find $\mathbf{u}_h \in \mathbf{V}_h$ and $p_h \in P_h$ such that

$$B(\mathbf{u}_h, p_h; \mathbf{v}, q) = F(\mathbf{v}, q), \qquad \forall (\mathbf{v}, q) \in \mathbf{V}_h \times P_h, \tag{43}$$

with

$$\begin{aligned} B(\mathbf{u}, p; \mathbf{v}, q) &= ((\nabla \mathbf{u})\mathbf{u}, \mathbf{v}) + (2\nu\boldsymbol{\varepsilon}(\mathbf{u}), \boldsymbol{\varepsilon}(\mathbf{v})) - (\nabla \cdot \mathbf{v}, p) + (\nabla \cdot \mathbf{u}, q) \\ &+ (\nabla \cdot \mathbf{u}, \delta \nabla \cdot \mathbf{v}) \\ &+ \sum_{K \in C_h} \left((\nabla \mathbf{u})\mathbf{u} + \nabla p - 2\nu \nabla \cdot \boldsymbol{\varepsilon}(\mathbf{u}), \tau((\nabla \mathbf{v})\mathbf{u} + \nabla q - 2\nu \nabla \cdot \boldsymbol{\varepsilon}(\mathbf{v}))\right)_K \end{aligned} \tag{44}$$

and

$$F(\mathbf{v}, q) = (\mathbf{f}, \mathbf{v}) + \sum_{K \in C_h} \left(\mathbf{f}, \tau((\nabla \mathbf{v})\mathbf{u} + \nabla q - 2\nu \nabla \cdot \boldsymbol{\varepsilon}(\mathbf{v}))\right)_K. \tag{45}$$

Similarly to the previous sections, we consider two alternative designs of the stability parameters $\tau$ and $\delta$. The first proposed in [11] is given by:

$$\delta = \lambda |\mathbf{u}(\mathbf{x})|_p h_K \xi(\mathrm{Re}_K(\mathbf{x})) \tag{46}$$

$$\tau(\mathbf{x}, \mathrm{Re}_K(\mathbf{x})) = \frac{h_K}{2|\mathbf{u}(\mathbf{x})|_p} \xi(\mathrm{Re}_K(\mathbf{x})) \tag{47}$$

$$\text{Re}_K(\mathbf{x}) = \frac{m_k |\mathbf{u}(\mathbf{x})|_p h_K}{4\nu(\mathbf{x})} \tag{48}$$

$$\xi(\text{Re}_K(\mathbf{x})) = \begin{cases} \text{Re}_K(\mathbf{x}) & , \; 0 \le \text{Re}_K(\mathbf{x}) < 1 \\ 1 & , \; \text{Re}_K(\mathbf{x}) \ge 1 \end{cases} \tag{49}$$

$$|\mathbf{u}(\mathbf{x})|_p = \begin{cases} \left( \sum_{i=1}^{N} |u_i(\mathbf{x})|^p \right)^{1/p} & , \quad 1 \le p < \infty \\ \max_{i=1,N} |u_i(\mathbf{x})| & , \quad p = \infty \end{cases} \tag{50}$$

$$m_k = \min \left\{ \frac{1}{3}, \; 2C_k \right\} \tag{51}$$

$$C_k \sum_{K \in \mathcal{C}_h} h_K^2 \| \nabla \cdot \boldsymbol{\varepsilon}(\mathbf{v}) \|_{0,K}^2 \le \| \boldsymbol{\varepsilon}(\mathbf{v}) \|_0^2 \qquad \mathbf{v} \in \mathbf{V}_h \, . \tag{52}$$

The alternative design for $k \ge 2$ proposed in [15] is given by:

$$\delta = \frac{|\mathbf{u}(\mathbf{x})|_p}{\sqrt{\lambda_K}} \xi(\text{Re}_K(\mathbf{x})) \tag{53}$$

$$\tau = \frac{\xi(\text{Re}_K(\mathbf{x}))}{\sqrt{\lambda_K} |\mathbf{u}(\mathbf{x})|_p} \tag{54}$$

$$\text{Re}_K(\mathbf{x}) = \frac{|\mathbf{u}(\mathbf{x})|_p}{4\sqrt{\lambda_K}\nu(\mathbf{x})} \tag{55}$$

$$\xi(\text{Re}_K(\mathbf{x})) = \begin{cases} \text{Re}_K(\mathbf{x}) & , \; 0 \le \text{Re}_K(\mathbf{x}) < 1 \\ 1 & , \; \text{Re}_K(\mathbf{x}) \ge 1 \end{cases} \tag{56}$$

$$\lambda_K = \max_{0 \ne \mathbf{v} \in (R_k(K)/R)^N} \frac{\| \nabla \cdot \boldsymbol{\varepsilon}(\mathbf{v}) \|_{0,K}^2}{\| \boldsymbol{\varepsilon}(\mathbf{v}) \|_{0,K}^2} \qquad , K \in \mathcal{C}_h \tag{57}$$

$$|\mathbf{u}(\mathbf{x})|_p = \begin{cases} \left( \sum_{i=1}^{N} |u_i(\mathbf{x})|^p \right)^{1/p} & , \quad 1 \le p < \infty \\ \max_{i=1,N} |u_i(\mathbf{x})| & , \quad p = \infty \, . \end{cases} \tag{58}$$

*Remarks*

1. The stabilized formulation given was introduced in [11], where a convergence analysis is given for a linearized form of the incompressible Navier-Stokes equations and a few numerical simulations are presented. A related form is also proposed in [22]. For low order interpolations, see [8,17,26]. Herein we wish to emphasize the definitions of the stability parameters $\tau$ and $\delta$. In particular, using larger values of $\tau$ than the ones presented above, may produce spurious oscillations when we employ high order interpolations (see [11] for a detailed discussion).

2. The design given by (53)-(58) excludes linear velocity interpolations. However, pressures may be linearly interpolated or higher, i.e., $l \ge 1$.

3. Similarly to the previous sections, the parameter $\lambda_K$ is computed as the largest eigenvalue of the generalized eigenvalue problem given in equation (13). Again, by definition of $\lambda_K$ in (57) combined with (13) and (52) yields

$$\lambda_K^{-1} = C_k h_K^2 \ . \tag{59}$$

The same link of $\lambda_K$ with $C_k$ and $h_K$ holds as in the Stokes model.

4. Convergence analysis taking into account the design in (53)-(58) of the stability parameters can be performed for a linearized model similarly as in [11]. As pointed out before, inverse estimates are no longer needed to establish stability and carry out the entire analysis, up to the point where interpolation estimates results are needed to characterize the rates of convergence. The analysis considerations for this case are similar to what is described in the advective-diffusive model section, which combined with the analysis presented in [11] yields a similar convergence result.

5. The design of the stability parameter $\tau$ for the Stokes flow given by (11)-(12) can be obtained by taking the limit as $\mathrm{Re}_K \to 0$ in (53)-(58). It can be viewed as the diffusive limit of the general situation of advective-diffusive incompressible flows governed by the Navier-Stokes equations.

6. Discretization of (43) is carried out by expanding the trial functions $\mathbf{u}_h, p_h$ and the test functions in terms of their finite element basis or shape functions. This leads to a set of nonlinear algebraic equations, parametrized, in particular, by the Reynolds number. To solve this set of equations for a fixed Reynolds number, we employ the standard Newton-Raphson method, combined with a Quasi-Newton strategy. The Quasi-Newton part consists of "freezing" the matrices if the norm of the residuals of the algebraic equations are monotonically decreasing. Otherwise, the matrices are updated with the latest increment, and the strategy is restarted. On the top of this algorithm we also make a continuation on the Reynolds number, which roughly consists in: solve the problem for a low Reynolds number, and use this solution as the initial guess for a larger Reynolds number problem, until the Reynolds number we wish to calculate is reached.

7. The matrix system for each Newton iteration is solved by a direct method (Gaussian elimination) for small two-dimension benchmark flows, and with GMRES for large two-dimensional and three-dimensional flows. We refer to [11,23] for numerical experiments employing this methodology.


## ACKNOWLEDGMENTS

The word *sacrilège* was borrowed from O. Pironneau's talk in Heidelberg. The author is grateful to the kind invitation of the organizers of this conference and acknowledges the partial support by the National Science Foundation under Grant ASC-9217394.

REFERENCES

[1] ARNOLD, D.N., BREZZI, F., FORTIN, M.: "A stable finite element for the Stokes equations", Calcolo, **23/4** (1984) pp. 337-344.

[2] BABUSKA, I.: "The finite element method with Lagrangian multipliers", Numer. Math., **20** (1973) pp. 179-192.

[3] BAIOCCHI, C., BREZZI, F., FRANCA, L.P.: "Virtual bubbles and the Galerkin-least-squares method", Comp. Meth. Appl. Mech. Engrg., **105** (1993) pp. 125-141.

[4] BREZZI, F.: "On the existence, uniqueness and approximation of saddle-point problems arising from Lagrange multipliers", RAIRO Ser. Rouge, **8** (1974) pp. 129-151.

[5] BREZZI, F., BRISTEAU, M.O., FRANCA, L.P., MALLET, M., ROGE, G.: "A relationship between stabilized finite element methods and the Galerkin method with bubble functions", Comp. Meth. Appl. Mech. Engrg., **96** (1992) pp. 117-129.

[6] BREZZI, F., FORTIN, M.: "Mixed and hybrid finite element methods", Springer Series in Computational Mathematics, Vol. 15, Springer-Verlag, Berlin, New-York 1991.

[7] BRISTEAU, M.O., MALLET, M., PERIAUX, J., ROGE, G.: "Development of finite element methods for compressible Navier-Stokes flow simulations in aerospace design", AIAA paper 90-0403, 28th Aerospace Meeting, Reno, Nevada (1990).

[8] BROOKS, A.N., HUGHES, T.J.R.: "Streamline upwind/Petrov-Galerkin formulations for convective dominated flows with particular emphasis on the incompressible Navier-Stokes equations ", Comp. Meth. Appl. Mech. Engrg., **32** (1982) pp. 199-259.

[9] CIARLET, P.G.: "Basic error estimates for elliptic problems", Handbook of Numerical Analysis, Vol. II, Finite Element Methods (Part 1), P.G. Ciarlet and J.L. Lions eds., pp. 17-351, Elsevier Science Publishers B.V., North Holland, 1991.

[10] DOUGLAS, J., WANG, J.: "An absolutely stabilized finite element method for the Stokes problem", Math. Comp. **52** (1989) pp. 495-508.

[11] FRANCA, L.P., FREY, S.L.: "Stabilized finite element methods: II. The incompressible Navier-Stokes equations", Comp. Meth. Appl. Mech. Engrg., **99** (1992) pp. 209-233.

[12] FRANCA, L.P., FREY, S.L., HUGHES, T.J.R.: "Stabilized finite element methods: I. Application to the advective-diffusive model", Comp. Meth. Appl. Mech. Engrg., **95** (1992) pp. 253-276.

[13] FRANCA, L.P., HUGHES, T.J.R.: "Two classes of mixed finite element methods", Comp. Meth. Appl. Mech. Engrg., **69** (1988) pp. 89-129.

[14] FRANCA, L.P., HUGHES, T.J.R., STENBERG, R.: "Stabilized finite element methods for the Stokes problem", Chap. 4 of Incompressible Computational Fluid

Dynamics-Trends and Advances, M.D. Gunzburger and R.A. Nicolaides eds., Cambridge University Press, 1993 pp. 87-108.

[15] FRANCA, L.P., MADUREIRA, A.L.: "Element diameter free stability parameters for stabilized methods applied to fluids", Comp. Meth. Appl. Mech. Engrg., **105** (1993) pp. 395-403.

[16] FRANCA, L.P., STENBERG, R.: "Error analysis of some Galerkin least squares methods for the elasticity equations", SIAM J. Numer. Anal., **28** (1991) pp. 1680-1697.

[17] HANSBO, P., SZEPESSY, A.: "A velocity-pressure streamline diffusion finite element methodfor the incompressible Navier-Stokes equations ", Comp. Meth. Appl. Mech. Engrg., **84** (1990) pp. 175-192.

[18] HARARI, I., HUGHES, T.J.R.: "What are $C$ and $h$?: Inequalities for the analysis and design of finite element methods", Comp. Meth. Appl. Mech. Engrg., **97** (1992) pp. 157-192.

[19] HUGHES, T.J.R., FRANCA, L.P.: "A new finite element formulation for computational fluid dynamics: VII. The Stokes problem with various well-posed boundary conditions: symmetric formulations that converge for all velocity/pressure spaces", Comp. Meth. Appl. Mech. Engrg., **65** (1987) pp. 85-96.

[20] HUGHES, T.J.R., FRANCA, L.P., BALESTRA, M.: "A new finite element formulation for computational fluid dynamics: V. Circumventing the Babuška-Brezzi condition: A stable Petrov-Galerkin formulation of the Stokes problem accommodating equal-order interpolations", Comp. Meth. Appl. Mech. Engrg., **59** (1986) pp. 85-99.

[21] HUGHES, T.J.R., FRANCA, L.P., HULBERT, G.M.: "A new finite element formulation for computational fluid dynamics: VIII. The Galerkin-least-squares method for advective-diffusive equations", Comp. Meth. Appl. Mech. Engrg., **73** (1989) pp. 173-189.

[22] LUBE, G., AUGE, A.: "Regularized mixed finite element approximations of incompressible flow problems. II. Navier-Stokes flow", Otto von Guericke University, Preprint 15/91, June 1991.

[23] MADUREIRA, A.L.: "On the development and analysis of stabilized finite element methods for fluids", M.Sc. Thesis, UFRJ, Brazil, 1992.

[24] PIERRE, R.: "Simple $C^0$ approximations for the computation of incompressible flows", Comp. Meth. Appl. Mech. Engrg., **68** (1988) pp. 205-227.

[25] PIRONNEAU, O., RAPPAZ, J.: "Numerical analysis for compressible viscous isentropic stationary flows", Publication du Laboratoire d'Analyse Numerique de Paris 6, 1988.

[26] TEZDUYAR, T.E., MITTAL, S., RAY, S.E., SHIH, R.: "Incompressible flow computations using stabilized bilinear and linear equal-order-interpolation velocity-pressure elements", Comp. Meth. Appl. Mech. Engrg., **95** (1992) pp. 221-242.

# ADAPTIVE COMPUTATION OF COMPRESSIBLE FLOW FIELDS WITH THE DLR-τ-CODE


VOLKER HANNEMANN, DANIEL HEMPEL, THOMAS SONAR

Institut für Strömungsmechanik

DLR Göttingen

Bunsenstraße 10

D-37073 Göttingen



ABSTRACT. The DLR-τ-Code is the implementation of a finite volume method for the numerical solution of compressible Navier-Stokes equations. The code uses TVD- or ENO-recovery procedures to achieve high resolution properties and works on control volumes dual to a triangulation. Mesh refinement and coarsening is done automatically according to residual based refinement indicators in order to use the full flexibility as provided by an approach based on triangular meshes. The choice of refinement indicators as well as their behaviour in boundary layers are still research topics.


## 1. GOVERNING EQUATIONS

We consider the two-dimensional compressible Navier-Stokes equations describing the time-dependent motion of a compressible ideal fluid in a bounded domain $\Omega \subset \mathbb{R}^2$, i.e.

$$\partial_t u + \sum_{j=1}^{2} \partial_{x_j} f_j^c(u) = \frac{1}{\mathrm{Re}} \sum_{j=1}^{2} \partial_{x_j} f_j^v(u) \tag{1.1}$$

where $u := (\rho, \rho v_1, \rho v_2, \rho E)^\mathsf{T}$ is the vector of conservative variables and the convective and viscous fluxes are given by

$$f_j^c(u) := \begin{bmatrix} \rho v_j \\ \rho v_1 v_j + p\delta_{1j} \\ \rho v_2 v_j + p\delta_{2j} \\ \rho H v_j \end{bmatrix} \text{ and } f_j^v(u) := \begin{bmatrix} 0 \\ \tau_{1j} \\ \tau_{j2} \\ v_1 \tau_{1j} + v_2 \tau_{j2} + \frac{\mu\kappa}{\mathrm{Pr}}\partial_{x_j}\epsilon \end{bmatrix},$$

respectively. The quantities $\rho, v_1, v_2, p, E, H$ are density, velocity component in $x_1$-direction, velocity component in $x_2$-direction, pressure, total energy and enthalpy of the fluid. The enthalpy is defined to be $H = E + \frac{p}{\rho}$. The elements of the shear stress tensor are given by

$$\tau_{ij} = \mu \left( \partial_{x_j} v_i + \partial_{x_i} v_j \right) + \delta_{ij}\lambda \left( \partial_{x_1} v_1 + \partial_{x_2} v_2 \right).$$

The Stokes' hypothesis defines $\lambda$ to be $\lambda = -\frac{2}{3}\mu$ and $\mu$ denotes the viscosity coefficient. The quantity $\epsilon$ denotes the specific internal energy and is defined by $\epsilon = E - \frac{1}{2}|v|^2$ where $v = (v_1, v_2)^\mathsf{T}$. The Reynolds and Prandtl numbers are denoted by Re and Pr respectively and $\kappa$ denotes the ratio of specific heats. The equation

---

[1] International Workshop on **Numerical Methods for the Navier-Stokes Equations**, Heidelberg, 25th-28th October 1993

101

of state for a perfect gas is given by $p = (\kappa - 1)\rho\left(E - \frac{|v|^2}{2}\right)$, the temperature $T$ is given by $T = \kappa(\kappa - 1)\mathrm{Ma}^2\epsilon$ and the viscosity coefficient is given by the Sutherland law reading as $\mu = T^{1.5}\frac{1+S}{T+S}$ where $S := 110°K/\overline{T}_\infty$ and $\overline{T}_\infty$ denotes the temperature at infinity measured in degree Kelvin.

## 2. The DLR-$\tau$-Code

In order to discretise the Navier-Stokes equations on the domain $\Omega$ we introduce a conforming triangulation $\mathcal{T}_h$ consisting of triangles $K_k, k = 1, \dots, \#K$, and nodes $i, i = 1, \dots, \#I$. The triangulation will be called the primary mesh while the union of boxes $\sigma_i, i = 1, \dots, \#I$, build from the straight lines $l_{ij}^m, m = 1, 2$, connecting the midpoints of the edges with the barycenters of the triangles will be called secondary mesh. See figure 1 for the geometry between two boxes. We are looking for weak

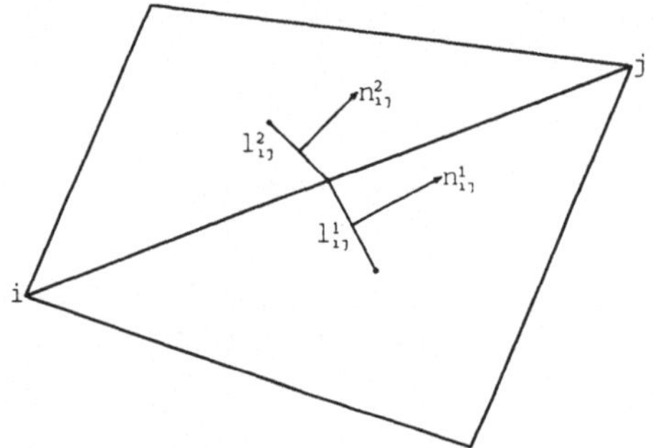


FIGURE 1. Geometry between two boxes

solutions of the Navier-Stokes equations, i.e. functions $u$ for which

$$\frac{d}{dt}\int_\sigma u\,dx + \oint_{\partial\sigma}\sum_{k=1}^{2} f_k^c(u)n_k\,ds = \frac{1}{Re}\oint_{\partial\sigma}\sum_{k=1}^{2} f_k^v(u)n_k\,ds \qquad (2.1)$$

holds on every bounded simply connected domain $\sigma \subset \mathbb{R}^2$ with Lipschitz boundary $\partial\sigma$. Here, $n = (n_1, n_2)$ denotes the outer unit normal vector at $\partial\sigma$. Introducing the cell average operator

$$u \xrightarrow{\mathbf{A}} \mathbf{A}(\sigma)u := \frac{1}{|\sigma|}\int_\sigma u\,dx \qquad (2.2)$$

we see that the definition of a weak solution immediately leads to an evolution equation on the boxes of the secondary mesh, i.e.

$$\frac{d}{dt}\mathbf{A}(\sigma_i)u(t) = -\frac{1}{|\sigma|}\left\{\oint_{\partial\sigma_i}\sum_{k=1}^{2}\left(f_k^c(u) - \frac{1}{Re}f_k^v(u)\right)n_k\,ds\right\} \qquad (2.3)$$

holds on any box $\sigma_i$. Introducing the cell average of the weak solution also into the right hand side leads into trouble since the line integrals are then no longer defined. Therefore a numerical flux function $\mathbb{R}^4 \times \mathbb{R}^4 \times \mathbb{R}^2 \ni (u, v, n) \xrightarrow{H} H(u, v; n) \in \mathbb{R}^4$ consistent in the sense $H(u, u, ; n) = \sum_{k=1}^{2} f_k^c(u)n_k$ is used. Denoting an approximation of the cell average on box $\sigma_i$ by $\overline{u}_i$ an intermediate stage in the development of a finite volume method can be written in the form

$$\frac{d}{dt}\overline{u}_i(t) = -\frac{1}{|\sigma|}\left\{\sum_{j\in N(i)}\sum_{m=1}^{2} H(\overline{u}_i,\overline{u}_j;n_{ij}^m)|l_{ij}^m| - \frac{1}{Re}\sum_{j\in N(i)}\sum_{m=1}^{2}\int_{l_{ij}^m}\sum_{k=1}^{2} f_k^v(u)n_k\,ds\right\}$$

in which $N(i)$ denotes the index set of neighbours of node (box) $i$. The remaining notation corresponds to figure 1.

Before describing the discretisation of the viscous fluxes we briefly outline recovery techniques to increase the order of approximation in the convective fluxes which is not yet sufficient for practical purposes. A linear polynomial recovery constructs on each box $\sigma_i$ for each fixed time a linear function $w_i(x) = w_{00} + \begin{pmatrix} w_{10} \\ w_{01} \end{pmatrix}\cdot(x - x_i)$, where $x_i$ denotes the barycenter of the box. Linear recovery does not make sense for the conservative variables $u$ since they contain products of their components. Therefore the recovery is done in the variables $\rho, v_1, v_2$ and $p$ separately. In our notation $w \in \{\rho, v_1, v_2, p\}$ is one of the variables to be recovered.

In order to keep cell averages conserved the linear recovery function has to satisfy $A(\sigma_i)w_i = \overline{w}_i$. This condition immediately determines the constant coefficient $w_{00} = \overline{w}_i$ and leaves the computation of an approximation to the gradient. We choose

$$\begin{pmatrix} w_{10} \\ w_{01} \end{pmatrix} = \frac{1}{|\sigma_i|}\int_{\sigma_i} \nabla_x \pi^K(w)\,dx \tag{2.4}$$

in which $\pi^K(w)$ denotes the $\Pi_1$-interpolant of $w$ on triangle $K$. Since the gradients of the linear interpolant are constant on each triangle the gradient computation is actually done by the sum $\begin{pmatrix} w_{10} \\ w_{01} \end{pmatrix} = \frac{1}{|\sigma_i|}\sum_K \nabla_x\pi^K(w)|K\cap\sigma_i|$ . As is well-known schemes using this type of recovery tend to produce spurious oscillations around shocks and a slope limiter is needed to control monotonicity. Denoting the limiter on $\sigma_i$ by $\Phi_i$ the final linear recovery polynomial reads as

$$w_i(x) = \overline{w}_i + \Phi_i \begin{pmatrix} w_{10} \\ w_{01} \end{pmatrix}\cdot(x - x_i). \tag{2.5}$$

The choice of the limiter is mostly based on personal taste. We use a device described by Barth and Jespersen [1] which is given by

$$\Phi_i \;=\; \min\left(1, \frac{\overline{w}_{ij}^{max} - \overline{w}_i}{w_i^{max} - \overline{w}_i}, \frac{\overline{w}_{ij}^{min} - \overline{w}_i}{w_i^{min} - \overline{w}_i}\right)$$

where $\overline{w}_{ij}^{max} := \max_{j\in N(i)}(\overline{w}_i, \overline{w}_j)$, $\overline{w}_{ij}^{min} := \min_{j\in N(i)}(\overline{w}_i, \overline{w}_j)$, $w_i^{max} := \max_{x\in\sigma_i} w_i(x)$, $w_i^{min} := \min_{x\in\sigma_i} w_i(x)$.

As an alternative the gradient can also be computed by means of an ENO-like algorithm. For each of the boxes $\sigma_i$ the gradient is taken to be $(w_{10}, w_{01}) = \nabla_x\pi^{K_m}(w)$ where $\pi^{K_m}(w)$ is the interpolant characterised by the criterion $|\nabla_x\pi^{K_m}(w)| = \min|\nabla_x\pi^K(w)|$ where the minimum is taken over all triangles $K$ which have node $i$ as a vertex, see also [4].

The linear recovery functions for $\rho, v_1, v_2$ and $p$ are used to compute values at the midpoints $x_{ij}^k$ of the edge segments $l_{ij}^k$. Instead of using the constant cell values $\overline{u}$ as arguments in the numerical flux function, values $u_i(x_{ij}^k)$ constructed from the $w_i(x_{ij}^k)$

are used in the form $H(u_i(x_{ij}^k, t), u_j(x_{ij}^k, t); n_{ij}^k)$, which is now a second order accurate approximation.

The discretisation of the viscous fluxes is done in a straight-forward central manner, i.e.

$$\int_{l_{ij}^m} \sum_{k=1}^2 f_k^v(u) n_k \, ds \approx \left( \sum_{k=1}^2 f_k^v(q_{l_{ij}^m}) n_{ij,k}^m \right) |l_{ij}^m| \tag{2.6}$$

where the value $q_{l_{ij}^m}$ is given by $q_{l_{ij}^m} := \frac{1}{2}\left(u_i(x_{ij}^m, t) + u_j(x_{ij}^m, t)\right)$ and derivatives occuring in the viscous fluxes are analogously averaged. The final finite volume scheme then reads as:

Find $\overline{u}_i$ for every box $\sigma_i$ as solution of the ODE system

$$\frac{d}{dt}\overline{u}_i(t) = -\frac{1}{|\sigma_i|} \left\{ \sum_{j \in N(i)} \sum_{m=1}^2 H(\overline{u}_i, \overline{u}_j; n_{ij}^m)|l_{ij}^m| - \right.$$

$$\left. \frac{1}{Re} \sum_{j \in N(i)} \sum_{m=1}^2 \left( \sum_{k=1}^2 f_k^v(q_{l_{ij}^m}) n_{ij,k}^m \right) |l_{ij}^m| \right\} \tag{2.7}$$

$$\overline{u}_i(0) = \mathbf{A}(\sigma_i)u(0). \tag{2.8}$$

At the time of this writing experiments with a $k - \epsilon$ turbulence model on triangular grids are underway but no conclusions can be drawn yet. The $\tau$-code itself is flexible enough to allow a hybrid approach in which boundary layers are discretised on quadrilateral meshes. In that case experiences from other DLR-codes can be used to a large extent and standard turbulence model implementations are possible.

The DLR-$\tau$-code uses an explicit Runge-Kutta scheme to advance the ODE-system (2.7) in time. Since care has to be taken not to spoil the spatial recovery process the TVD-Runge-Kutta schemes developed by Shu and Osher [3] were adopted. Since the CFL number is bounded by 1 there is no speed-up in using these type of methods. We are currently investigating the possibilities of implicit time stepping schemes. A domain decomposition approach to increase the efficiency of the explicit time stepping was developed and is currently tested.


## 3. Local Adaption

In this section we describe the geometrical part of local grid refinement and recoarsening. Note that other algorithms as those described here were used with the $\tau$-Code in the past. For a review, see [5] and [6].

The adaption-procedure starts with a conforming triangulation $\mathcal{T}_h$, a subset $\mathcal{R} \subseteq \mathcal{T}$ of triangles which have to be refined, and a subset $\mathcal{C} \subseteq \mathcal{T}, \mathcal{R} \cap \mathcal{C} = \emptyset$ of triangles which have to be coarsened. Initially the adaption routine inserts new points into the triangles of $\mathcal{R}$ as follows: Let $K \in \mathcal{R}$ be a triangle that has to be refined. If $K$ has an obtuse angle and the opposite edge of this angle is a boundary of the triangulation, we insert two new points symmetrically at this edge and devide the triangle into three subtriangles. Otherwise we insert the barycenter of $K$ as a new point and divide $K$ also into three triangles. Figure 2 illustrates these rules.

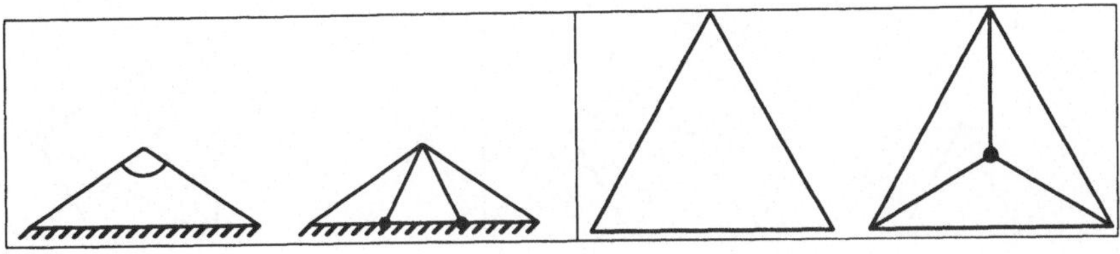

FIGURE 2.

After inserting new points into the triangles of $\mathcal{R}$ as described above, vertices of the triangles of $\mathcal{C}$ have to be removed. To delete points from a triangulation note that not all boundary points are removable—otherwise the geometry of the triangulated boundary would be destroyed. To avoid this we restrict the removement algorithm to inner points and those boundary points which are located on a straight line of the boundary. We call those points of the triangulation *removable* points. According to the following rules we determine a set $\mathcal{P}_C$ of vertices of triangles $K \in \mathcal{C}$, which have to be removed:

(1) If $K \in \mathcal{C}$ has a removable vertex, remove at least one vertex of K.
(2) Remove not more than two vertices of a triangle and not more than one vertex of a boundary edge of a triangle.

To remove a point $i \in \mathcal{P}_C$ the surrounding domain has to be remeshed. This domain consists of the adjacent triangles of $i$ and it is a star-shaped polygon. To remesh this domain it is usefull to simplify it first by *swapping edges* to minimise the number of edges leading from the point $i$: two neighbouring triangles with a common edge generally form a quadrilateral. The meaning of *swapping edges* is to replace this edge with the other diagonal of the quadrilateral, see figure 3. This is possible if and only if the domain is a convex, non-degenerated quadrilateral, see figure 3 for exceptions.

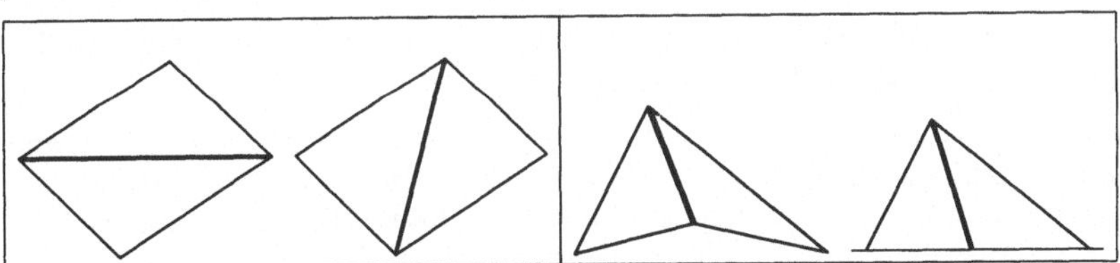

FIGURE 3. Swapping of diagonals. Only if the domain is a convex, non-degenerated quadrilateral, an edge can be swapped.

We use the procedure of *swapping edges* to simplify the domain around $i \in \mathcal{P}_C$. Whenever an edge leading from $i$ can be swapped, swapping will reduce the number of edges around the point as shown in figure 4. We apply this procedure up to the point when no more edge can be swapped.

This algorithm terminates with a very simple geometry of the edges leading from $i$: If $i$ is an inner point either four edges are left and there are two colinear edges or three edges around $i$ are left; otherwise if $i$ is a boundary point it is located on a straight line of the boundary and only one inner edge leading from $i$ is left. In each situation the point may be removed by uniting the adjacent triangles as shown in figure 5. This procedure will be applied for all points of $\mathcal{P}_C$.

Generally this procedure and the point insertion routine produce a lot of flat triangles. Therefore we apply a smoothing algorithm which eliminates such obtuse and

105

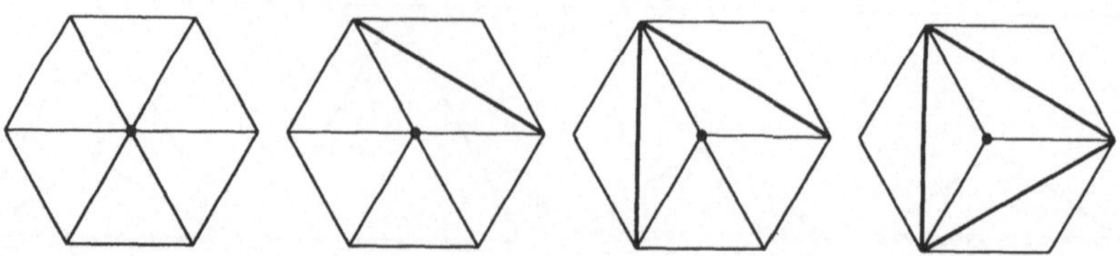

FIGURE 4. Minimization of the number of edges leading from a point by successive edge swappings.

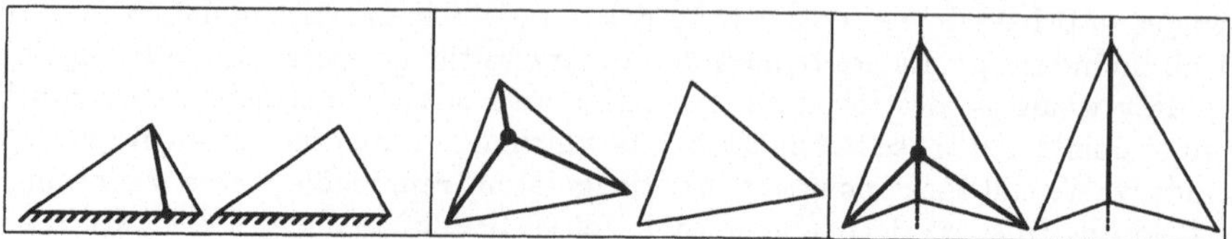

FIGURE 5. The minimisation of edges leading from a point terminates with one of three standard situations and the point may be removed by uniting the adjacent triangles.

accute angled triangles as a final step of the adaption procedure: We take a triangle's smallest angle as a suitable quantity for it's flatness. Now we will swap an edge of two neighbouring triangles, if swapping maximises the triangles minimal angle (max-min), see figure 6. To smooth a whole triangulation or a locally bounded region we

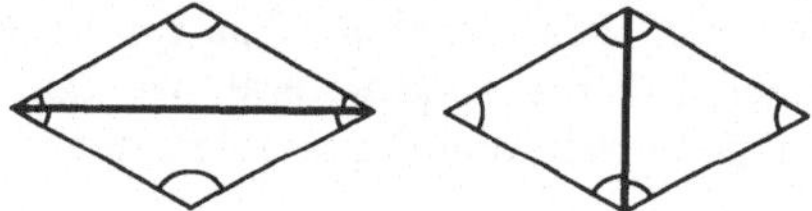

FIGURE 6. An edge will be swapped, if that maximises the minimal angle of the adjacent triangles.

apply this procedure successively up to the point when swapping of edges won't be an improvement according to the principle of maximisation of minimal angles.
A further part of the adaption procedure is the interpolation of conservative data while manipulating the grid. This is done by interpolation formulae for each elementary grid operation: swapping an edge, inserting two points at the boundary, inserting a triangles barycenter and removing a point in three standard situations as shown in figure 5.

## 4. REFINEMENT INDICATORS

In order to decide wether a certain region of $\mathcal{T}_h$ has to be refined or coarsened a reliable refinement indicator is a necessary ingredient for any adaptive algorithm. Traditional criteria often used in computational fluid mechanics are based on gradients of flow variables which is an approach unable to capture true errors in the numerical solution. In [5] an alternative approach was advocated using a finite element-like residual. If $\mathcal{L}u = 0$ denotes an abstract partial differential equation and $u^h$ the solution of a discretization, then $e^h := u^h - u$ denotes the error and $r^h := \mathcal{L}u^h$ the residual. Since $u = u^h - e$ it follows that $r^h - \mathcal{L}e = 0$. If now $\mathcal{L}$ is a linear, invertible differential operator, then it follows that

$$\|e\| \leq \|\mathcal{L}^{-1}\|\|r^h\| \tag{4.1}$$

where we leave the spaces and norms unspecified. Thus, the true error is bounded by the residual of the numerical solution which - in contrast to the error - is a computable quantity.

In view of the Navier-Stokes equations it is clear that the inherent nonlinearity none of the assumptions leading to the above inequality holds true. However, results obtained for the streamline diffusion finite element method [2] encourage the use of residual based refinement indicators for compressible flow computations.

In the case of our finite volume method the residual can be easily computed by means of the linear interpolant $u^h\big|_K := \pi^K(\overline{u})$ which is the linear polynomial interpolating the values $\overline{u}_{i_1}, \overline{u}_{i_2}$ and $\overline{u}_{i_3}$ at the nodes $i_1, i_2$ and $i_3$ of each triangle K. The (unsteady residual) can then be defined on K as

$$r^h := \frac{1}{3} \sum_{k=1}^{3} \frac{d}{dt}\overline{u}_{i_k} + \sum_{j=1}^{2} \partial_{x_j} f_j^c(u^h) - \frac{1}{Re} \sum_{j=1}^{2} \partial_{x_j} f_j^v(u^h), \tag{4.2}$$

where the first term is the average of the right hand side of (2.7) on the three boxes $\sigma_{i_1}, \sigma_{i_2}, \sigma_{i_3}$ lying at the vertices of K. This residual is a four-component vector. We compute $\|r^h\| := \sum_{i=1}^{4} \|r_i^h\|$ to get a number characterising the residual on each triangle.

As was remarked in [5] the $L^2$-norm of the residual behaves like $\mathcal{O}(h^{d/2-1})$ at shocks where d denotes space dimension. Thus, in 2-d an artificial power of h is needed to construct a convergent refinement indicator. It became clear from numerical experiments that

$$R^h(K) := h\|r^h\|_{L^2(K)} \tag{4.3}$$

is a useful refinement indicator for compressible flow computations. Süli [7] suggested the use of the weak norm $\|r^h\|_{H^{-1}(K)}$ which is essentially equivalent to the indicator (4.3) but harder to compute. Both indicators are well suited for inviscid flow computations but problems remain in the adaption within a boundary layer. Süli and Sonar are currently working on indicators in certain dual graph norms. This type of indicators are located somehow between $H^0$ and $H^{-1}$, are theoretically sound and hopefully provide better scaling properties with respect to boundary layers, see [8]. In all numerical computations presented the refinement indicator (4.3) was used in the form

$$h\|r^h\|_{L^2(K)} \leq TOL_1 \quad \Rightarrow \quad \mathcal{R} \leftarrow \mathcal{R} \cup K$$
$$h\|r^h\|_{L^2(K)} > TOL_2 \quad \Rightarrow \quad \mathcal{C} \leftarrow \mathcal{C} \cup K$$

with two user-specifies constants $TOL_1$ and $TOL_2$.


### 5. Numerical Examples


As a numerical test case the flow over the inlet lip of a scramjet was simulated. The situation is shown schematically in figure 7 where a typical shock configuration around an engine inlet in hypersonic flight is shown. In front of the lip of the intake cowl the bow shock of the intake ramp impinges on the bow shock around the cowl. According to the angle at which the impinging shock meets the bow shock of the lip six different types of shock-shock interactions can be characterised as shown in the right half of figure 7. In the test case the onflow Mach number was chosen to

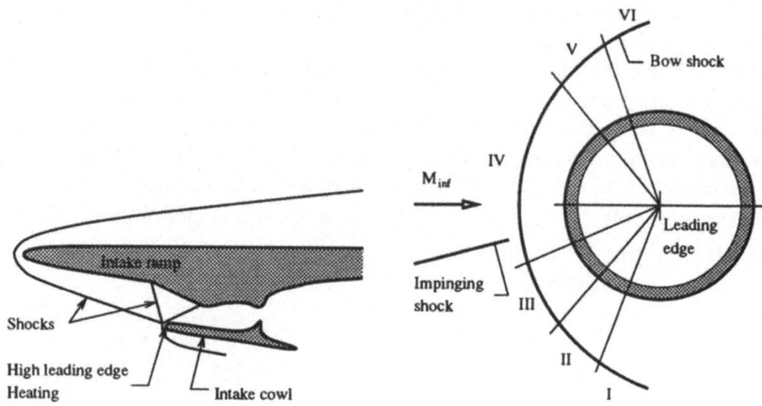


FIGURE 7. Scramjet inlet and classification of the flow cases occuring at the inlet lip.

be $Ma_\infty = 8.03$, the temperature at infinity $T_\infty = 122.1°K$, the Reynolds-Number $Re = 387500$ computed with the diameter of the cylinder $d = 76.2mm$. The wall temperature was given to be $T_w = 294.4°K$ and the angle of the impinging shock was $\delta = 18.11°$. These conditions correspond to a case measured by Wieting and Holden et. al. in 1987 and 1988 [9], [10].

The adapted grid for this test case is shown in figure 8 together with the Mach number distribution. As can be seen the refinement indicator was able to detect the overall phenomena of this flow field quite well. The shocks are sharply captured as is the supersonic jet just behind the interaction of bow shock and impinging shock. Note that a detachement bubble has formed on the upper side of the cylinder.

A closer look at the supersonic jet is plotted in figure 9. As can be seen the jet is terminated by a compression shock just in front of the cylinder's surface. All overall flow phenomena are in good agreement with measurements and other computational results.

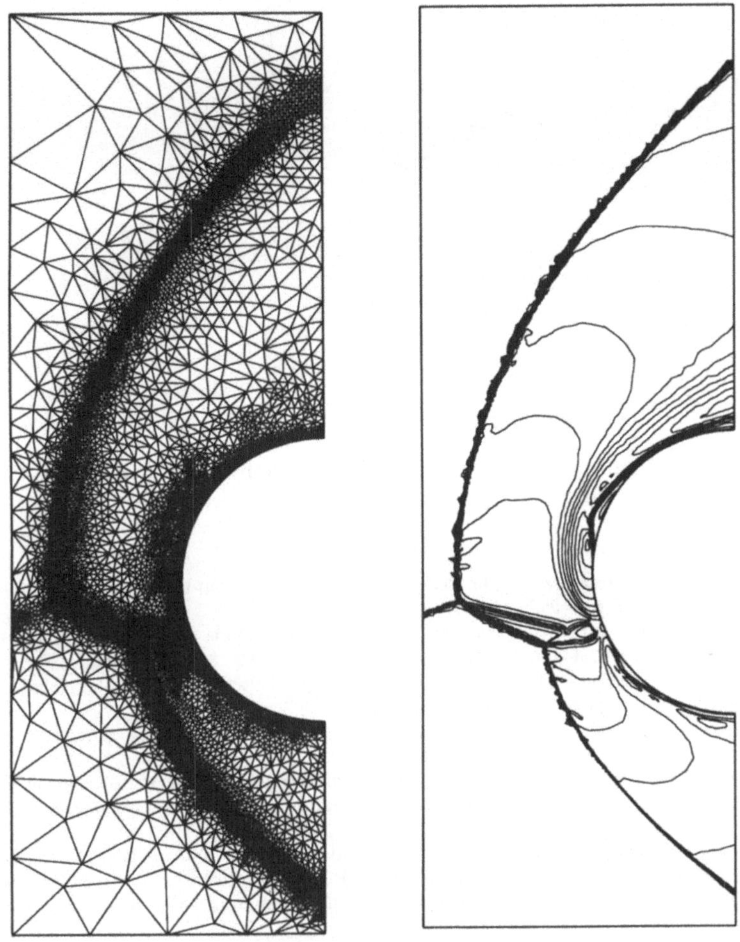

FIGURE 8. Adapted grid and Mach number distribution.

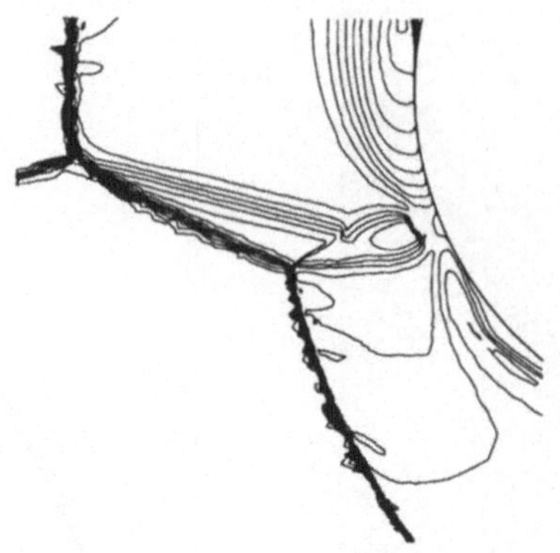

FIGURE 9. Zoom of the supersonic jet.

ACKNOWLEDGEMENT The authors would like to thank their colleague Stefan Brück who contributed figure 7.

## REFERENCES

1. T.J. Barth, D.C. Jespersen – The design and application of upwind schemes on unstructured meshes. *AIAA paper 89-0366 (1989)*.

2. P. Hansbo, C. Johnson – Adaptive streamline diffusion methods for compressible flow using conservative variables. *Comput. Methods Appl. Mech. Engrg. 87, 267-280, (1991)*.

3. C.-W. Shu, S. Osher – Efficient implementation of essentially nonoscillatory shock-capturing schemes. *J. Comp. Phys. 77, 439-471, (1988)*.

4. Th. Sonar – On the design of an upwind scheme for compressible flow on general triangulations. *Numerical Algorithms 4, 135-149, (1993)*.

5. Th. Sonar – Strong and weak norm refinement indicators based on the finite element residual for compressible flow computations. *Impact of Computing in Science and Engineering 5, 111-127, (1993)*.

6. Th. Sonar, V. Hannemann, D. Hempel – Dynamic adaptivity and residual control in unsteady compressible flow computation. *to be published: Mathematical and Computer Modelling, (1994)*.

7. E. Süli – private communication. *Oxford University Computing Laboratory, Numerical Analysis Group, (1992)*.

8. E. Süli, Th. Sonar – Weak norm error estimators for the adaptive computation of compressible Flow Fields. *in preparation*.

9. A.R. Wieting, M.S. Holden – Experimental Study of Shock Wave Interference Heating on a Cylindrical Leading Edge at Mach 6 and 8. *AIAA paper 87-1511, (1987)*.

10. A.R. Wieting, M.S. Holden, J.R. Moselle, C. Glass – Studies of Aerothermal Loads Generated in Regions of Shock-Shock Interaction in Hypersonic Flow. *AIAA paper 88-0477, (1988)*.

# DEFECT CORRECTION
## FOR
## CONVECTION DOMINATED FLOW

by


**Wilhelm Heinrichs**
*Mathematisches Institut der Heinrich–Heine–Universität Düsseldorf,*
*Universitätsstr. 1, D-40225 Düsseldorf, Germany*


## SUMMARY


A defect correction method for the convection–diffusion equation is presented. The discretization is performed by $2nd$ order finite difference schemes ($\beta$-schemes) where the $2nd$ order upstream scheme is combined with the standard central scheme. Furthermore higher order discretizations with spectral methods are considered. For preconditioning the usual first order upstream scheme is employed. The defect correction iteration is used for relaxation inside a multigrid procedure. For the spectral scheme GMRES is used for the outer iteration. The method is applied to the Boussinesq flow problem in vorticity-streamfunction formulation with high Rayleigh numbers.


## INTRODUCTION

A defect correction method for the convection–diffusion equation is presented. The discretization is performed by $2nd$ order finite difference schemes ($\beta$-schemes) where the $2nd$ order upstream scheme is combined with the standard central scheme. Furthermore higher order discretizations with spectral methods are considered. For preconditioning the usual first order upstream scheme is employed. The defect correction iteration is used for relaxation inside a multigrid procedure. It is shown that the smoothing analysis yields rather pessimistic results. In the practical computation the discretization error is reached in two V-cycles. Numerical results are presented which demonstrate the high efficiency of our treatment. The convection–diffusion equation yields a good model for the numerical solution of the Navier-Stokes equations with high Reynolds numbers. One is interested in finding a stable method which is of at least $2nd$ order accuracy. Unfortunately, the standard stable schemes as the usual upstream scheme or the artificial viscosity discretization lead to only $1st$ order methods. But in many applications the $2nd$ order accuracy is necessary in order to get a realistic impression of the flow.

The $\beta$ - schemes were already analyzed by Desideri & Hemker [2] and Luh [10]. For $\beta = 1$ we obtain the standard $2nd$ order upstream scheme and for $\beta = 0$ the central scheme. $\beta = \frac{1}{2}$ results in the *Fromm's* scheme. For $\beta = \frac{1}{3}$ we obtain the *upwind biased* scheme which is of $3rd$ order accuracy. Since an iterative solver for these higher order schemes yields bad convergence factors we propose a defect correction procedure. We already made good experience with this method for spectral discretizations. Here the higher order scheme is also preconditioned by the standard $1st$ order upwind scheme. We investigate the smoothing properties of this procedure for the convection equation. This method is used for relaxation in a multigrid cycle. In the spectral scheme the solution is approximated by Chebyshev polynomials. By a Fourier analysis it can be shown that the eigenvalues of the preconditioned operator are bounded but complex. Hence one has to employ a nonsymmetric matrix iteration for the solution. Here we recommend the GMRES iteration which belongs to the residual minimization methods. Clearly, for the general convection–diffusion problem the first derivatives have to be approximated according to the sign of the coefficients. Therefore for the iterative solution we recommend flow directed schemes. Since the Chebyshev nodes are dense near the boundary it is necessary to use line Gauss–Seidel relaxation (in an alternating manner). Finally this iterative solver is applied to the Boussinesq flow problem in vorticity–streamfunction formulation with high Rayleigh numbers.

## CONVECTION-DIFFUSION PROBLEM

Here we consider convection-diffusion problems which can in its most general form be written as

$$-\epsilon \Delta u + a u_x + b u_y \;=\; f \;\text{ in } \Omega = (-1,1)^2, \tag{1}$$

$$u \;=\; g \;\text{ on } \partial\Omega, \tag{2}$$

where $\epsilon = \frac{1}{Re}$ and $Re$ denotes the Reynolds number. $f$ is defined in $\Omega$, and $g$ is defined on $\partial\Omega$. $a$ and $b$ denote given constants. Such problems arise after a linearization of the Navier–Stokes equations (or Boussinesq flow problems where $\epsilon = \frac{1}{Ra}$ and $Ra$ denotes the Rayleigh number). It is well known that $Re$ is about the square root of $Ra$. The part $-\epsilon \Delta u$ denotes the diffusive part and $a u_x + b u_y$ denotes the convective part of the above equation. Here we are mainly interested in convection dominated flows where $\epsilon << h$ or $\epsilon << N^{-2}$. Here $h$ denotes the step size of the finite difference (FD) scheme and $N$ the maximal degree of the polynomials in a spectral scheme. It is well known that discretizations for this type of problem are in general unstable. One possibility to avoid the phenomenon of instability is to use upstream discretization for $u'$. Clearly, an obvious disadvantage of this scheme lies in the fact that the method now becomes only first order accurate. Hence it makes sense to use the first order upstream scheme only as a preconditioner for a higher order scheme. We analyze the preconditioning properties of this method for the following higher order schemes:

- $\beta$-schemes, $\beta \in [0, 1]$,

- spectral methods.

The first order upstream scheme is explicitly given by the upstream operator $L_h^1$ (here in 1D):

$$L_h^1 \cong \frac{1}{h}\left[-1\ \underline{1}\ 0\right].$$

The second order hybrid scheme is the $\beta$-scheme $L_h^\beta$ which is a combination of the standard $2nd$ order upstream scheme $L_h^{su,2}$ and the $2nd$ order central scheme $L_h^{ce,2}$:

$$L_h^\beta = \beta L_h^{su,2} + (1-\beta)L_h^{ce,2}, \quad \beta \in [0, 1]. \tag{3}$$

For all $\beta$ we obtain at least $2nd$ order discretizations. Clearly, for $\beta = 0$ the method becomes unstable. Especially, for $\beta = \frac{1}{3}$ we obtain a $3rd$ order scheme. The following special cases of the hybrid schemes are of importance (here in 1D):

$\beta = 1$    standard upstream, $2nd$ order     $\frac{1}{h}\left[\frac{1}{2}\ -2\ \underline{\frac{3}{2}}\ 0\right]$,

$\beta = \frac{1}{2}$    Fromm's scheme, $2nd$ order     $\frac{1}{h}\left[\frac{1}{4}\ -\frac{5}{4}\ \underline{\frac{3}{4}}\ \frac{1}{4}\right]$,

$\beta = \frac{1}{3}$    upwind biased scheme, $3rd$ order     $\frac{1}{h}\left[\frac{1}{6}\ -1\ \underline{\frac{1}{2}}\ \frac{1}{3}\right]$,

$\beta = 0$    central scheme, $2nd$ order, unstable     $\frac{1}{h}\left[0\ -\frac{1}{2}\ \underline{0}\ \frac{1}{2}\right]$.

Here we study the preconditioning properties of $L_h^1$ for the 2D convection operator

$$au_x + bu_y.$$

The defect correction iteration is defined by the operator

$$M_h = I_h - \omega\left(L_h^1\right)^{-1}L_h^\beta, \quad \beta \in [0, 1].$$

Here $\omega$ denotes a relaxation parameter which should accellerate the convergence speed. By a Fourier analysis for the Fourier components $\underline{\theta}$ ( $\underline{\theta} = \theta_1$ in $1D$, $\underline{\theta} = (\theta_1, \theta_2)$ in $2D$ ) the above operator $M_h$ leads to the *amplification factor*

$$\mu(\underline{\theta}) = 1 - \omega\frac{\lambda^\beta(\underline{\theta})}{\lambda^1(\underline{\theta})},$$

where $\lambda^\beta(\underline{\theta})$ and $\lambda^1(\underline{\theta})$ denote the factors of the Fourier analysis for the operators $L_h^\beta$ and $L_h^1$. Now the convergence factor $\rho$ of the defect correction procedure is defined as the supremum of $\mu(\underline{\theta})$ taken over all frequencies $\underline{\theta} \in (-\pi, \pi]$ in 1D and $\underline{\theta} \in (-\pi, \pi]^2$ in 2D:

$$\rho(M_h) \;:=\; \sup_{\underline{\theta} \in (-\pi,\pi]^2} |\mu(\underline{\theta})|.$$

Furthermore we were interested in defect correction as a smoother in a multigrid procedure. The efficiency of a smoother can be measured by means of the *smoothing rate* $\mu_\beta$. This rate can be obtained by taking the above supremum only for the high frequencies $|\underline{\theta}| := \max(|\theta_1|, |\theta_2|) \in (\frac{\pi}{2}, \pi]$ :

$$\mu_\beta \;:=\; \sup_{\underline{\theta} \in (\frac{\pi}{2},\pi]} |\mu(\underline{\theta})|.$$

From the Fourier analysis of Yvonne Luh [10] it became clear that for all $\beta \in [0, 1]$, $\omega$ and independent of the *alignment* $\frac{b}{a}$ the prediction

$$\rho(M_h) \;=\; 1$$

holds.

However, in the multigrid procedure we are more interested in the damping only of the high frequencies. Here we consider the special cases $\beta = 1$ and $\beta = \frac{1}{2}$. From the analysis in [10] the following was seen:

- $\beta = 1$, $\omega_{opt} = 0.68$: $\mu_1 \leq 0.77$

- $\beta = \frac{1}{2}$, $\omega_{opt} = 1.00$: $\mu_{\frac{1}{2}} \leq 0.72$.

For $\beta = 1$ the optimal parameter $\omega$ is about 0.68 for all alignments $\frac{b}{a}$. The corresponding smoothing rate $\mu_1$ is decreasing for increasing alignment. The maximum is attained at $\frac{b}{a} = 0.1$ where $\mu_1 = 0.77$. For $\beta = \frac{1}{2}$ the value $\omega_{opt} = 1$ yields a quite good choice for all alignments $\frac{b}{a}$. For $\beta = 0$ the smoothing rate $\mu_0$ is always equal to 1 (independent of the parameter choice of $\omega$). From these considerations it becomes clear that the smoothing rates are rather bad compared to the usual rates for symmetric problems (e.g., the Poisson problem).

Here we apply a standard multigrid method. The transfer operators are given by *full weighting* restriction and *bilinear* interpolation. For relaxation we choose the already mentioned *Richardson* iteration with *defect correction*. It is explicitly defined as follows:

$$u_h^{j+1} \;=\; u_h^j \,-\, \omega \left(L_h^1\right)^{-1} \left(L_h^\beta u_h^j - f_h\right)$$

for $j = 0, 1, 2, \ldots$. $u_h^0$ denotes a start approximation which is chosen to be identically zero. $\omega$ denotes the relaxation parameter. Optimal choices are given by the smoothing analysis. Instead of solving the first order problem relative to $L_h^1$ exactly

we employ a *lexikographic* Gauß-Seidel step which nearly yields an exact solver if
$a$, $b > 0$. If $a$ and $b$ have different sign then after a renumbering of the grid points
the same effect can be achieved. In case of variable coefficients $a$ and $b$ which change
sign one has to use the *flow directed point relaxation*.

Here we consider a multigrid method with 7 grids and corresponding step sizes
$h_\nu = \frac{1}{2^\nu}$, $\nu = 1, \ldots, 7$. In general, we employ a *V-cycle*. Other cycle structures as
*W-cycle*, *F-cycle* or the *full multigrid* technique could not improve the convergence
speed significantly. We employ two relaxations before and one after the coarse grid
correction. The results are compared with the pure Richardson iteration without
using multigrid. The absolute errors between the exact solution and the $IT.-$ iterate are measured in the discrete $L^1$ and $L^2$ norms: $L1$, $L2$. By $Q(L1)$, $Q(L2)$ we
denote the quotient of the errors for two successive iterates. Numerical results are
provided for the example where the exact solution is given by

$$u(x,y) \;=\; \sin\left(8\pi(y - \frac{b}{a}x)\right),$$

where $a = 4$, $b = 1$ and $\epsilon = 10^{-6}$ (see Y. Luh [10]). The right hand side $f$ and the
Dirichlet boundary conditions are determined by $u$. We present results for $\beta = 1$.
The corresponding numerical results are given in table I.

Table I. $\beta = 1$, V–cycle

| IT | $L1$ | $L2$ | $Q(L1)$ | $Q(L2)$ |
|----|------|------|---------|---------|
| 1 | $4.28 \cdot 10^{-2}$ | $5.42 \cdot 10^{-2}$ | 0.067 | 0.076 |
| 2 | $2.44 \cdot 10^{-2}$ | $3.04 \cdot 10^{-2}$ | 0.569 | 0.560 |
| 3 | $2.42 \cdot 10^{-2}$ | $3.01 \cdot 10^{-2}$ | 0.994 | 0.989 |

The numerical results show the highly improved efficiency of the V-cycle. There
is nearly no improvement by using other cycle structures. In particular, the first
rate of the multigrid scheme is very small and yields already an error which is nearly
equal to the discretization error. Hence in general it can be seen that 2 or 3 V-cycles
are enough to reach the truncation error. So the smoothing analysis gives rather
pessimistic results. In practice, the discretization error is reached very fast. By
comparing the results for different step sizes we also confirmed the at least second
order accuracy. For $\beta = \frac{1}{3}$ we obtain a $3rd$ order method. Here we obtain the most
precise results. For increasing $\beta$ the results become less accurate.

Let us now consider the case of variable coefficients $a$, $b$, i.e., $a = a(x,y)$, $b = b(x,y)$.
For the iterative solution we recommend *flow directed schemes*. For smoothing it is

recommended to use alternate iterations of FDHI (Flow Directed Horizontal Iterations) and FDVI (Flow Directed Vertical Iterations). In the literature this combination is called FDHVI (see [4], [5]). The iterative scheme FDHI is a variant of line Gauss-Seidel relaxation. Let $P_i$ denote the mesh points on the vertical line $x = x_i$. We divide $P_i$ into two subsets:

$$P_{i,E} := \{(i,j) : a(x_i, y_j) \geq 0\},$$
$$P_{i,W} := \{(i,j) : a(x_i, y_j) < 0\}.$$

The FDHI partitioning and ordering of the unknowns consists of the subsets $P_{i,E}$ arranged in order of increasing $i$, followed by the subsets $P_{i,W}$, arranged in order of decreasing $i$. The difference equations on each of the subsets $P_{i,E}$ or $P_{i,W}$ are a collection of tridiagonal systems. By considering the mesh points $P_j$ on a horizontal line $y = y_j$ and dividing $P_j$ into subsets $P_{j,N}$, $P_{j,S}$ we may construct the iterative scheme FDVI. Finally one alternates between FDHI and FDVI, resulting in FDHVI. For a more detailed description including numerical results we refer to [4] and [5]. Han et al. [5] describe a procedure based on directed graphs to partition and order the unknowns of the Gauss-Seidel process. This is performed by inspection of the coefficient matrix. Nevertheless, this algorithm is expensive for nonlinear problems, like those coming from the Navier-Stokes or Boussinesq equations, when the coefficients are solution dependent and require the reconstruction of the directed graph several times. The penalty for such a choice is proportional to the number of mesh points. Here the FDHVI scheme is applied to the Boussinesq flow problem.

## THE BOUSSINESQ FLOW PROBLEM

The problem specifically considered here is that of the two-dimensional flow of a Boussinesq fluid of Prandtl number $Pr = 0.71$ (i.e., air) in an upright square cavity (see [1], [3]). The walls are non-slip and impermeable. The horizontal walls are adiabatic and the vertical sides are at fixed temperatures. In addition to the Navier-Stokes equations we have one further equation for the temperature $T$. By $Ra$ we denote the Rayleigh number. The Boussinesq flow problem in vorticity-streamfunction formulation reads as follows:

$$4\Delta\psi + \omega \quad = \quad 0 \ \text{ in } \Omega = (-1,1)^2, \tag{4}$$

$$-2Pr\Delta\omega + \frac{\partial}{\partial x}(v_1\omega) + \frac{\partial}{\partial y}(v_2\omega) \quad = \quad RaPr\frac{\partial T}{\partial x} \ \text{ in } \Omega, \tag{5}$$

$$-2\Delta T + \frac{\partial}{\partial x}(v_1 T) + \frac{\partial}{\partial y}(v_2 T) \quad = \quad 0 \ \text{ in } \Omega. \tag{6}$$

As usual $(v_1, v_2)^t$ denotes the velocity. The scalar factors 2 in equations (5), (6) and 4 in equation (4) are due to the fact that we here define the problem in $(-1,1)^2$ instead of the original square cavity $(0,1)^2$. $\psi$ fulfills homogeneous Dirichlet boundary conditions, i.e.,

$$\psi = \frac{\partial \psi}{\partial \nu} = 0 \ \text{on} \ \partial \Omega$$

and $T$ fulfills mixed Dirichlet/Neumann boundary conditions

$$T(-1, y) = 1, \ T(1, y) = 0 \ \text{for} \ y \in (-1, 1),$$
$$\frac{\partial T}{\partial y}(x, -1) = \frac{\partial T}{\partial y}(x, 1) = 0 \ \text{for} \ x \in [-1, 1].$$

The homogenous Neumann boundary conditions correspond to the fact that the horizontal walls are adiabatic.

Now the equations (4)–(6) are linearized by a Quasi-Newton method, where the velocity from the previous iteration is employed. Hence we have to solve the linear problem:

$$4\Delta\psi^{n+1} + \omega^{n+1} = 0 \ \text{in} \ \Omega, \tag{7}$$

$$-2Pr\Delta\omega^{n+1} + \frac{\partial}{\partial x}(v_1^n \omega^{n+1}) + \frac{\partial}{\partial y}(v_2^n \omega^{n+1}) = RaPr\frac{\partial T^{n+1}}{\partial x} \ \text{in} \ \Omega, \tag{8}$$

$$-2\Delta T^{n+1} + \frac{\partial}{\partial x}(v_1^n T^{n+1}) + \frac{\partial}{\partial y}(v_2^n T^{n+1}) = 0 \ \text{in} \ \Omega, \tag{9}$$

where the index $n$ denotes the $n$-th iterate.

Clearly, the velocity is related to the streamfunction by

$$v_1^n = \frac{\partial \psi^n}{\partial y}, \ \ v_2^n = -\frac{\partial \psi^n}{\partial x}.$$

The linear system (7)–(9) is now approximately solved by a spectral multigrid (SMG) method (see [6], [7], [8], [9]). In the spectral scheme the solution is approximated by polynomials in $\mathbf{P}_N$, $N \in \mathbf{N}$ where $\mathbf{P}_N$ denotes the space of polynomials of degree $\leq N$. The discretization is performed by a pseudo spectral (or collocation) method in the Chebyshev-Gauss-Lobatto nodes

$$(x_i, y_j) = (\cos\frac{i\pi}{N}, \cos\frac{j\pi}{N}), \ i, j = 0, \ldots, N.$$

We first describe the pseudo spectral discretization of the system (7)–(9). The functions $\psi^{n+1}$ and $\omega^{n+1}$ are spectrally approximated by polynomials $u_{N+2} \in \mathbf{P}_{N+2}$, $v_N \in \mathbf{P}_N$ (see [8], [9]). $T^{n+1}$ is approximated by a polynomial $w_N \in \mathbf{P}_N$. Here the index $n + 1$ is omitted. Hence $v_1^n$, $v_2^n$ are approximated by the polynomials $v_{1,N}$, $v_{2,N} \in \mathbf{P}_{N+2}$, where

$$v_{1,N} = \frac{\partial \overline{u}_{N+2}}{\partial y}, \ v_{2,N} = -\frac{\partial \overline{u}_{N+2}}{\partial x},$$

and $\bar{u}_{N+2} \in \mathbf{P}_{N+2}$ corresponds to the polynomial $u_{N+2}$ from the previous iteration. Now the pseudo spectral problem related to (7)–(9) reads as follows:
Find $u_{N+2} \in \mathbf{P}_{N+2}$, $v_N \in \mathbf{P}_N$, $w_N \in \mathbf{P}_N$ such that

$$[4\Delta u_{N+2} + v_N](x_i, y_j) \;=\; 0 \tag{10}$$

for $i, j = 0, \dots, N$ and

$$[-2Pr\Delta v_N + \frac{\partial}{\partial x}(v_{1,N}v_N) + \frac{\partial}{\partial y}(v_{2,N}v_N)](x_i, y_j) \;=\; RaPr\frac{\partial w_N}{\partial x}(x_i, y_j), \tag{11}$$

$$[-2\Delta w_N + \frac{\partial}{\partial x}(v_{1,N}w_N) + \frac{\partial}{\partial y}(v_{2,N}w_N)](x_i, y_j) \;=\; 0 \tag{12}$$

for $i, j = 1, \dots, N - 1$. Since $u$ has to fulfill two types of boundary conditions we choose $u_{N+2} \in \mathbf{P}_{N+2}$ fulfilling the homogeneous Dirichlet boundary conditions. For $v_N$ there are no boundary conditions. The pseudo spectral boundary conditions for $w_N \in \mathbf{P}_N$ are given by

$$w_N(-1, y_j) = 1, \; w_N(1, y_j) = 0 \;\text{ for } j = 1, \dots, N - 1, \tag{13}$$

$$\frac{\partial w_N}{\partial y}(x_i, -1) = \frac{\partial w_N}{\partial y}(x_i, 1) = 0 \;\text{ for } i = 0, \dots, N. \tag{14}$$

The system (10)–(14) completely determines the spectral approximations $u_{N+2}$, $v_N$, $w_N$. It is clear that the polynomials $u_{N+2}$, $v_N$ are uniquely determined by the equations (10), (11). Furthermore the equation (12) together with the boundary conditions (13), (14) uniquely determine the polynomial $w_N$. Hence in the linearized version the systems (10), (11) for determining $u_{N+2}$, $v_N$ and (12), (13), (14) for determining $w_N$ can be handled separately. First, one solves the system (12), (13), (14) for $w_N$ by a SMG method, then one calculates $\frac{\partial w_N}{\partial x}(x_i, y_j)$, $i, j = 1, \dots, N - 1$, and finally one solves the system (10), (11) by the SMG method introduced in [8], [9]. Here we employed 6 V-cycles of SMG in order to get a nearly exact solution of the linear systems.

Now we turn to a more precise description of the SMG method. We use the same components as already introduced. A somewhat different treatment results from the fact that the diffusive part is now perturbed by the first order derivatives $\frac{\partial}{\partial x}$, $\frac{\partial}{\partial y}$. For an increasing Rayleigh number the convective part becomes dominant. Hence in the defect correction step one has to use a FD approximation which remains stable also for an increasing Rayleigh number. Furthermore the FD problem has to be solved approximately by a suitable iterative method which also works for convection dominated flows. Here we employed the FDHVI iteration for preconditioning of the spectral system resulting from the equations (11), (12). In order to handle the complex eigenvalues of the preconditioned spectral operator we employ nonsymmetric matrix iterations. Here we choose the GMRES iteration.

By using these components we numerically calculated for various Rayleigh numbers and mesh sizes the following quantities:

$|\psi|_{mid}$ : absolute value of the streamfunction at the midpoint of the cavity,
$|\psi|_{max}$ : maximum absolute value of the streamfunction.

The local heat flux in a horizontal direction at any point in the cavity is given by

$$Q := v_1 T - 2\frac{\partial T}{\partial x}.$$

Let us further introduce the following Nusselt numbers:

$\overline{Nu} := \frac{1}{4}\int_{-1}^{1}\int_{-1}^{1}Q(x,y)dxdy$ : average Nusselt number throughout the cavity,

$Nu_{\frac{1}{2}} := \frac{1}{2}\int_{-1}^{1}Q(0,y)dy$ : average Nusselt number on the vertical mid-plane,

$Nu_0 := \frac{1}{2}\int_{-1}^{1}Q(-1,y)dy$ : average Nusselt number on the vertical boundary,

$Nu_{max} := \max\{|Q(-1,y)| : y \in [-1,1]\}$ : maximum value of the Nusselt number,

$Nu_{min} := \min\{|Q(-1,y)| : y \in [-1,1]\}$ : minimum value of the Nusselt number.

The above integrals in the definition of $\overline{Nu}$, $Nu_{\frac{1}{2}}$ and $Nu_0$ are evaluated by the Clenshaw-Curtis quadrature. In the tables II - III we present the numerical results for the Rayleigh numbers $Ra = 10^4, 10^5$. The numerical results are in good accordance with the results obtained in [3]. However, for a larger Rayleigh number or increasing $N$ the above SMG method is no more convergent. The reason is that upstream preconditioning is not good enough. Here one has to find some better ways of preconditioning. At the moment we try to find improved preconditioners where the finite difference discretization is performed on staggered grids.

**Table II.** Results for $Ra = 10^4$.

| N | $|\psi|_{mid}$ | $|\psi|_{max}$ | $\overline{Nu}$ | $Nu_{\frac{1}{2}}$ | $Nu_0$ | $Nu_{max}$ | $Nu_{min}$ |
|---|---|---|---|---|---|---|---|
| 8 | 5.0713 | 5.0713 | 2.2474 | 2.1946 | 2.1870 | 3.6170 | 0.5067 |
| 16 | 5.0736 | 5.0736 | 2.2448 | 2.1946 | 2.1870 | 3.5314 | 0.5853 |
| 24 | 5.0981 | 5.0980 | 2.2340 | 2.2350 | 2.2420 | 3.5450 | 0.5920 |

**Table III.** Results for $Ra = 10^5$.

| N | $|\psi|_{mid}$ | $|\psi|_{max}$ | $\overline{Nu}$ | $Nu_{\frac{1}{2}}$ | $Nu_0$ | $Nu_{max}$ | $Nu_{min}$ |
|---|---|---|---|---|---|---|---|
| 8 | 14.3409 | 18.8519 | 4.4140 | 4.7345 | 4.7590 | 10.4740 | 0.3438 |
| 16 | 11.3720 | 12.3330 | 4.5030 | 4.5061 | 4.5313 | 7.9010 | 0.7551 |
| 24 | 9.1600 | 9.6530 | 4.5100 | 4.5120 | 4.5231 | 7.7700 | 0.7361 |

# REFERENCES

1. Behnia, M., Wolfstein, M., and De Vahl Davis, G.: A stable fast marching scheme for computational fluid mechanics. Int. J. for Num. Meth. in Fluids **10**, 607-621(1990).

2. Desideri, J.A. and Hemker, P.W.: Analysis of the convergence of iterative implicit and defect-correction algorithms for hyperbolic systems. Centre for Mathematics and Computer Science, Report NM-R9004, 1992 (unpublished).

3. De Vahl Davis, G.: Natural convection of air in a square cavity: a bench mark numerical solution. Int. J. for Numer. Meth. in Fluids **3**, 249-264(1983).

4. Deville, M.D. and Mund, E.H.: Finite element preconditioning of collocation schemes for advection-diffusion equations. Proceedings of the IMACS International Symposium on Iterative Methods in the Linear Algebra, edited by R. Beauwens and P. de Groen, Brüssel, 181-189, 1992.

5. Han, H., Il'in, V.P., and Kellogg, R.B.: Flow directed iterations for advection dominated flow, *in* BAIL V, Proceedings of the Fifth Int. Conf. on Boundary and Interior Layers - Computational and Asymptotic Methods, edited by G. Ben-yu, J.J.H. Miller and S. Zhong-ci, Shanghai, China, 1988.

6. Heinrichs, W.: Line relaxation for spectral multigrid methods. J. Comp. Phys. **77**, 166-182(1988).

7. Heinrichs, W.: Multigrid methods for combined finite difference and Fourier problems. J. Comp. Phys. **78**, 424-436(1988).

8. Heinrichs, W.: Spectral multigrid techniques for the Stokes problem in streamfunction formulation. J. Comput. Phys. **102**, 310-318(1992).

9. Heinrichs, W.: Spectral multigrid techniques for the Navier-Stokes equations. Comput. Meth. Appl. Mech. Eng. **106**, 297-314(1993).

10. Luh, Y.: Diskretisierungen und Mehrgitteralgorithmen zur Lösung hyperbolischer Differentialgleichungen, am Beispiel der Wellengleichung, der Advektionsgleichung und der verallgemeinerten Stokes-Gleichungen. Ph. D. Thesis, Bonn 1992. GMD-Bericht Nr. 205, Oldenbourg-Verlag, Oldenbourg 1992.

# AN OPERATOR SPLITTING APPROACH FOR COMPUTING COMPRESSIBLE FLOWS IN ASTROPHYSICS


Ahmad Hujeirat and Rolf Rannacher

Institut für Angewandte Mathematik
Universität Heidelberg
D–69120 Heidelberg, Germany


## Summary


We consider a proto-type flow problem in astrophysics: the formation of a viscous boundary layer in the Keplerian accretion disk surrounding a central compact star which rotates with sub-Keplerian velocity. The underlying mathematical model are the compressible Navier-Stokes equations in axi-symmetric approximation. We present a numerical method for solving this problem based on operator decomposition and dimensional splitting. This approach has proven to be successful in dealing with the various inherent difficulties in this problem: the stiff source terms, the extreme variations in density and velocity and the different relevant time scales.


# 1 Introduction

Most of the astrophysical fluid flows are highly turbulent, compressible and time dependent. Since friction exists everywhere, they are also viscous. Therefore, part of their kinetic energy is converted into heat and part of this heat is transported away via radiation which is then astronomically observed. For simulating the thermo- and the hydro-structure of the occurring phenomena, very efficient multi-dimensional radiation-hydrodynamical codes are needed. A proto-type problem is the formation of steady state structures in the boundary layer (BL) between the accretion disk rotating with Keplerian velocity around a compact central star which rotates with sub-Keplerian velocity. This phenomenon is of great importance for understanding the short and long time evolution of the disk and possibly novae explosions.

An implicit, robust and highly accurate multi-dimensional solver is needed to follow the evolution of the BL on the viscous time scale. For that, one has to solve the full 3-D axi-symmetric compressible Navier-Stokes equations. We describe a method based on the concept of operator decomposition and dimensional splitting by which one can compute steady state or quasi-steady state solutions to this problem. The accuracy and stability

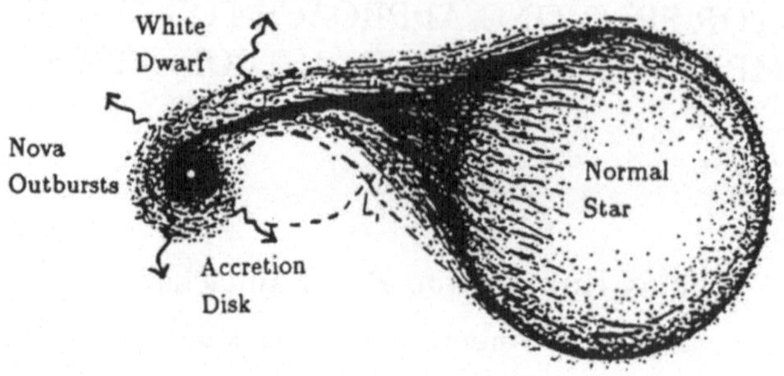


Figure 1: The accretion phenomenon around a compact star

of this method has been tested at the shock tube problem with satisfactory results. Then, the 2-D and the 3-D axi-symmetric boundary layer problem in the accretion disk was treated and physically consistent results where obtained. The developed method turned out to be efficient enough for carrying these long-time computations out on starndard workstations. Among the astrophyically interesting results obtained through these simulations is that the flow in the BL is governed by shock waves which are mainly depending on the viscosity.

## 2  The Astrophysical Problem

Let us describe the astrophysical problem to be solved in somewhat more detail. We consider a binary system consisting of a compact star (White Dwarf) and a mass-losing normal star which is in an expansion stage of evolution. However, this expansion is limited by the equipotential surface that overlaps the two stars. As the matter flows through the saddle point - $L_1$ - with almost sound speed (see Fig. 1), it undergoes successive momentum interactions which ends in a nearly circular motion and, consequently, an accretion disk (AD) around the compact star is formed (for a numerical simulation of the evolution of the AD, see Hensler [3]). The viscosity and the gravity in the disk act to decouple the specific angular momentum from the matter such that, while a large amount of matter flows inwards due to gravity, the viscous forces transport only some of the angular momentum outwards. The structure of the boundary layer (BL), i.e., the region connecting the disk and the central star (see Fig. 2), is of great importance for the following two reasons:
- Energy conservation indicates that for a slowly rotating star, the energy that is liberated in the BL is of the same order as that of the whole disk. This may explain the soft and hard X-rays observed in some cataclysmic variables (see, e.g., Pringle [8], Hujeirat [5] and the references therein).
- Resolving the hydrodynamical structure of the BL may provide some insight into the time evolution of novae explosions that are observed in cataclysmic variables on the time scale of several dozens of years (see, e.g., Shara [10], Hujeirat [6] and the references therein).

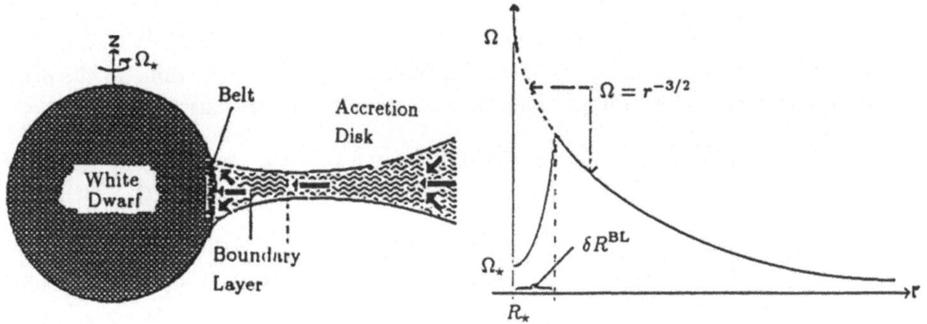


Figure 2: The geometrical location of the BL

# 3 The Mathematical Problem

The full physical problem described above is by far too complicated to be solved with present-day computers. Hence, some simplifications are necessary:
- The 3-D configuration of the flow in the BL is assumed to possess axi-symmetry.
- The magnetic field of the central star as well as any chemical interactions are ignored.
- Radiation transfer is treated in the limit of diffusion approximation.
In this case, using cylindrical coordinates, the non-dimensional compressible 3-D axi-symmetric time dependent Navier-Stokes equations take the following form:

1. Continuity equation:
$$\partial_t \rho + \epsilon r^{-1} \partial_r (r\rho u) + \epsilon \partial_z (\rho v) = 0\,.$$

2. Radial momentum equation:
$$\partial_t(\rho u) + \epsilon r^{-1}\partial_r(r\rho u^2) + \epsilon\partial_z(\rho uv) = -\epsilon\partial_r P + \epsilon^{-1}\rho r(\Omega^2 - \bar{r}^{-3})$$
$$+\epsilon r^{-1}\partial_r r\sigma_{rr} + \epsilon\partial_z \sigma_{rz} - \epsilon r^{-1}\sigma_{\varphi\varphi}\,.$$

3. Vertical momentum equation:
$$\partial_t(\rho v) + \epsilon r^{-1}\partial_r(r\rho uv) + \epsilon\partial_z(\rho v^2) = -\epsilon\partial_z P - \epsilon^{-1}\rho z\bar{r}^{-3}$$
$$+\epsilon r^{-1}\partial_r r\sigma_{rz} + \epsilon\partial_z \sigma_{zz}\,.$$

4. Angular momentum equation:
$$\partial_t(r\rho v_\varphi) + \epsilon r^{-1}\partial_r(r^2 \rho u v_\varphi) + \epsilon\partial_z(r\rho v v_\varphi) = \epsilon r^{-1}\partial_r(r^2 \sigma_{r\varphi}) + \epsilon\partial_z(r\sigma_{\varphi z})\,.$$

5. Energy equation:
$$\partial_t(\rho\mathcal{E}) + \epsilon r^{-1}\partial_r(r\rho u\mathcal{E}) + \epsilon\partial_z(\rho v\mathcal{E}) = -\epsilon P(r^{-1}\partial_r(ru) + \partial_z v)$$
$$+\epsilon f_{diff} r^{-1}\partial_r(r\check{\kappa}\partial_r T) + \epsilon f_{diff}\partial_z(\check{\kappa}\partial_z T) + \epsilon f_\Phi \Phi\,.$$

Here, $\mathcal{E} = \frac{C_V T}{\gamma}$, $P = \frac{\varrho T}{\gamma}$, $\check{\kappa} = \lambda_f \xi \frac{T^3}{(\kappa+\sigma)\rho}$, $\lambda^f = \frac{6}{6+3R_f+R_f^2}$, $R_f = 2.12 \times 10^{-4} \frac{1}{(\kappa+\sigma)\rho} \frac{|\nabla T^4|}{T^4}$, and $f_{diff}$, $f_\Phi$ are switch on-off coefficients. Further, $\bar{r} = (r^2 + z^2)^{1/2}$, while all the other symbols have their usual meanings. The components of the stress tensor are:

$$\sigma_{rr} = 2\eta \partial_r u - \tfrac{2}{3}\eta \nabla \cdot \vec{V}, \qquad \sigma_{zz} = 2\eta \partial_z v - \tfrac{2}{3}\eta \nabla \cdot \vec{V}, \qquad \sigma_{\phi\phi} = 2\eta r^{-1} u - \tfrac{2}{3}\eta \nabla \cdot \vec{V},$$

$$\sigma_{r\phi} = \eta r \partial_r \Omega, \qquad\qquad\qquad \sigma_{rz} = \eta(\partial_r v + \partial_z u), \qquad\qquad \sigma_{z\phi} = \eta r \partial_z \Omega,$$

and $\Phi$ has the form $\Phi = 2\eta \langle D_{rr}^2 + D_{zz}^2 + D_{\phi\phi}^2 + 2D_{rz}^2 + 2D_{r\phi}^2 + 2D_{z\phi}^2 \rangle - (2/3)\eta(\nabla \cdot \vec{V})^2$, where $\eta = \rho\nu = \rho\alpha V_s H_h$ and $H_h$ is the harmonic mean of the disk thickness and the radial pressure scale height. Further u, v and $v_\varphi\,(= r\Omega)$ are the r, z and $\varphi$ velocity components, respectively, of the velocity vector $\vec{V}$. The elements of the deformation tensor are:

$$D_{rr} = \partial_r u, \qquad\qquad D_{zz} = \partial_z v, \qquad D_{\phi\phi} = r^{-1} u,$$

$$D_{rz} = \tfrac{1}{2}(\partial_r v + \partial_z u), \qquad D_{r\phi} = \tfrac{1}{2}r\partial_r\Omega, \qquad D_{z\phi} = \tfrac{1}{2}r\partial_r\Omega.$$

The mathematical aim is to determine whether the above set of equations together with their corresponding boundary conditions possess a stationary or quasi-stationary solution. Before we describe the numerical algorithm for attacking this problem, let us discuss some hydrodynamical aspects of the expected flow in the boundary layer:

- Assuming the accretion disk to be geometrically thin, the Mach number can vary in the range $0 \leq \mathcal{M} \leq 100$.

- With $\epsilon = $ [disk thickness/star radius] $= \mathcal{O}(10^{-2})$, the sources terms are about $\epsilon^{-2} = \mathcal{O}(10^{-4})$ times larger than the advection and viscous terms. Therefore, the flow is highly turbulent and perturbations may be easily amplified generating a dynamically unstable flow and the density waves may steepen into shock waves.

- The internal energy in the outer flow region is about four orders of magnitudes less than the kinetic energy, i.e., the internal energy could be of the same order or less than the truncation error of the numerical method and negative temperatures may occur.

- The order of the non-dimensional viscous time scale is $\mathcal{T}_{visc} \sim 10^4$, while the dynamical time scale is around $\mathcal{T}_{dyn} \sim 10^{-2}$. Therefore, in order to reach steady state with a reasonable number of time steps, the size of the time step should be limited only by accuracy requirements, i.e., the method used must be implicit.

- The flow in the regions close to the BL may undergo rapid changes in its macro-structure, i.e., from incompressible to pressure free, to highly compressed by shocks and to flow in hydrostatic equilibrium. Therefore, the numerical method is expected to deal with four types of fluids: highly viscous flow, highly supersonic flow with shock formation, corona type flow with circulations, and quasi-hydrostatic flow.

- Consistent initial conditions are not known, and also some of the boundary conditions have to be artificially prescribed.

In view of these facts, even an only qualitatively correct simulation of the flow in the BL is a challange for numerical methods.


# 4  The Numerical Algorithm

A numerical algorithm for solving the problem described above should meet the following requirements:

- Strongly refined grid with grid sizes varying by four to six order of magnitude.
- Conservative treatment of the equations.
- Second-order discretization in space and time with low numerical diffusion.
- Highly robust and efficient solution method.

We now describe the key steps of our approach towards satisfying these conditions (for additional details see Hujeirat [5] and Hujeirat&Rannacher [7]).

## 4.1  Physical splitting

The key step is the decomposition of the physical process into a series of events which represent the main steps of the scheme. We try to decouple the set of equations as much as possible to avoid full interaction between them within the time step. This is based on a careful analysis of the main physical processes in the problem, mainly their evolution and their relevant length and time scales. If the full set of equations is solved simultaneously the order of the equations, of course, does not play any role. However, it plays a very important role within such a splitting process. Test calculations show that inconsistency in the order of solving the equations may excite numerical instabilities which enforce the reduction of the time step as far down as required for an explicit scheme. Different flows require different physical splittings according to the particular forces controling the flow. Further, in some cases the continuity equation or the energy equation have to be solved twice in each time step in the course of the so-called Strang-splitting (see Strang [11]), so that a set of 6 or 8 equations has to be solved instead of the original 5 equations. In the 3-D axi-symmetric case, the following splitting is proposed:

1. Angular momentum equation

$$q^\Omega + L_r q_r^\Omega + L_z q_z^\Omega + L_{rr} q_r^\Omega + L_{zz} q_z^\Omega = 0.$$

2. Radial momentum equation

$$q^u + L_r q_r^u + L_z q_z^u + L_{rr} q_r^u + L_{zz} q_z^u + L_{rz} q_{rz}^u + f^u = 0.$$

3. Sub-splitting of the continuity equation if needed

4. Vertical momentum equation

$$q^v + L_r q_r^v + L_z q_z^v + L_{rr} q_r^v + L_{zz} q_z^v + L_{rz} q_{rz}^v + f^v = 0.$$

5. Continuity equation

$$q^\rho + L_r q_r^\rho + L_z q_z^\rho = 0.$$

6. Energy equation

$$q^T + L_r q_r^T + L_z q_z^T + L_{rr} q_r^T + L_{zz} q_z^T + f^T = 0.$$

Here, $L_x$ and $L_{xx}$ are first- and second-order spatial difference operators, respectively.

## 4.2 Dimensional Splitting

Going further in our simplification of the solution method, we use a dimensional factorization of ADI-type (see, e.g., Beam&Worming [1]) to convert the penta-diagonal solution step normally required for each scalar equation (in the 3-D axi-symmetric case) into two successive tri-diagonal solution steps. The underlying idea here is that the above physical splitting indirectly generates a preferable direction for the space splitting (see Gourlay [2]). In our case, test calculations have shown that the order in which an equation is solved is strongly related to the direction of the net force in the right hand side of each equation which controls the rate of changes of the momentum. Such a splitting enhances convergence as well as stability. For example, the effective force in the radial momentum equation (gravity and centrifugal forces) is radially directed and the order of the splitting is accordingly as follows:

$$\{\frac{\rho^{ex}}{\delta t} + L_r(\rho u)^{ex} + L_{rr}(\rho \nu)^{ex}\}u^{\star} = \frac{(\rho u)^n}{\delta t} + L_{rz}(\rho \nu v)^{ex} + f_{ex}^u(\vec{r}),$$

$$\{\frac{\rho^{ex}}{\delta t} + L_z(\rho u)^{ex} + L_{zz}(\rho \nu)^{ex}\}u^{n+1} = \frac{(\rho u)^{\star}}{\delta t} + L_{rz}(\rho \nu v)^{ex} + f_{ex}^u(\vec{r}),$$

where the superscript "ex" refers to extrapolated values from the preceding time level.

On the other hand, from the theory of radiation transfer, we know that radiation energy transport is inversely proportional to the optical depth which in turn is also inversely proportional to the mean free path. From accretion disk point of view, this means that most of the energy produced in the disk is liberated away from its two sides, and thus, the preferable order of splitting in the energy equation is correlated to the preferable direction of the energy transfer. Of course, the splitting in equations with zero effective forces in their right hand side is problem dependent (for a detailed discussion, see Hujeirat [4]).

## 4.3 Time Discretization

Following Rannacher [9] and Heywood&Rannacher [7], we apply a modified Crank-Nicolson scheme in the following form:

$$q_j^{n+1} + \theta\langle L_x q_j^x + L_y q_j^y + L_{xx} q_j^{xx} + L_{yy} q_j^y + L_{xy} q_j^{xy}\rangle^{n+1} = \theta f_j^{n+1} + (1 - \theta) f_j^n$$

$$-(1 - \theta)\langle L_x q_j^x + L_y q_j^y + L_{xx} q_j^{xx} + L_{yy} q_j^y + L_{xy} q_j^{xy}\rangle^n,$$

where $\theta = (1 + \alpha \delta t)/2$, with $\alpha = O(1)$. This method is of also of second-order accurate but due to the shift $\alpha \delta t/2$ more robust than the usual Crank-Nicolson scheme.

## 4.4 Spatial Discretization

In order to keep the calculated numerical solution consistent with the conservative form of the Navier-Stokes equations, we adapt the finite volume formulation. Let the difference form of the divergence theorem be written as

$$\frac{\delta(vol\, q)}{\delta t} + \Delta F = 0,$$

where F represents the flux. To be consistent with this formulation, we define the scalar quantities $q$ at the cell center and the flux $F$ at the cell surface. In our case, density, pressure, temperature, viscosity and any combination of them are defined at the cell center, while u, v and $v_\varphi$ are defined at the cell surface. Additionally, to avoid numerical instabilities that may be generated due to non-consistent interpolation of the pressure in steep gradient regions, the cell shifting method (see, e.g., Tscharnuter [12]) is used in the evaluation of the momentum equation. The convective terms are approximated by using first- or second-order upwinding.

In computing the approximate solution, we use a defect correction strategy. To clarify the idea, consider the following 1-D equation:

$$q_t^{n+1} + L_x F(q) + L_{xx} G(q) = f(q),$$

where $L_x$ is a convection operator and $L_{xx}$ a central viscous operator. We want to solve the system

$$q_t^{n+1} + L_x^c F(q) + L_{xx} G(q) = f(q),$$

or in short $A^c q^c = b$, where $L_x^c$ is a second-order approximation to $L_x$. Solving this system directly leads to unstable behaviour of the numerical solution (numerical oscillations, over- and undershooting, etc.). Instead we solve it step-wise in a predictor-corrector mode. First, we solve the system $A^{up} q^{up} = b \implies q^{up}$, where $A^{up}$ is the matrix resulting from the first-order upwind discretization of $L_x$, and $q^{up}$ is the corresponding upwind solution. Next, we compute the defect $d = b - A^c q^{up}$ with the matrix $A^c$ corresponding to a second-order discretization of $L_x$. Then, we use the same upwind matrix to solve the defect equation $A^{up} \mu = d$ and set $q^{\tilde{c}} = q^{up} + \mu$. This process may be repeated until no further improvement in accuracy is observed. In the limit one gets $q^{\tilde{c}} \longrightarrow q^c$.


## 4.5  The Computational Grid

Since large gradients are expected to occur in the BL, we use a strongly refined grid. The distribution of grid points is implicitly determined through the four parameters $dr_{min}$, $R_{out}$, $R_{in}$, $JR$, where $dr_{min}$ is the required finest grid size at the inner boundary, $R_{out}$ is the outer radius, $R_{in}$ is the inner radius, and $JR$ is the pre-fixed number of grid points. Using the relation $\xi = dr_j/dr_{j+1} > 1$, we obtain an equation of the form

$$\xi^{JR-1} + (\xi - 1)\frac{\Delta R_{in}^{out}}{dr_{min}} - 1 = 0,$$

where $\Delta R_{in}^{out} = R_{out} - R_{in}$. A fixed-point iteration may be used to compute the parameter $\xi$. Once $\xi$ is found, all grid spacings are known.


## 4.6  Artificial Viscosity

Since the flow considered is naturally viscous, the information about local compression is carried through its natural viscosity. Therefore, unless the natural viscosity is droped down beyond the numerical diffusion of the method, the importance of the artificial viscosity is heavily reduced. We allow information about real shock fronts to be carried through an additional coefficient of the artificial viscosity, while the structures of the viscous terms of the Navier-Stokes equations are kept in their original form. Therefore, the coefficient of the artificial viscosity has the form (shock capturing):

$$\nu_{art} = \begin{cases} \alpha l^2 (\nabla \cdot \vec{V}) & \text{if } \nabla \cdot \vec{V} \leq 0 \\ 0 & \text{otherwise,} \end{cases}$$

where $l = (dr(j)^\beta$, $\beta = 0.75 \to 1.0$, $0 < \alpha = O(1)$. Thus, following Tscharnuter [12], the total viscosity coefficient consists of a really physical and an artificial viscosity coefficient, i.e., $\nu_{tot} = \nu_{real} + \nu_{art}$.

# 5　Applications

Using the algorithm described, we performed several numerical calculations. First, the method was validated at the common shock tube problem. Then, the 1-D polytropic boundary layer problem was solved on a strongly refined mesh consisting of $N.P. = 1100 - 1500$ points and using more than $10^5$ time steps to follow the viscous development of the BL over several thousands of orbital periods. Typical CPU-times on a workstation (SUN Sparc 10/20) ranged between 9 and 39 hours. The corresponding results are plotted in Fig. 4. They show the development of a strong steady state or slightly oscillatory shock depending on the viscosity parameter $\alpha$. Next, the 2-D isothermal adiabatic and optically

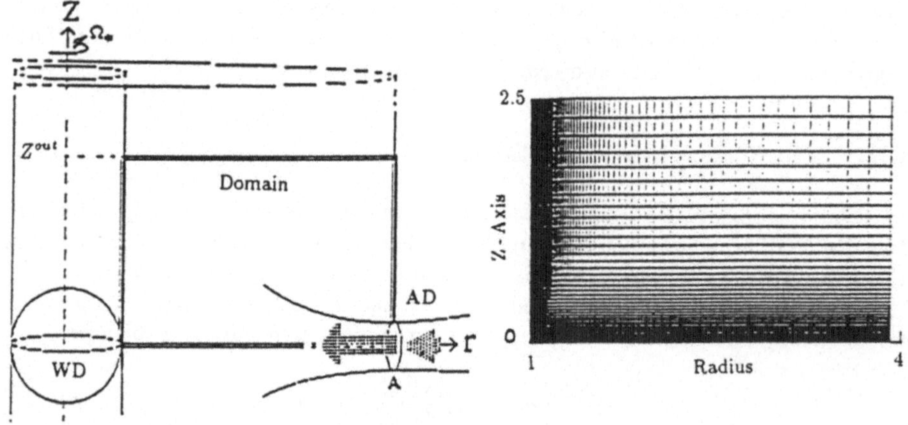


Figure 3: The physical and numerical domain of the 2-D calculations

thick BL in cylindrical coordinates was simulated. In this case, a strongly refined tensor-product mesh with 15000 points was used (Fig. 3). The calculation was carried over the viscous time scale with about 50000 time steps. Typical CPU-times on a workstation ranged between 28 and 48 hours. A selection of results is presented in Fig. 5. Additionally, the calculations were repeated using spherical coordinates to examine the results and their sensitivity to the geometrical compression. For the physical interpretation of the obtained results, we refer to Hujeirat [5] and [6].

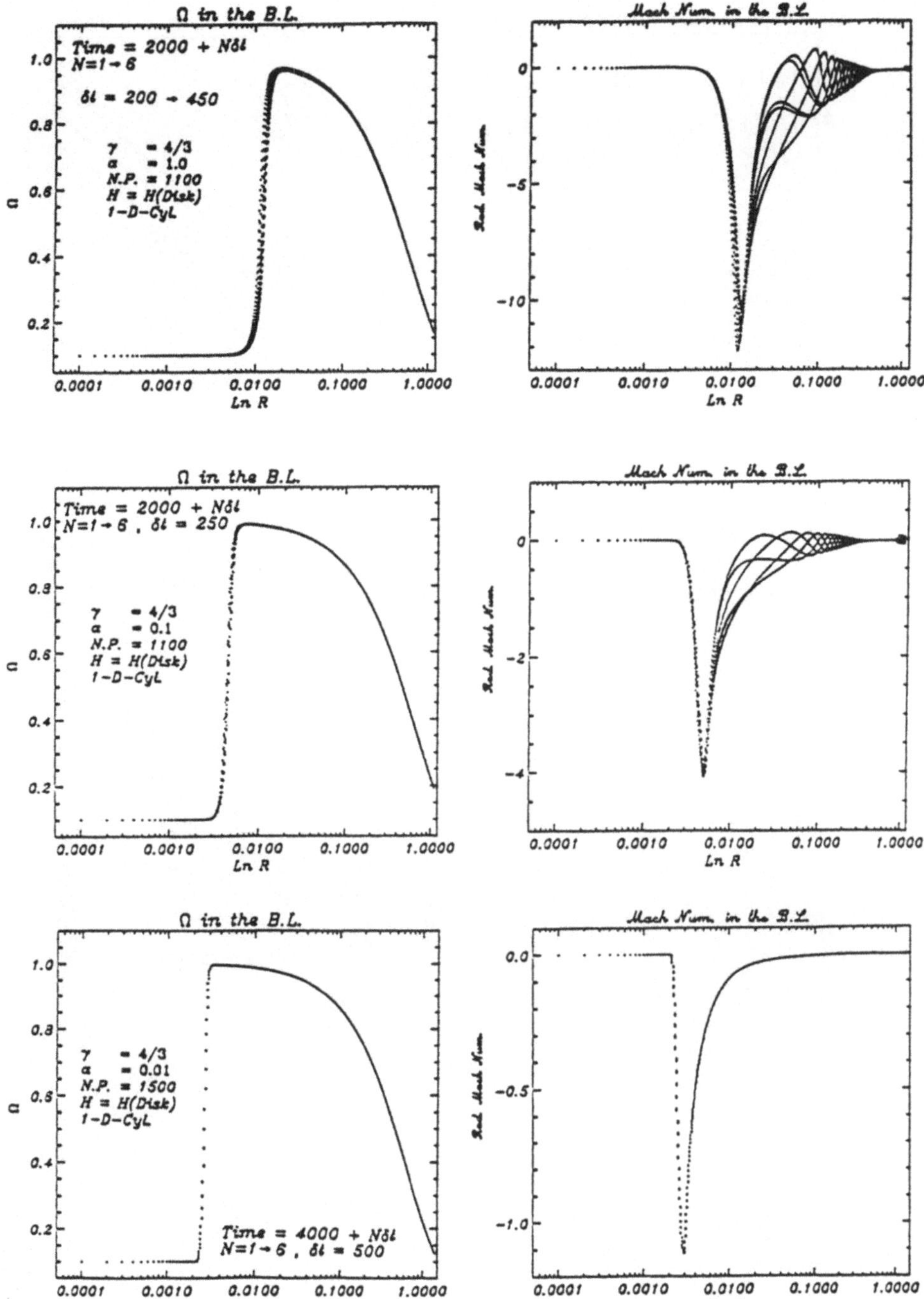

Figure 4: The 1-D distributions of the angular velocity and the radial Mach number for different values of the viscosity

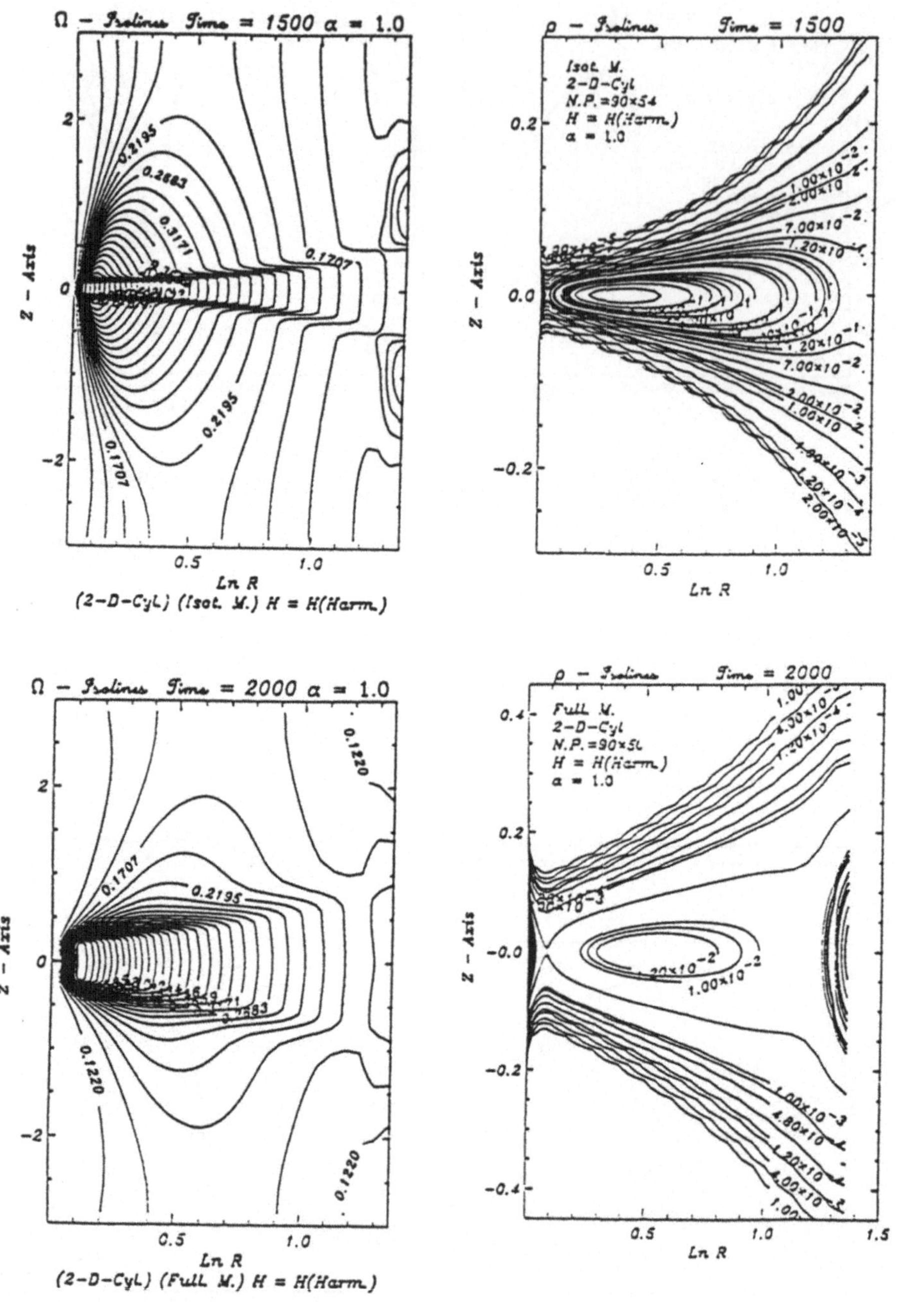

Figure 5: The 2-D angular velocity and density distributions in the BL-AD

# 6 Discussion

We have presented a numerical algorithm for solving the compressible Navier-Stokes equations over long time-intervals and its application to a complicated astrophysical problem. The algorithm is capable to provide steady and quasi-steady solutions in one and two dimensional models efficiently and independent of initial conditions. The robustness and accuracy of the method has been proved by various tests. Many thousands of time steps can be performed on a standard workstation and within several hours of CPU time. The algorithm seams to capture accurately strong time-dependent solutions like moving shocks and corona type circulations.

A weak point of this algorithm is the undesirable step size restriction due to the nonlinerity even when a steady state is reached. Further, the CPU-times required for a transient calculation is still to high in order to perform systematic parameter studies or to include more realistic models for energy transfer. A solution to these problems may be obtained by a modification of the operator splitting and by incorporating multi-level concepts into the solution process. Results in this direction will be reported in Hujeirat&Rannacher [7].

**Acknowledgment**
The authors thank the Deutsche Forschungsgemeinschaft (DFG) for supporting this work through the SFB 359 "Reaktive Strömungen, Diffusion und Transport" at the University of Heidelberg.

# References


[1] Beam, R. M., and Warming, R. F., An implicit factored scheme for the compressible Navier-Stokes equations, AIAA J. 16 (1978), 393-402.

[2] Gourlay, A. R., Splitting methods for time dependent PDEs, in The State of the Art in Numerical Analysis (D. Jacobs, edt.), pp. 757-791, Academic Press, London, 1977.

[3] Hensler, G., Hydrodynamical calculations of accretion disks in close binary systems, I, II, Astronomy and Astrophysics, 114 (1982), 309-327.

[4] Heywood, J. G., and Rannacher, R., Finite-element approximation of the nonstationary Navier-Stokes problem, Part IV: Error analysis for second-order time descretization, SIAM J.Numer.Anal. 27 (1990), 353-384.

[5] Hujeirat, A., A Numerical Approach Towards Steady State Solutions in Boundary Layers in Astrophysics, Dissertation, Universität Heidelberg, 1993.

[6] Hujeirat, A., Hydrodynamical calculations towards steady state structures in boundary layers in accretion disks, Part I and II, J. Astronomy and Astrophysics, 1994, submitted.

[7] Hujeirat, A., and Rannacher, R., A numerical method for solving the compressible Navier-Stokes equations modeling viscous boundary layers in accretion disks, in preparation.

[8] Pringle, J. E., Soft X-ray emission from dwarf novae, M.N.R.A.S. 178 (1977), 195-202.

[9] Rannacher, R., On the stabilization of the Crank-Nicolson scheme for long-time calculations, Technical Report, University of Saarbrücken, 1986.

[10] Shara, M., Resent progress in understanding the eruptions of classical novae, Puplications of the Astronomical Society of the Pacific, 101 (1989), 5-31.

[11] Strang, G., On the construction and comparison of difference schemes, SIAM J.Numer.Anal. 5 (1968), 506-517.

[12] Tscharnuter, W., and Winkler, K. H. A., A method for computing selfgravitating gas flow with radiation, Comp. Phys. Commun. 18 (1978), 171 -199.

# ON ERROR CONTROL IN CFD


Claes Johnson
Mathematics Department
Chalmers University of Technology
S–41296 Göteborg, Sweden

Rolf Rannacher
Institut für Angewandte Mathematik
Universität Heidelberg
D–69120 Heidelberg, Germany


## SUMMARY


We present in a simple generic situation a new approach towards quantitative error control in computational fluid mechanics. Combining so-called strong stability and Galerkin orthogonality, we derive sharp a posteriori and a priori $L_2$ error estimates for stationary nearly parallel pipe flow governed by the incompressible Navier–Stokes equations. These estimates state explicitly the dependence on the Reynolds number $Re$ in the form of a multiplicative constant proportional to $Re$.


# 1 Introduction

The available error analysis for numerical methods for the incompressible Navier-Stokes equations contains stability constants $C^s$ multiplying truncation errors or residuals, which are very large even for moderate Reynolds numbers. For instance, for smooth laminar non-stationary flow on relevant time scales, typically $C^s \sim \exp(Re)$ where $Re$ is the Reynolds number, see [9]. In cases of interest from fluid dynamics point of view, we usually have $50 \le Re \le 10^9$, and in these cases the existing error estimates for non-stationary problems thus appear to be void of content from any practical point of view. Similarly, the available estimates for stationary problems contain unspecified constants (except in the small data case) related to assuming that the exact solution is "isolated" or "on a regular branch of solutions", and thus are void of content as long as the size of the stability constants is left completely open (see e.g. [6],[13],[12]). In [9] we pointed out this unsatisfactory state of affairs and gave an a posteriori and a priori error analysis for time-discretization of a model case of non-stationary pipe flow with multiplicative stability constants $C^s$ proportional to $Re$, thus realizing an improvement from $\exp(Re)$ to $Re$. As far as we know, these are the first meaningful error estimates for the non-stationary incompressible Navier-Stokes equations in a case with moderately large Reynolds number.

In this note we give a similar error analysis for space-discretization of a stationary model problem of nearly parallel pipe flow, again containing a stability constant proportional to $Re$. The present paper is part of a long-term project to realize reliable and efficient quantitative adaptive error control for finite element methods for partial differential equations in general, and the Navier-Stokes equations in particular. For an introduction and references into this work, we refer to [11],[3]. Our adaptive algorithms are based on a posteriori error estimates guaranteeing reliability and the efficiency of the algorithms is evaluated through sharp a priori estimates. The theoretical foundation thus has two "legs", an a posteriori leg connected with reliability, and an a priori leg connected with efficiency. Each leg is built from combination of so called "strong stability" coupled with Galerkin orthogonality. For the a posteriori error estimates the strong stability concerns a linearized continuous dual problem, and for the a priori error estimates a corresponding discrete problem.

The purpose of this paper is to isolate an essential feature of the general problem of (adaptive) error control in CFD in the context of a model problem related to stationary nearly parallel flow. Nearly parallel flow represents a basic flow pattern met in e.g. pipe flow and boundary layer flow.

The stability constants $C^s$ measure the stability properties of the (continuous) linearized Navier-Stokes equations coupling residuals of approximate solutions to the corresponding perturbations of the exact solution. The residuals may be related to perturbations of data such as applied forces, boundary conditions or geometry, or to the numerical discretization. If $C^s$ is very large, then the Navier-Stokes equations are "unstable" and "non-computable" in the pointwise sense; a situation typically corresponding to turbulent flow. If the stability constant is not too large, then the flow is "computable". It appears that the stability constant $C^s$ for "organized" laminar flow in 3d may be of order $O(Re)$, and for complex flows possibly much larger. With the computer power of today, laminar flows in 3d with $Re$ in the range $10^2 - 10^3$ may be computable. For flows, with larger Reynolds numbers, turbulence modelling of small scales will have to be used reducing the effective Reynolds number to computable ranges (so called large eddy simulation LES). The set of computable flows is of course growing with increasing computer power.

Real flows are almost always three dimensional and thus only 3d computations could be directly meaningful. A rapid development towards CFD in 3d is taking place today: Flexible general 3d mesh generators exist and massive parallelism is beginning to give the required computer power. CFD in 2d requires significantly less computational work, not only because of dimensional reduction but also because stability constants in 2d may be much smaller than in 3d.

With 3d capabilities in CFD it also appears to be possible to compute the crucial stability constants $C^s$ for real flows by solving the linearized Navier-Stokes equations numerically. In this way it should be possible to realize quantitative adaptive error control in CFD for "computable" flows, cf [9]. The results obtained so far indicate e.g. that for Pouiseuille flow between two parallel plates, roughly $C^s \sim \frac{Re}{50}$, see [4]. We plan to perform extensive computations to determine the $C^s$ for different flows in 3d.

Note that since at best we expect to have $C^s \sim Re$, it follows that the limit of the 3d Navier-Stokes equations (2.1) as $Re$ tends to infinity, that is the Euler equations, in general are "infinitely unstable" and thus cannot in general be expected to have physically pointwise meaningful solutions. It thus appears that the meaning of a numerical solution of the Euler equations in 3d in general is very unclear; a strong mesh dependence could be expected in many cases and no convergence as the mesh size tends to zero may be observed. The numerical results for 3d Euler now starting to appear should shed light on this fundamental question.

An outline of the paper is as follows. In Section 2 we recall the Navier-Stokes equations for incompressible flow and in Section 3 we define certain stability constants $C_0^s$ and $C_1^s$ measuring different aspects of hydrodynamic stability. In Section 4 we consider the model problem of stationary nearly parallel streamwise constant pipe flow, and prove in this case estimates for the stability constants with explicit dependence on the Reynolds number. In Section 5 we formulate the finite element method and prove a posteriori and a priori error estimates with a linear dependence on the Reynolds number made explicit in the model case.

## 2  Navier-Stokes Equations

We recall the stationary Navier-Stokes equations: Find $(u, p)$ such that

$$\begin{aligned}
(u \cdot \nabla)u - \nu \Delta u + \nabla p &= f &&\text{in} \quad \Omega, \\
\nabla \cdot u &= 0 &&\text{in} \quad \Omega, \\
u &= 0 &&\text{on} \quad \partial\Omega,
\end{aligned} \tag{2.1}$$

where $u = (u_1, u_2, u_3)$ and $p$ are the velocity and the pressure of a Newtonian fluid with viscosity $\nu > 0$, enclosed in the volume $\Omega$ in $\mathbf{R}^3$, and $f$ is a given driving force. We assume that (2.1) is normalized, without loss of generality, so that the "reference" velocity $U$ and length scale $L$ are both equal to one. The Reynolds number $Re$ is then by definition given by $Re = ULv^{-1} = \nu^{-1}$ (where $\nu$ is to be understood as a dimensionless quantity). We consider here the case with $Re$ moderately large, say in the range $10 - 10^4$, in which case (2.1) may be expected to have a physically significant stationary solution under suitable conditions on the data. For Reynolds numbers larger than $10^4$, the chance of observing a stationary flow appears to be small.

## 3  Hydrodynamical Stability

Hydrodynamical stability is concerned with the stability properties of the Navier-Stokes equations and is of fundamental importance e.g. in the study of transition into turbulence and also in the analysis of numerical methods. The objective is to estimate the perturbation in the solution due to a perturbation of data. More precisely, let $(u, p)$ be a given solution of the Navier-Stokes equations (2.1) corresponding to the right hand side $f$, and let $(u + \varphi, p + \chi)$ be a second solution corresponding to the right hand side $f + g$. Subtracting the equations for $(u, p)$ and $(u + \varphi, p + \chi)$, we get the following equation for the perturbation $(\varphi, \chi)$:

$$\begin{aligned}
(u \cdot \nabla)\varphi + (\varphi \cdot \nabla)u + (\varphi \cdot \nabla)\varphi - \nu \Delta \varphi + \nabla \chi &= g &&\text{in } \Omega, \\
\nabla \cdot \varphi &= 0 &&\text{in } \Omega, \\
\varphi &= 0 &&\text{on } \partial\Omega.
\end{aligned} \tag{3.1}$$

If the quadratic term in $\varphi$ is omitted, we get the corresponding linearized perturbation equations (the linearized Navier-Stokes equations) linearized at the solution $(u, p)$:

$$\begin{aligned}
(u \cdot \nabla)\varphi + (\varphi \cdot \nabla)u - \nu \Delta \varphi + \nabla \chi &= g &&\text{in } \Omega, \\
\nabla \cdot \varphi &= 0 &&\text{in } \Omega, \\
\varphi &= 0 &&\text{on } \partial\Omega.
\end{aligned} \tag{3.2}$$

The basic question in quantitative hydrodynamic stability is to give bounds for the solution $(\varphi, \chi)$ of the "full" perturbation equation (3.1) or the linearized perturbation equation (3.2) in certain norms in terms of certain norms of the data $g$. A typical such bound could read as follows:

$$\|\varphi\| \leq C_0^s(u)\|g\|, \tag{3.3}$$

where $\| \cdot \|$ denotes the $L_2(\Omega)$-norm, and $C_0^s(u)$ is a stability constant depending on $u$, defined by

$$C_0^s(u) = \sup_{g \in L_2} \frac{\|\varphi\|}{\|g\|}, \tag{3.4}$$

where $\varphi$ is the solution of (3.1) or (3.2) corresponding to $g$. The stability constant $C_0^s(u)$ measures the $L_2$-norm of the perturbation of the solution in terms of the $L_2$-norm of the data. A large stability constant $C_0^s(u)$ corresponding to large perturbation growth, reflects the presence of energy transfer from the base flow to the perturbations. Typically, the stability constant will increase with increasing Reynolds number. A basic question in quantitative hydrodynamic stability related to transition to turbulence is to determine the dependence of the stability constant $C_0^s(u)$ on the Reynolds number for some laminar flows of interest. Below we prove for a model problem of nearly parallel streamwise constant pipe flow that $C_0^s(u) \sim Re^2$, an estimate which is sharp in the depencence on $Re$. The proof is based on simple "energy" estimates using the decoupled nature of the model problem. A result of similar nature is given in [7], where it is also showed by numerical experiments that a perturbation of size $Re^2$ may be sufficient to initiate turbulence (which from an intutive point of view fits with the estimate $C_0^s(u) \sim Re^2$).

In the error analysis below of Galerkin methods we shall need quantitative stability estimates for the dual of (3.2) defined by: Find $(\varphi, \chi)$ such that

$$\begin{aligned}
-(u \cdot \nabla)\varphi - \varphi \cdot \nabla v - \nu\Delta\varphi + \nabla\chi &= g && \text{in } \Omega, \\
\nabla \cdot \varphi &= 0 && \text{in } \Omega, \\
\varphi &= 0 && \text{on } \partial\Omega,
\end{aligned} \tag{3.5}$$

where the dependence on $u$ has also been extended to a dependence on the two functions $u$ and $v$, and where $(\varphi \cdot \nabla v)_i = \sum_{j=1}^{3} \varphi_j u_{j,i}$. Below, the function $v$ will play the role of the approximate solution $u_h$, and we thus should think of $v$ as being close to $u$. In addition to the stability constant $C_0^s(u, v)$ defined in analogy to (3.4) with (3.2) replaced by (3.5) with now dependence on both $u$ and $v$, it is in the error analysis of the finite element method natural to use another stability constant $C_1^s(u, v)$, measuring a weighted norm involving derivatives of the solution $(\varphi, \chi)$ of (3.5). Typically, $C_1^s(u, v)$ may be defined to be the smallest constant satisfying:

$$\|\nabla\chi\| + \nu\|D^2\varphi\| \leq C_1^s(u, v)\|g\| \qquad \forall g \in L_2(\Omega), \tag{3.6}$$

where $(\varphi, \chi)$ is the solution of (3.5), and $D^m v = \max_{|\alpha|=m} |D^\alpha v|$ with the usual multi-index notation for derivatives $D^\alpha v$ of order $\alpha$. We refer to the stability constant $C_0^s(u, v)$ not involving any derivative of $\varphi$ as the "weak" stability constant and the constant $C_1^s(u, v)$ as the "strong" stability constant. Note the normalization used in the definition of $C_1^s$ in (3.6) with a $\nu$-dependent norm, which is motivated by the analysis given below. Variants of this normalization may also be appropriate, as will be seen below.

We shall prove that for the pipe flow model problem, $C_1^s \sim Re$, which will be used to prove error estimates for finite element methods with multiplicative stability constants proportional to $Re$. For more aspects on the advantages in error analysis of using strong stability and Galerkin orthogonality, we refer to [9],[3].

Below, $C$ will denote a positive constant of moderate size independent of the mesh size $h$ and the Reynolds number $Re$.

# 4 Stationary $x_1$-independent Pipe Flow

As a basic model problem we shall now consider the Navier-Stokes equations (2.1) in the case of stationary flow in an infinitely long straight pipe $\Omega = \mathbf{R} \times \omega$, where $\omega$ in the $(x_2, x_3)$-plane is the crosssection (with smooth boundary) of the pipe, and the axis of the pipe is oriented along the $x_1$-axis. We shall first consider the case when both the base flow and the perturbations are $x_1$-independent. We shall below consider extensions to weak $x_1$-dependence of a certain form. Our basic assumption is that the base flow $(u, p)$ is nearly parallel in the sense that the transversal velocity $(u_2, u_3)$ is small in comparison with $u_1$. More precisely, we shall assume that

$$\|u_1\|_{1,\infty} = 1 \qquad \|u_i\|_{1,\infty} \le c\nu, \qquad i = 2, 3, \tag{4.1}$$

where $c$ is a sufficiently small constant and $\| \cdot \|_{1,\infty}$ denotes the $W^{1,\infty}(\omega)$-norm. In the $x_1$-independent case considered first, the Navier-Stokes equations take the form:

$$\begin{aligned}
(\bar{u} \cdot \bar{\nabla})u_1 - \nu\Delta u_1 &= f_1 && \text{in } \omega, & (4.2.\text{a}) \\
(\bar{u} \cdot \bar{\nabla})\bar{u} + \bar{\nabla}p - \nu\Delta\bar{u} &= \bar{f} && \text{in } \omega, & (4.2.\text{b}) \\
\bar{\nabla} \cdot \bar{u} &= 0 && \text{in } \omega, & (4.2.\text{c}) \\
u &= 0 && \text{on } \partial\omega, & (4.2.\text{d})
\end{aligned}$$

where $\bar{v} = (v_2, v_3)$ and $\bar{\nabla} = (\partial/\partial x_2, \partial/\partial x_3)$. We note the structure of (4.2) with the equations for $\bar{u}$ decoupled from the first equation for $u_1$. Using this decoupling and the standard small data result for the equations for $\bar{u}$, we see that (4.2) admits a unique solution if $\|\bar{f}\| \le c\nu^2$ with $c$ sufficiently small, and $f_1 \in L_2(\omega)$. Note that because of the normalization contained in (4.1), we typically expect to have $\|f_1\| \le C\nu$, but we are not limited to the small data case with $\|f\| \le c\nu^2$ corresponding to $Re \sim 1$. In our case $Re \sim \nu^{-1} \gg 1$.

The linearized perturbation equations (3.2) linearized at the solution $(u, p)$ just given take the following form assuming that also the perturbation $(\varphi, q)$ is $x_1$-independent:

$$\begin{aligned}
(\bar{u} \cdot \bar{\nabla})\varphi_1 + (\bar{\varphi} \cdot \bar{\nabla})u_1 - \nu\Delta\varphi_1 &= g_1 && \text{in } \omega, & (4.3.\text{a}) \\
(\bar{u} \cdot \bar{\nabla})\bar{\varphi} + (\bar{\varphi} \cdot \bar{\nabla})\bar{u} + \bar{\nabla}q - \nu\Delta\bar{\varphi} &= \bar{g} && \text{in } \omega, & (4.3.\text{b}) \\
\bar{\nabla} \cdot \bar{\varphi} &= 0 && \text{in } \omega, & (4.3.\text{c}) \\
\varphi &= 0 && \text{on } \partial\omega. & (4.3.\text{d})
\end{aligned}$$

The dual of (4.3) corresponding to (3.5) with a dependence on both $u$ and $v$, takes the form:

$$\begin{aligned}
-\bar{u} \cdot \bar{\nabla}\varphi_1 - \nu\Delta\varphi_1 &= g_1 && \text{in } \omega, \\
-(\bar{u} \cdot \bar{\nabla}\bar{\varphi} - \nu\Delta\bar{\varphi} + \bar{\nabla}v \cdot \varphi + \bar{\nabla}\chi &= \bar{g} && \text{in } \omega, \\
\bar{\nabla} \cdot \bar{\varphi} = 0 \text{ in } \omega, \varphi = 0 && \text{in } \partial\omega,
\end{aligned} \tag{4.4}$$

where $\bar{\nabla} v \cdot \varphi = (v_{1,j}\varphi_1 + \bar{v}_{,j} \cdot \varphi)^2_{j=1}$. We note the structure of the linearized problems (4.3) and (4.4) with $\bar{\varphi}$ and $\varphi_1$ being decoupled. Further, zero order terms with coeffients of order one involving $\varphi_1$ occur in the equations for $\bar{\varphi}$ and vice versa. This means that the coupling through zero order terms in (4.4) has a special form, which opens the possibility of analytically proving quantitative stability estimates. The assumption of $x_1$-independence makes the problems (4.2), (4.3) and (4.4) effectively two-dimensional, but as opposed to the fully two-dimensional case usually studied with independence with respect to the spanwise coordinate $x_3$ and with spanwise velocity $u_3 = 0$, we retain here important properties of three-dimensional flow , cf. [9].

For the model problem, it is natural to modify the definition of the stability constant $C_1^s(u,v)$ related to the dual (4.4) to be the smallest constant such that

$$\|\bar{\nabla}\chi\| + \nu\|D^2\bar{\varphi}\| + \|D^2\varphi_1\| \le C_1^s(u,v)(\|g_1\| + \nu\|\bar{g}\|) \qquad \forall g \in L_2(\omega), \qquad (4.5)$$

where $\|\cdot\|$ denotes the $L_2(\omega)$-norm and $(\varphi, \chi)$ is the solution of (4.4) with data $g \in L_2(\omega)$.

We now state and prove for the dual problem (4.4) basic estimates of the stability constant $C_0^s(u,v)$ defined by the analog of (3.4), and of the stability constant $C_1^s(u,v)$ defined by (4.5).

**Theorem 4.1** *Suppose that*

$$\|\bar{u}\|_\infty \le C\nu \qquad \|v_1\|_{1,\infty} \le C \qquad \|\bar{v}\|_{1,\infty} \le c\nu, \qquad (4.6)$$

*with the constant $c$ small enough. Then there is a constant $C$ such that the stability constants $C_i^s(u,v)$ for (4.4) in the context of $x_1$-independent nearly parallel pipe flow, satisfy the following bounds:*

$$C_0^s(u,v) \le C\nu^{-2}, \qquad C_1^s(u,v) \le C\nu^{-1}. \qquad (4.7)$$

**Proof** First, multiplying the first equation in (4.4) by $\varphi_1$, and integrating over $\omega$ using the fact that $\bar{\nabla}\cdot\bar{u} = 0$, gives

$$\nu\|\bar{\nabla}\varphi_1\|^2 \le \|g_1\|\,\|\varphi_1\|,$$

which proves by Poincares inequality that

$$\|\varphi_1\| + \|\nabla\varphi_1\| \le C\nu^{-1}\|g_1\|. \qquad (4.8)$$

Next, multiplying the momentum equation for $\bar{\varphi}$ by $\bar{\varphi}$ and using again the fact that $\bar{\nabla}\cdot\bar{u} = 0$ and that $c$ is small enough, we find easily that

$$\nu\|\bar{\nabla}\bar{\varphi}\|^2 \le C\nu^{-1}(\|\bar{g}\|^2 + \|\varphi_1\|^2) \le C\nu^{-3}\|g\|^2,$$

from which the desired estimate for $C_0^s(u,v)$ follows. Inserting this estimate into (4.4) and using elliptic regularity, we finally obtain the desired estimate for $C_1^s(u,v)$.

# 5 Finite Element Approximation

## 5.1 The Finite Element Method

We shall now prove a posteriori and a priori error estimates for a finite element method for (4.2). More precisely, we first prove an "abstract" a posteriori error estimate for the full three-dimensional problem (2.1) involving an unspecified stability constant $C_1^s(u, u_h)$. We then make this estimate concrete in the model case of nearly parallel $x_1$-independent pipe flow by providing the dependence of $C_1^s(u, u_h)$ on the Reynolds number. We also give in this case an a priori error estimate showing that the a posteriori error estimate is sharp up to a constant.

Introducing the space $\hat{V} = V \times H$, where $V = H_0^1[(\Omega)]^3$ and $H = L_2(\Omega)/\mathbf{R}$, we give the problem (2.1) the following variational formulation: Find $\hat{u} \equiv (u, p) \in \hat{V}$ such that

$$A(u; \hat{u}, \hat{v}) = L(v) \qquad \forall \hat{v} \equiv (v, q) \in \hat{V}, \tag{5.1}$$

where, with $\hat{w} = (w, r)$,

$$A(u; \hat{w}, \hat{v}) = a(w, v) + b(u; w, v) - (r, \nabla \cdot v) + (q, \nabla \cdot w),$$

$$b(u; w, v) = \int_\omega (u \cdot \nabla) w \cdot v \, d\bar{x},$$

$$a(w, v) = \nu \int_\omega \sum_{i=1}^3 \nabla w_i \cdot \nabla v_i \, d\bar{x}, \qquad L(v) = (f, v) \equiv \int_\Omega f \cdot v \, dx.$$

To formulate the finite element method, let $V_h$ and $H_h$ be finite dimensional subspaces of $V$ and $H$ consisting of continuous piecewise polynomials on a triangulation $T_h = \{K\}$ of $\Omega$ into elements $K$ of diameter $h_K$ given by the "mesh function" $h(x) = h_K$ for $x \in K$. Our results are stated with $h(x)$ variable, but for simplicity we assume in the proofs that $h(x) \sim h$ where $h$ is a constant "mesh parameter", corresponding to quasi-uniform meshes, see [2]. We further assume that $T_h$ satisfies the usual "minimal angle condition", i.e., the quotient of the inscribed and circumscribed sphere for each element is bounded below by a constant $\theta$ independent of $h$. We also assume that the pair $(V_h, H_h)$ satisfies the usual "inf-sup" condition:

$$\inf_{q \in H_h} \sup_{v \in V_h} \frac{(q, \nabla \cdot v)}{\|\nabla v\|} \geq \alpha \|q\|, \tag{5.2}$$

where $\alpha$ is a positive constant independent of $h$. For definiteness we assume that $V_h$ contains piecewise linears and $H_h$ piecewise constants. The assumption that $H_h$ consists of continuous functions is not essential. We may also consider stabilized Galerkin methods with equal order (continuous) approximation for which an infsup-condition is built in through the stabilizing terms.

Below we shall by $C$ denote "interpolation" constants of moderate size, which only depend on $\theta$, and the degree and the type of piecewise polynomials used (i.e. the $C$ may also depend on the constant $\alpha$ in the infsup-condition).

We consider the following standard Galerkin finite element method for (2.1): Find $\hat{u}_h \equiv (u_h, p_h) \in \hat{V}_h$, such that

$$A(u_h; \hat{u}_h, \hat{v}) = F(v), \qquad \forall \hat{v} \in \hat{V}_h. \tag{5.3}$$

In general, further stabilizations of the convective terms corresponding to the weighted least squares control of the Galerkin-least squares (or streamline diffusion) method, are necessary. In the concrete cases of nearly parallel flow studied below, this is not necessary because the convective terms are small in these cases, and we thus do not consider these modifications in detail in this note.

We now proceed to prove first the a posteriori and then the a priori error estimates. Before entering into the proof, the reader may take a quick look at Remark 6.1 below containing a simple model problem of relevance.

## 5.2 A Posteriori Error Estimates

The proof of the a posteriori error estimate for the finite element method (5.3) for the Navier-Stokes equations (2.1) has the following structure:

1. Error representation via a linearized dual continuous problem.

2. Use of the Galerkin orthogonality.

3. Interpolation error estimates for the dual solution.

4. Strong stability for the dual continuous problem.

We first represent the error $e = u - u_h$, through the solution $\hat{\varphi} \equiv (\varphi, \chi)$ of the dual linearized problem (3.5), with $g = e$, and $v = u_h$. In variational form this problem reads as follows: Find $\hat{\varphi} \in \hat{V}$, such that

$$L(u, u_h; \hat{v}, \hat{\varphi}) = (v, e) \qquad \forall \hat{v} \in V, \tag{5.4}$$

where

$$L(u, u_h; \hat{v}, \hat{\varphi}) = -((u \cdot \nabla)\varphi - \nabla u_h \cdot \varphi, v) + a(\varphi, v) + (\nabla \chi, v) - (\nabla q, \varphi). \tag{5.5}$$

Note that $L(u, u_h; \cdot, \cdot)$ represents the Frechet derivative of $A(u; \cdot, \cdot)$ evaluated "between" $u$ and $u_h$ so that

$$A(u; \hat{u}, \hat{\varphi}) - A(u_h; \hat{u}_h, \hat{\varphi}) = L(u, u_h; \hat{e}, \hat{\varphi}).$$

Choosing now $v = e$ in (5.4) and integrating by parts gives

$$\|e\|^2 = L(u, u_h; \hat{e}, \hat{\varphi}) = A(u; \hat{u}, \hat{\varphi}) - A(u_h; \hat{u}_h, \hat{\varphi})$$

$$= F(\hat{\varphi}) - A(u_h; \hat{u}_h, \hat{\varphi}) = F(\hat{\varphi} - \hat{\varphi}^h) - A(u_h; \hat{u}_h, \hat{\varphi} - \hat{\varphi}^h)$$

$$= ((u_h \cdot \nabla)u_h - f + \nabla p_h, \varphi - \varphi^h) + (\nu \nabla u_h, \nabla(\varphi - \varphi^h)) + (\nabla \cdot u_h, \chi - \chi^h),$$

where $\hat{\varphi}^h = (\varphi^h, \chi^h) \in \hat{V}$ is a nodal interpolant of $\hat{\varphi}$.

Estimating now the interpolation error $\varphi - \varphi^h$ and recalling the definition (3.6) of the strong stability constant $C_1^s(u, u_h)$ in the general case, we obtain the following "abstract" result (cf.[2]) :

**Theorem 5.1** *Let $u$ and $u_h$ be the solutions of the exact and discrete Navier-Stokes equations (2.1) and (5.3). Then there is a constant $\bar{C}$ such that*

$$\|u - u_h\| \leq \bar{C} C_1^s(u, u_h)(\nu^{-1}\|h^2 R_1(u_h)\| + \|h R_2(u_h)\|), \tag{5.6}$$

*where*

$$R_1(u_h)|_K = |(u_h \cdot \nabla)u_h + \nabla p_h - \nu\Delta u_h - f| + |\nu D_h^2 u_h| \quad \text{on } K,$$

$$R_2(u_h) = |\nabla \cdot u_h|.$$

*Here*

$$D_h^2 v = \frac{1}{2} \max_{S \subset \partial K} |[\frac{\partial v}{\partial n_S}]| \, h_K^{-1} \quad \text{on } K,$$

*where $[\frac{\partial v}{\partial n_S}]$ is the jump in normal derivative $\frac{\partial v}{\partial n_S}$ across the side $S$ of the element $K \in T_h$.*

We next give a concrete form of this estimate in the case of $x_1$-independent nearly parallel pipe flow by using the stability estimate of Theorem 4.1 and the definition (4.5) of $C_1^s(u, u_h)$.

**Theorem 5.2** *Suppose that the hypothesis (4.6) in Theorem 4.1 is satisfied with $v = u_h$. Then there is a constant $C$ such that*

$$\|u_1 - u_{1h}\| \leq \nu^{-1} C \bar{C}(\nu^{-1}\|h^2 \bar{R}_1(u_h)\| + \|h^2 R_{11}(u_h)\| + \|h \bar{R}_2(u_h)\|), \tag{5.7}$$

$$\|\bar{u} - \bar{u}_h\| \leq \bar{C}(\nu^{-1}\|h^2 \bar{R}_1(u_h)\| + \|h \bar{R}_2(u_h)\|), \tag{5.8}$$

*where $\| \cdot \| = \| \cdot \|_{L_2(\omega)}$ and*

$$\bar{R}_1(u_h) = |(\bar{u}_h \cdot \bar{\nabla})\bar{u}_h + \bar{\nabla} p_h - \nu\bar{\Delta}\bar{u}_h - \bar{f}| + \nu|D_h^2 \bar{u}_h|,$$

$$R_{11}(u_h) = |(\bar{u}_h \cdot \bar{\nabla})u_{1h} - f_1| + \nu|D_h^2 u_{1h}|,$$

$$\bar{R}_2(u_h) = |\bar{\nabla} \cdot u_h|.$$

## 5.3 A Priori Error Estimates

We next prove an a priori error estimate for the finite element method (5.3) for the Navier-Stokes equations (4.2). The proof uses the decoupled nature of (4.2) allowing an optimal estimate for $\bar{u} - \bar{u}_h$ to be proved by the usual techniques in the small data case, and then using this estimate in a standard way to estimate $u_1 - u_{1h}$. In both cases we may renormalize to the case $\nu = 1$. To emphasize the principal aspects, we shall here not write the proof this way, but instead structure the proof as follows paralleling the approach used above for the a posteriori error estimates:

1. Error representation via a linearized dual discrete problem.

2. Use of the Galerkin orthogonality.

3. Interpolation error estimates for the exact solution.

4. Strong stability for the dual discrete problem.

We represent the error $e^h = u^h - u_h$, where $u^h \in V_h$ is the nodal interpolant of $u$, through the solution $\hat{\varphi}_h \in \hat{V}_h$ of the discrete anlogue of the dual linearized problem (3.5), with $g = e^h$, and $v = u_h$. In variational form this problem reads as follows: Find $\hat{\varphi}_h \in \hat{V}_h$, such that

$$L(u, u_h; \hat{v}, \hat{\varphi}_h) = (v, u^h - u_h) \qquad \forall \hat{v} \in \hat{V}_h, \tag{5.9}$$

Choosing here $\hat{v} = \hat{e}^h$ gives

$$\|e^h\|^2 = L(u, u_h; \hat{e}^h, \hat{\varphi}_h)$$

$$= L(u, u_h; \hat{u} - \hat{u}_h, \hat{\varphi}_h) + L(u, u_h; \hat{u}^h - \hat{u}, \hat{\varphi}_h) = L(u, u_h; \hat{u}^h - \hat{u}, \hat{\varphi}_h).$$

We now define the discrete strong stability constant $C_{1h}^s(u, u_h)$ to be the smallest constant satisfying

$$\|\bar{\nabla}\chi_h\| + \nu\|\hat{D}_h^2\bar{\varphi}_h\| + \|\hat{D}_h^2\varphi_{1h}\| + \nu^{-1}\|h\bar{\nabla}\cdot\bar{\varphi}_h\| \leq C_{1h}^s(u, u_h)\|g\| \qquad \forall g \in L_2(\omega), \tag{5.10}$$

where $\hat{\varphi}_h = (\varphi_h, \chi_h) \in \hat{V}_h$ is the solution of (5.9) with $e^h$ replaced by $g$, and $\hat{D}_h^2 v = D_h^2 v + \Delta v$ in $K$.

For the model case of nearly parallel $x_1$-independent flow, we have the following estimate of the stability constant $C_{1h}^s(u, u_h)$:

**Theorem 5.3** *Suppose that the exact velocity $u$ satisfies(4.1). Then there is a constant $C$ such that*

$$C_{1h}^s(u, u_h) \leq C\nu^{-1}. \tag{5.11}$$

**Proof** The proof follows using the stability estimates of Theorem 4.1 for the continuous dual problem (4.4) with solution $\varphi$, together with standard error estimates for $\|\varphi - \varphi_h\|$ derived using the decoupled nature of (4.4). In paricular it follows from these estimates that the assumptions of Theorem 4.1 with $v = u_h$ are satisfied. We omit the details.

Combining this estimate with the above representation formula, we get the following a priori error estimate (estimating also $\|\bar{\nabla}p - \bar{\nabla}p_h\|$ in an analogous way):

**Theorem 5.4** *Let $u$ and $u_h$ be the solutions of the exact and discrete Navier-Stokes equations in the context of nearly parallel $x_1$-independent flow satisfying 4.1. Then there is a constant $\bar{C}$ such that*

$$\nu^{-1}\|h(\bar{\nabla}p - \bar{\nabla}p_h))\| + \|\bar{u} - \bar{u}_h\| \leq \bar{C}(\|h^2 D^2\bar{u}\| + \nu^{-1}\|h^2\nabla p\|), \tag{5.12}$$

$$\|u_1 - u_{1h}\| \leq \bar{C}\|h^2 D^2 u_1\| + \nu^{-1}\bar{C}(\|h^2 D^2\bar{u}\| + \nu^{-1}\|h^2\nabla p\|). \tag{5.13}$$

We note the equilibration of approximation with respect to $u$ and $p$; if $H_h$ contains higher order polynomials, we get corresponding increase in the power of $h$ in the $p$-term.

Concerning the efficiency of adaptive algorithms based on the a posteriori error estimate of Theorem 5.2 we note that the right hand sides of the a posteriori error estimate of Theorem 5.2 are bounded by the corresponding right hand sides of the a priori error estimates of Theorem 5.4 up to a constant $\bar{C}$. This follows as in [2] bounding the a posteriori quantities using inverse estimates and the a priori error estimates.

# 6 Extension to Weak $x_1$-dependence

In this section we indicate an extension of the above results for $x_1$-independent flows to flows with weak $x_1$-dependence of the form:

$$v(x_1, \bar{x}) = v(\bar{x}) \exp(i\alpha x_1), \tag{6.1}$$

where

$$\alpha \leq c\nu, \tag{6.2}$$

with $c$ sufficiently small. We note the convention used in (6.1): All functions $v(x)$ occuring are assumed to have a dependence on $x_1$ according to (6.1), where we denote the $\bar{x}$-dependent multiplicative factor again by $v(\bar{x})$.

The dual linearized problem (3.5) now takes the following form:

$$
\begin{aligned}
-i\alpha u_1 \varphi_1 - \bar{u} \cdot \bar{\nabla}\varphi_1 - \nu\bar{\Delta}\varphi_1 + \nu\alpha^2\varphi_1 + v_{1,1}\varphi_1 + \bar{v}_{,1} \cdot \bar{\varphi} + i\alpha\chi &= g_1 && \text{in } \omega, \\
-i\alpha u_1 \bar{\varphi} - (\bar{u} \cdot \bar{\nabla}\bar{\varphi} - \nu\Delta\bar{\varphi} + +\nu\alpha^2 + \bar{\nabla}v \cdot \varphi + \bar{\nabla}\chi &= \bar{g} && \text{in } \omega, \\
i\alpha\varphi_1 + \bar{\nabla} \cdot \bar{\varphi} &= 0 && \text{in } \omega, \\
\varphi &= 0 && \text{in } \partial\omega.
\end{aligned}
\tag{6.3}
$$

The extension of Theorem 4.1 to the present situation of slow $x_1$-dependence reads as follows:

**Theorem 6.1** *Suppose that (6.2) holds and*

$$\|\bar{v}\|_{1,\infty} + \nu^{-1}\|\bar{v}\|_{1,\infty} \leq c\nu, \qquad \|v_1\|_{1,\infty} + \nu^{-1}\|v_{1,1}\|_{\infty} \leq C, \tag{6.4}$$

$$\|\bar{u}\|_{\infty} \leq C\nu, \tag{6.5}$$

*where the constant $c$ in (6.4) and (6.2) are small enough. Then the stability constants $C_i^s(u,v)$ for (6.3) in the context of slowly $x_1$-dependent nearly parallel pipe flow, satisfy the following bounds:*

$$C_0^s(u,v) \leq C\nu^{-2}, \qquad C_1^s(u,v) \leq C\nu^{-1}. \tag{6.6}$$

**Proof**    The proof follows estimating first $\varphi_1$ in terms of $\bar{\varphi}$ through the first equation, and inserting this result into the equation for $\bar{\varphi}$.

We can now prove a posteriori and priori error estimates analogous for slowly $x_1$-dependent flow according to (6.1)-(6.2) which are analogous to the estimates stated above for $x_1$-independent flow. These results could probably be applied to more general slow $x_1$-dependence by using Fourier transformation in $x_1$. We leave the details for future publication.

**Remark 6.1** A basic feature of the model problem may be exhibited through the linear system

$$
\begin{aligned}
-\nu\Delta\varphi_1 + \varphi_2 &= g_1 && \text{in } \omega, \\
-\nu\Delta\varphi_2 &= g_2 && \text{in } \omega, \\
\varphi_1 = \varphi_2 &= 0 && \text{on } \partial\omega.
\end{aligned}
\tag{6.7}
$$

A finite standard piecewise linear finite element approximation $u_h = (u_{1h}, u_{2h})$ of this system satisfies the following bounds (if $\partial\omega$ is smooth or $\omega$ is convex):

$$
\begin{aligned}
\|u_2 - u_{2h}\| &\leq C\|h^2 D^2 u_2\|, \\
\|u_1 - u_{1h}\| &\leq \nu^{-1} C\|h^2 D^2 u_2\| + C\|h^2 D^2 u_1\|.
\end{aligned}
\tag{6.8}
$$

We note the analogy with the estimates of Theorem 5.4 with a stability constant depending linearly on $\nu^{-1}$.

# References


[1] K. M. Butler and B. F. Farrell, Three-dimensional optimal perturbations in viscous shear flow, Phys. Fluids, 8(1992), 1637-1650.

[2] K. Eriksson and C. Johnson, Adaptive finite element methods for parabolic problems: A linear model problem, SIAM J. Numer. Anal., 28(1991), 43-77.

[3] K. Eriksson, D. Estep, P. Hansbo and C. Johnson, Adaptive Finite Element Methods, North Holland, to appear.

[4] N. Eriksson, On the stability of pipe flow, Master of Science Thesis, Mathematics Department, Chalmers University of Technology, 1993.

[5] H. Gustavsson, Energy growth of three-dimensional disturbances in plane Poiseuille flow, J. Fluid Mech., 224(1991), 241-260.

[6] V. Girault and P.A. Raviart, Finite Element Methods for the Navier-Stokes Equations, Lecture Notes in Math. 749, Springer, Berlin, 1979.

[7] G. Kreiss, Anders Lundblad and D. Henningson, Bounds for treshold amplituides in subcritical shear flows, Preprint Trita-NAS-9307, Department of Numerical Analysis and Computing Science, Royal Institute of Technology, Stockholm.

[8] C. Johnson and A. Szepessy, Adaptive finite element methods for conservation laws, to appear in CPAM.

[9] C. Johnson, R. Rannacher and M. Boman, Numerics and hydrodynamic stability: Towards error control in CFD, Preprint 1993-13, Mathematics Department, Chalmers University of Technology, 1993, to appear in SINUM.

[10] C. Johnson and R. Rannacher, A quantitative stability analysis of nearly parallel flow, to appear.

[11] C. Johnson, A new paradigm for adaptive finite element methods, Proc MAFELAP Conf Brunel Univ. 93, Wiley, to appear.

[12] L. Tobiska and R. Verfürth, Analysis of a streamline diffusion finite element method for the Stokes and the Navier-Stokes equations, SIAM J. Numer. Anal., to appear.

[13] R. Verfürth, Adaptive finite element methods, preprint, Zürich Univ 1993.

[14] L. N. Trefethen, A. E. Trefethen, S. C. Reddy and T. A. Driscoll, A new direction in hydodynamic stablity: Beyond eigenvalues, Technical Report CTC92TR115 12/92, Cornell Theory Center, Cornell University, 1992.

# Parallel Grid Adaptation


Y. Kallinderis*

Dept. of Aerospace Engineering and Engineering Mechanics
The University of Texas at Austin
Austin, TX 78712


## Abstract


A unified parallel algorithm for grid adaptation by local refinement/coarsening is presented. It is designed to have two unique features: independence from type of the grid, as well as from any particular parallel architecture. This is achieved by employing a generic data template which is configured to capture the data structures for any computational grid regardless of structure and dimensionality. The unified parallel algorithm is employed for dynamic adaptation of 3-D, unstructured tetrahedral grids on a partitioned memory MIMD architecture. Performance results are presented for the iPSC/860.


## 1 Introduction

Adaptive grid algorithms are employed extensively in computational fluid dynamics (CFD). They provide flexibility to adjust the grid during the solution procedure without intervention by the user. A popular method divides initial coarse grid-cells, thus creating locally embedded grids. Several levels of such finer grids are allowable, and they can be limited to those regions of the domain in which important features exist. Conversely, excessive resolution is removed by deleting grid-cells locally over regions in which the solution does not vary appreciably. Several such algorithms for two-dimensional grids have been developed [1, 2]. Furthermore, adaptive local refinement/coarsening of unstructured tetrahedral grids has been developed and implemented for complex, 3-D geometry flow simulations [3, 4]. Different types of grid topology have been been employed within the same domain in order to resolve the various types of flow features [5]. Very little work has been done on development of efficient parallel algorithms for grid adaptation [6].

There is a need to develop architecture-independent algorithms for grid adaptation with sustained high performance across a wide variety of parallel machines. It is also imperative that the data structure for such algorithms be decoupled from the specific topology and dimensionality of the computational grid. A generic parallel adaptive algorithm for solution of the Navier-Stokes equations on a 2-D, quadrilateral mesh, which is portable across SIMD and shared-memory MIMD architectures has been developed [6]. The main idea is to express the algorithm in terms of generic parallel "primitives" which are independent of the underlying system. The architecture-specific details are encapsulated in the implementation of these primitives on that architecture.

---

*Associate Professor

The current work presents a more "unified" approach to parallel grid adaptation. It consists of a unified parallel algorithm for grid adaptation which uses generic parallel primitives similar to those presented in [6]. This algorithm however, encompasses partitioned memory MIMD architectures in addition to the two discussed in [6]. It is efficiently implemented on the Intel iPSC/860 and manifests high performance and scalability. Also, the data structure it employs is in the form of a "data template" which can be configured to capture any computational grid regardless of dimensionality and topology. The universal characteristic of this template is illustrated by showing how it can be configured to represent a 3-D unstructured, tetrahedral grid as well as the computational grid employed in [6].

## 2 The Adaptive Grid Method

We consider a 3-D unstructured grid made up of tetrahedral cells. Adaptive embedding introduces finer cells in those regions of the field that need be resolved, while simultaneously removing cells from previously embedded regions that do not require extra resolution. The adaptive grid method considered in this work is discussed in detail in [3]. We present here, an overview of the same.

Grid embedding essentially involves dividing a grid cell by inserting new nodes. The new nodes are introduced in the middle of the edges of the tetrahedra to be refined. Each edge is divided into two edges and each face into four faces after introducing three edges in the interior of the divided face. This scheme thus results in the formation of eight children cells as shown in Figure 1(a).

Two special types of refinement are also employed. One type involves the division of only one face of a tetrahedral cell into four children. The cell itself is then divided into four children as shown in Figure 1(b). The second method divides only one out of the six original edges, and the cell is henceforth divided into two children as shown in Figure 1(c).

An important aspect of grid adaptation is the deletion of previously divided tetrahedra. Cells that are to be deleted are eliminated along with their sibling cells which were formed from the same parent cell. In this way, the parent cells are recovered after the deletion of their child cells.

### 2.1 The Postorder Representation

In order to achieve efficient parallel grid adaptation, it is essential to impose certain rules for addition and removal of entities from the data structure lists during the refinement/coarsening process [6]. For instance, when a grid gets refined, one or more cells get divided into multiple children. This results in the formation of "adaptation trees", one for each original grid cell as shown in Figure 2(a). The insertion of these new cells into the list of cells $C$ has to be done in a manner most efficient from the point of view of parallelisation. Furthermore, the resulting list of cells should have a structure which allows efficient deletion of these children cells if the need arises later. Since refinement and coarsening are both "local" operations, it is apparent that the most efficient way of insertion would be to insert new children corresponding to a particular cell in its immediate vicinity in the lists. This would in turn, allow efficient coarsening since a cell to be coarsened would find all its children next to it in the cell listing.

Due to the above reasons, the new cells to be created are inserted in such a manner that the final listing of cells corresponds exactly to the postorder listings of all the adaptation trees placed contiguously one after the other. This special ordering is illustrated in Figure 2(b). It is seen that the newly introduced children of a cell in the original grid are to the immediate left of that cell. This creates a "local access mechanism" for a parent cell which needs to access its children cells for coarsening.

Similarly, when a face gets divided into multiple faces, all its children are inserted in contiguous fashion on its immediate left in $F$. The same technique is also applied for division of edges into multiple edges.

The Prefix Primitive

The Prefix primitive performs a "prefix" operation on a given list of numbers. For each element in the initial list, its prefix is obtained by summing together all the elements on its left including itself [7]. A list L is said to be a prefix of list M ($L = Prefix(M)$) if and only if:

$$L(i) = \Sigma_{j=1}^{j=i} M(j), \forall i, 1 \le i \le |M|.$$

The prefix primitive is used to calculate the new positions of entities in a list after refinement/coarsening. For instance, in the division case, the input to the prefix operation is a list of integers of the same length as the list of entities with an integer in position $i$ holding the number of entities to be added as a result of division of the $i^{th}$ entity of the original entity list. For an entity not to be divided, this integer is 0. After the prefix operation, the $i^{th}$ integer indicates the number of positions through which the $i^{th}$ entity has to move in order to reach its correct position in the entity list after division.

## 3  The Unified Parallel Algorithm

The unified parallel algorithm for grid adaptation can be described in terms of the generic parallel primitives and the data template defined earlier. It employs a "hierarchical approach" wherein entities are divided and undivided in ascending and descending order of dimensionality respectively. In other words, edges, faces and cells are divided in that order during refinement, while unrefinement is performed in the reverse order. The division/deletion procedure is essentially the same for all entities algorithmically, with the only differences being in the number of new entities added and the assignment of attributes to the newly created entities. For instance, during division of cells in a 3-D tetrahedral grid, eight new cells are created for each cell divided, while during division of edges, only two new edges are created for each edge to be divided. As a result, this procedure can be described in an entity-independent fashion and the overall algorithm can then be expressed as repeated applications of this procedure to various entity lists.

### 3.1  The Entity Division/Deletion Procedure

The procedure takes in as inputs, an integer n and two lists I and M, where I is the list of entities and M is a list of integers of the same cardinality as list I. An element in M has value 1 if the corresponding element in I is to be divided, -1 if it is to be deleted and 0 if neither. The integer n specifies how many children entities are formed when an entity in I gets refined or how many are deleted when it is coarsened. A positive value of n denotes division while a negative value denotes deletion. The steps in the procedure are as follows:

147

1. $M_p = n * M$. This step multiplies all elements of the list $M$ by $n$ and creates a new list $M_p$. The list $M_p$ now holds the number of new entities that are to be introduced or removed for each entity. Thus $M_p$ contains 0 if the corresponding entity is neither to be divided nor deleted and $n$ otherwise.

2. $M_p = Prefix(M_p)$. The list $M_p$ is assigned the result of performing a prefix operation on it. It now holds for each entity, the "correction" that has to be applied to its position in the entity list as a result of the division/deletion.

3. $M_p = M_p + I_{pos}$, where $I_{pos}$ is simply a list of integers containing the integer $i$ in its $i^{th}$ position. As a result of this operation, the list $M_p$ now contains the new position for the $i^{th}$ entity in position $i$, with $i$ being an integer from 1 to $|L|$.

4. Setup the attributes for the newly created entities. This step is performed only in case of division. The particulars of this step depend on the type of entities in the entity array as well as their attributes. For example, in case of the 3-D unstructured grid, division of the list of cells would be followed by a step to fill in the cell-to-face, cell-to-edge and cell-to-node mappings for each newly created cell.

It can be seen that the above procedure uses only the generic primitives defined earlier as its building blocks.

## 3.2 The Hierarchical Adaptation Method

The grid adaptation algorithm consists of a refinement step and an unrefinement step. During the refinement step, entities that are marked for division are divided by application of the division/deletion procedure described earlier. The input to this step is a list of "flags" of cardinality equal to that of the highest dimensional entity list i.e. cells in case of a 3-D tetrahedral grid. An entity is flagged 1 if it is to be divided and zero otherwise. The refinement is performed as follows:

1. The flags for the highest entity list are scattered to all entities of lower dimensions. Thus, in case of a 3-D tetrahedral grid, a cell to be divided flags all its constituent faces and edges as candidates for division.

2. The division/deletion procedure is applied to all entity lists starting from the list corresponding to the entity lowest in dimensionality. This ensures that when an entity list of a particular dimensionality is being divided, all lists of lower dimensionality are already divided thereby enabling the division of higher entities. In case of a tetrahedral grid, the edges are divided first, followed by the faces and cells in that order.

In the unrefinement step, the division/deletion procedure is applied to the entity lists in the reverse order. The input is a list of flags similar to that in the case of refinement, except that an entity to be deleted is marked -1. The overall unrefinement is performed as follows:

1. The highest entity list scatters the flags to all lower entities. This step is exactly the same as the first step in refinement.

2. Those entities in the highest entity list that are *not* to be deleted now scatter their flags, which are all zeros, to all lower entities. This is termed as the "constraint imposition step".

3. The division/deletion procedure is now applied to all entity lists in descending order of dimensionality. Thus, in case of the 3-D tetrahedral grid, the cells are deleted first, followed by the faces and edges.

## 4  Application to a Partitioned Memory MIMD Architecture

A generic, parallel adaptive Navier-Stokes algorithm has been developed [6] for shared memory MIMD (CRAY-YMP) and SIMD (CM-2) architectures. The grid is a 2-D quadrilateral grid which is structured in two dimensions initially, but is rendered unstructured by adaptation. The adaptive Navier-Stokes algorithm itself was based upon generic parallel primitives which are similar to those used by the unified algorithm. In this section, we demonstrate the universal nature of the unified algorithm by presenting its implementation and performance results on a partitioned memory MIMD architecture, the Intel iPSC/860.

The same user program is executed on all the processors each with its own set of data. Coordination among processors is achieved through message passing for which "send" and "receive" primitives are provided. These can be either synchronous or asynchronous depending upon the requirement of the algorithm. The programming paradigm is essentially that of any ordinary sequential machine. That is, the actual structure of any program written for the iPSC has basically a sequential form with additional calls to the message passing routines for synchronisation among processors.

We consider the problem of local refinement/coarsening of a 3-D unstructured, tetrahedral grid. The grid is partitioned among the processors in such a way that any cell belongs to exactly one processor while faces, edges and nodes can be shared by more than one processors. The partitions are assigned one each to all the processors. Details regarding the partitioning algorithm are discussed in [8].

The data structure for the grid is derived from the generic template. It consists of all the components described earlier and some additional information regarding the entities that are shared with other processors. Each processor maintains this data structure for its portion of the computational grid. The additional components of a processor's data structure are as follows:

1. Considering the total number of processors to be P, $P - 1$ pairs of integers are added as attributes to the attribute lists of all faces, edges and nodes. For a given entity, there is one pair for each other processor in the system. The first integer of the pair is the processor number and the second integer gives the *id* of that entity on that processor. For an entity which belongs to only one processor, the second integer is -1 for all pairs. Thus the attributes for a node on processor 0 for a system consisting of 4 processors are : $< x, y, z, < 0, -1 >, < 1, i_1 >, < 2, i_2 >, < 3, i_3 >>$, where $i_1, i_2, i_3$ are "images" or *ids* of that node on processors 1,2 and 3 respectively.

2. Three lists $F_{adj}$, $E_{adj}$ and $N_{adj}$ which hold information regarding processors that share entities with the processor. The list $F_{adj}$ is a list of integers denoting the

numbers of all processors that share at least one face with this processor. Similarly, $E_{adj}$ and $N_{adj}$ contain the numbers of processors sharing at least one edge and node respectively.

The division part of the algorithm takes in as input the data structure for the computational grid and a list $F_c$ of integers of cardinality equal to that of the list of cells. A cell is to be divided if the corresponding element in $F_c$ is 1, and it is to be deleted if it is -1. The cell is unchanged by adaptation otherwise. Additional lists $F_f$ and $F_e$ are used to hold similar "flags" for the faces and edges respectively.

1. For each cell $i$, where $F_c(i) = 1$,

   (a) Set $F_f(C[j]) = 1$, for $j = 11, 14$
   (b) Set $F_e(C[j]) = 1$, for $j = 5, 10$.

2. Call procedure $divide(\text{E},F_e)$.

3. Call procedure $divide(\text{F},F_f)$.

4. Call procedure $divide(\text{C},C_f)$.

The attributes 11 through 14 for a cell in the 3-D tetrahedral grid data structure hold the *ids* of its faces while attributes 5 through 10 hold the edges.

The above steps illustrate the importance of the hierarchical approach to adaptive refinement employed by the unified algorithm. The edges, which are dimensionally the lowest, are divided first so that higher entities can be divided without consideration to the fact whether their constituent edges have been refined. The faces are divided next, followed by the cells.

The inputs to the deletion algorithm are the grid data structures, a list of flags for the cells an element of which contains a -1 if the corresponding cell is to be deleted and a 0 otherwise. Additional lists are used to hold similar flags for the faces and edges as well. The steps in the algorithm are as follows:

1. For each cell $i$, where $F_c(i) = -1$,

   (a) Set $F_f(C[j]) = -1$, for $j = 11, 14$
   (b) Set $F_e(C[j]) = -1$, for $j = 5, 10$.

2. For each cell $i$, where $F_c(i) = 0$,

   (a) Set $F_f(C[j]) = 0$, for $j = 11, 14$
   (b) Set $F_e(C[j]) = 0$, for $j = 5, 10$.

3. Call procedure $delete(\text{E},F_e)$.

4. Call procedure $delete(\text{F},F_f)$.

5. Call procedure $delete(\text{C},C_f)$.

The first scatter step above can be called an "intention to delete" step where every cell that is to be deleted marks all its associated entities with a -1 flag. The second scatter step is the constraint imposition step wherein all cells that are to be retained ensure that none of *their* entities are deleted by the deletion module. Thereafter, the deletion module is called once each for cells, faces and edges.

Some additions to the above manifestation of the unified algorithm are required for synchronisation across partition boundaries. During division, after the faces and edges of cells to be divided are marked with flags as per the unified algorithm, each processor performs a "logical OR" operation for each marked face/edge that is shared with other processors. This basically involves exchanging information regarding that entity with all processors that share it. This is due to the fact that if one processor divides its copy of the shared entity, then all processors sharing that entity have to divide their own copies as well.

The deletion part of the adaptive algorithm needs only one enhancement to the unified version. This is done in the constraint imposition step when cells that are not marked for deletion impose constraints against the deletion of their faces and edges. A logical OR operation, exactly the same as the one described above is performed on all shared faces and edges. This is because even if one processor does not delete its copy of a shared face or node, then no other processor can delete its copy of the same entity.

Figure 3 shows an application of the above algorithm for parallel adaptation of a 3-D channel grid with a moving blast wave front. The wave starts from the origin and propagates away from it, continuously growing in size. The adaptive algorithm monitors the position of the wave and places locally finer grids in its vicinity. Figure 3(b) shows the adapted grid with the wavefront having moved away from its previous position of Figure 3(a). It can be seen that the adaptation in the vicinity of the previous position is removed while additional cells are introduced around the new position of the wave. The amount of adaptation is seen to increase in proportion to the size of the wave.

## 4.1 Performance Results

We present performance results on the Intel iPSC/860 for a 3-D channel grid consisting of tetrahedra. The grid is adapted once by dividing a certain number of cells within each processor partition. These are then immediately deleted. The total execution time for this step is measured on each processor. The total time for execution of the algorithm is taken to be the maximum of all the individual processor timings. The different parameters to be considered are: (i) the number of processors into which the original grid is divided, (ii) the partitioning method used, and (iii) the number of cells divided/deleted within each partition.

We consider two different methods for grid partitioning for performance evaluation of the parallel adaptive algorithm. The "strip" partitioning method partitions the original grid among processors by using cutting planes along one of the three coordinate axes. The "all-round' partitioning method partitions the initial grid using cutting planes along all three coordinate axes. Details regarding these methods are discussed in [8]. These two methods differ significantly in terms of communication costs and as a result, it is insightful to evaluate the performance of the adaptive algorithm for these two partitioning techniques.

Two trends were observed vis-a-vis the execution times for both forms of grid partitioning. The total execution time reduced as additional processors were introduced in the system, with the number of cells being adapted remaining the same. This implies that the communication overhead caused by additional processors does not overwhelm the efficiency of the overall algorithm. The second trend was observed in case of increasing number of cells being adapted for a given number of processors. The execution time increased with increase in the amount of adaptation. This is to be expected as larger amount of adaptation implies additional work per processor in the system.

Figure 4 graphically illustrates the variation in execution time with number of processors with the number of cells being adapted held fixed at 200. The dashed line corresponds to a strip partitioning while the solid line corresponds to an all-round partitioning. The scale on the X-axis is logarithmic with base 2. It can be seen that strip partitioning results in lesser execution time as compared to the time for the corresponding instance with all-round partitioning. Furthermore, this discrepancy increases with increase in the number of processors involved in the execution. This is due to the startup overhead of messages during interprocessor communication. In case of the strip partitioning, a processor only needs to communicate with at most two other processors while in case of the all-round partitioning, the number of adjacent processors can be as high as eight. Since the communication takes place across interpartition boundaries which are essentially 2-D, the total communication overhead is dominated by the startup overhead of each message rather than the actual length of the message. As a result, this overhead is directly proportional to the number of messages being sent during adaptation. The number of messages in turn depends upon the number of adjacent processors that a given processor needs to communicate with. Consequently, the communication overhead tends to increase with increase in number of processors in the all-round partitioning case leading to the discrepancy in execution times.

Figure 5(a) shows the surface plot corresponding to an initial, unadapted grid for the ONERA M6 wing. The grid consists of 35008 tetrahedral cells which are partitioned among all 128 processors of the iPSC/860 using the all-round partitioning method. Four cutting planes are used along each of X and Z axes while 8 cutting planes are used along the Y-axis. The view for the surface plot is taken along the Y-axis. Transonic flow of Mach number 0.84 involves two shock waves on the upper surface of the wing: a fore shock close to the leading edge, and an aft shock which forms a lamda system with the former [3]. The parallel adaptive algorithm is employed to perform one step of adaptation on this initial grid. A total of 9000 cells from the initial grid were divided resulting in an adapted grid consisting of 103,580 cells. Figure 5(b) shows the surface plot for the one-level adapted grid. A local embedded grid has been placed by the algorithm in the regions of the leading edge and shock waves [3]. The adapted grid was obtained in a total execution time of *1.65 seconds.*

## 5   Acknowledgements

This work was supported by DARPA Grant DABT 63-92-0042, and by NSF Grant ECS-9023770. Computing time on the Intel iPSC/860 was provided by the NAS Division of NASA Ames Research Center, Moffett Field, CA.

# References

[1] Y. Kallinderis and J.R. Baron, "Adaptation Methods for a New Navier-Stokes Algorithm," *Journal of the American Institute of Aeronautics and Astronautics*, Vol. 27, pp 37-43, January 1989.

[2] Y. Kallinderis and J.R. Baron, "A New Adaptive Algorithm for Turbulent Flows," *Computers and Fluids Journal*, Vol. 21, No. 1, pp. 77-96, 1992.

[3] Y. Kallinderis, and V. Parthasarathy, "An Adaptive Refinement / Coarsening Scheme for 3-D Unstructured Meshes", AIAA Journal, Vol. 31, No. 8, pp 1440-1447, August 1993.

[4] R. Lohner, and J. Baum "Numerical Simulation of Shock Interaction with Complex Geometry Three-Dimensional Structures Using A New Adaptive H-Refinement Scheme on Unstructured Grids ," AIAA Paper 90-0700, 1990.

[5] S.Ward and Y.Kallinderis "Hybrid Prismatic/Tetrahedral Grid Generation for Complex 3-D geometries", AIAA Paper 93-0669, Reno, NV, 1993.

[6] Y. Kallinderis and A. Vidwans, "A Generic Parallel Adaptive-Grid Finite-Volume Navier-Stokes Algorithm", AIAA Journal, Vol. 32, No. 1, pp 54-61, January 1994.

[7] C. P. Kruskal, L. Rudolph and M. Snir "The Power of Parallel Prefix" *IEEE Transactions on Computers*, Vol 34, pp 965-968, 1984.

[8] A. Vidwans, Y. Kallinderis and V. Venkatakrishnan, "A Parallel Dynamic Load Balancing Algorithm for 3-D Unstructured Grids", AIAA Paper 93-3313, Proceedings of the 11$^{th}$ Computational Fluid Dynamics Conference, Orlando, FL, July 1993.

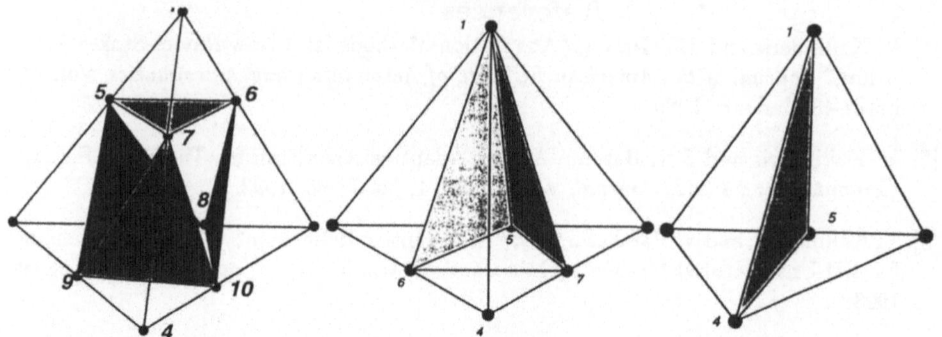

Figure 1: (a) Isotropic division into eight children cells. (b) Directional division into four children cells. (c) Directional division into two children cells.

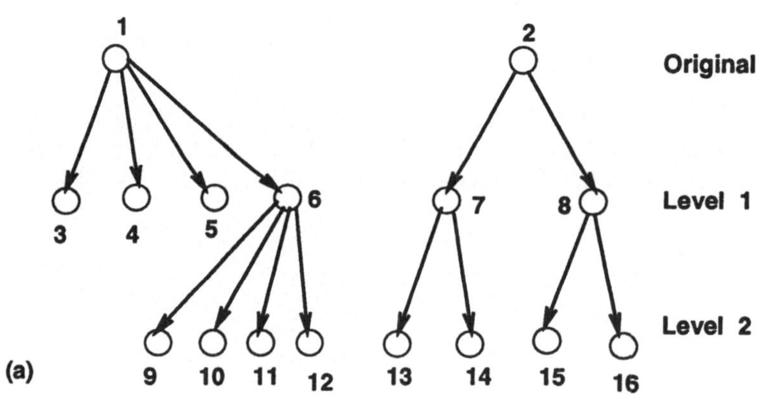



**Postorder Representation :**

Figure 2: (a) Formation of adaptation trees of cells. One tree can be formed for each cell of the original, unadapted grid. (b) The postorder representation.

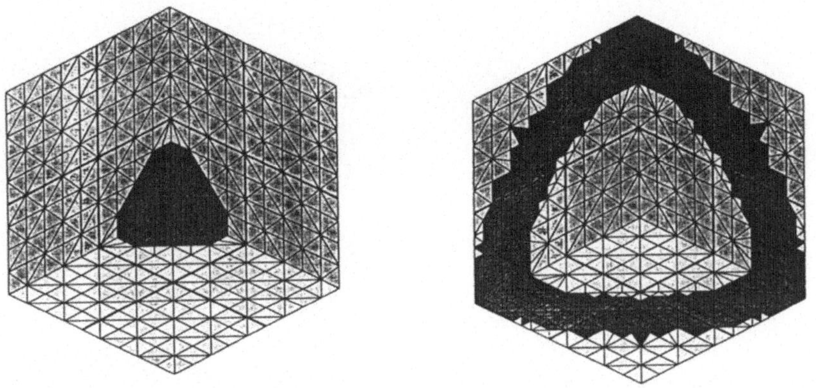

Figure 3: Illustration of the 3-D grid adaptation process involving both division and deletion of tetrahedra. A blast wave is shown moving through a channel grid. Dark area shows adaptation around the position of the wave. (a) Initial position of the wave. (b) Position of the wave as it is about to move out of the channel.

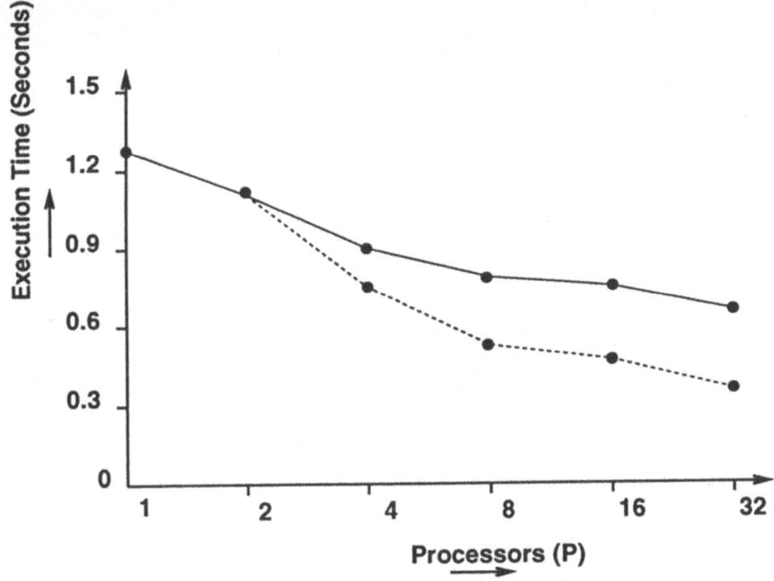

Figure 4: Reduction in execution times with increasing number of processors for one adaptation of a 3-D channel grid wherein 200 cells are divided and deleted. The scale on the X-axis is logarithmic. Dashed line corresponds to a *strip partitioning* while the solid line corresponds to an *all-round partitioning* of the initial computational grid.

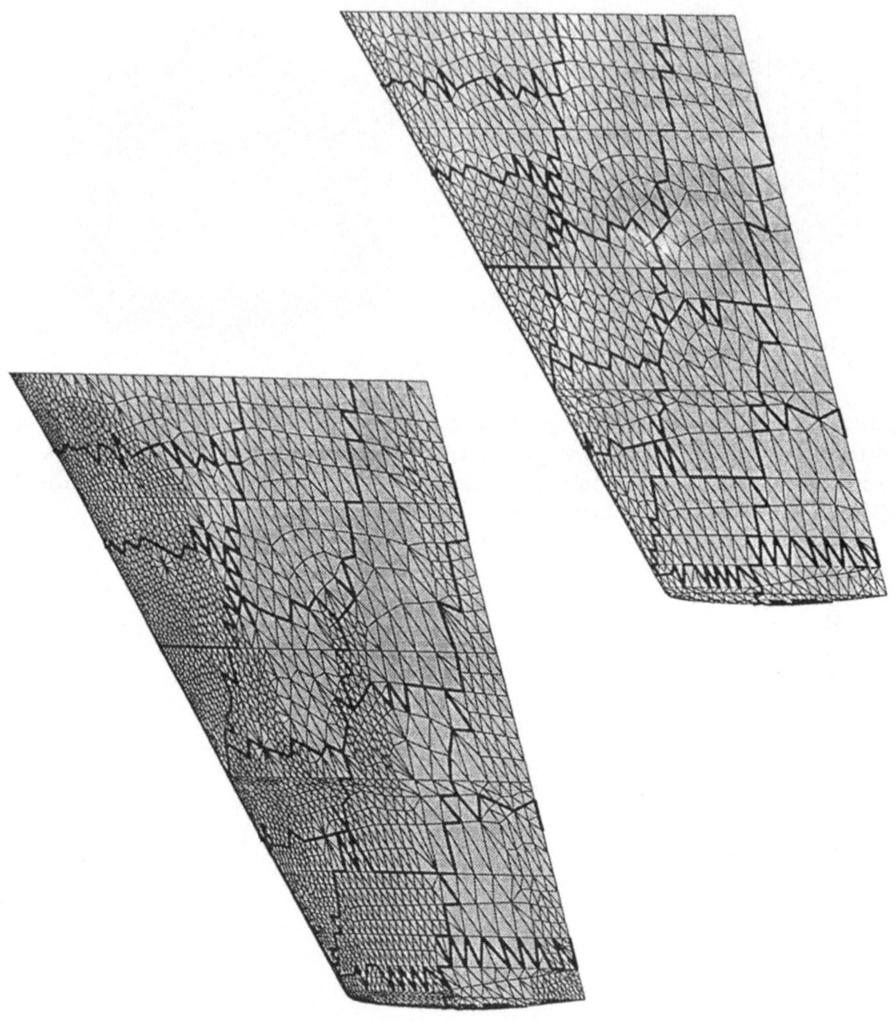

Figure 5: (a) Surface plot of the original grid for an Onera M6 wing consisting of 35008 tetrahedral cells partitioned among 128 processors of the iPSC/860. (b) Surface plot of a one-level adapted grid for the Onera M6 wing consisting of 103,580 cells obtained in a total execution time of 1.65 seconds on a 128-processor iPSC/860.
Thick lines denote inter-partition boundaries.

# NUMERICAL SIMULATION OF TURBULENT THREE-DIMENSIONAL FLOW PROBLEMS ON PARALLEL COMPUTING SYSTEMS


M. Kurreck, R. Koch, S. Wittig
Lehrstuhl und Institut für Thermische Strömungsmaschinen
Universität Karlsruhe (T.H.)
Kaiserstr. 12, 76128 Karlsruhe, Germany


## SUMMARY


The parallelized version of the finite-volume code EPOS (Elliptic Package on Shear Flows) for the prediction of turbulent three-dimensional flows is presented. EPOS was parallelized using the domain decomposition method. Three different methods for coupling the subdomains and two iterative methods for solving the system of algebraic equations are outlined. The two-dimensional flow in a model combustor and the flow in an experimental combustor were predicted with the parallelized version of EPOS. The calculations were performed on the Transputer Cluster GCel (PARSYTEC) using several hundred processors. Although good efficiencies were achieved, the comparision of the CPU-times of the parallel and the vectorized version of EPOS, running on a SNI S600/20 computer, shows, that current vector computers give the best performance.


## NOMENCLATURE

| | | | |
|---|---|---|---|
| $\vec{c}$ | velocity vector | Greek symbols | |
| $NP$ | number of processors | $\rho$ | density |
| $p'$ | pressure correction | $\phi$ | flow quantity |
| $S$ | source term | $\Gamma$ | diffusion coefficient |
| $T$ | total CPU-time | | |
| $\bar{t}$ | average CPU-time | Subscripts | |
| | for one outer iteration | $NP$ | number of processors |
| $u, v, w$ | velocity components | $num$ | numerical |
| $x, y, z$ | spatial coordinates | $par$ | parallel |
| | | $tot$ | total |
| | | 1 | one processor |

## INTRODUCTION

For the development of advanced gas turbine components, such as low emission combustors and improved blade cooling configurations, the numerical simulation of the flow becomes increasingly important. However, due to the limited performance of present computing systems, the numerical simulation is restricted to simplified models. The new generation of parallel computers with increasing performance offers a wide range of opportunities for the use of more advanced turbulence and combustion models on larger computational domains.

The drawback, however, is, that programming of parallel computers is more complicated compared to scalar or vector computers. New parallel algorithms have to be developed and evaluated for applying this new type of computing systems efficiently. Until now, relatively few attempts have been made to simulate complex flow problems on parallel systems.

Agarwal [1] used an explicit finite-volume scheme for the calculation of two-dimensional flows on a Connection Machine (CM2). Keyes [6] solved two-dimensional laminar flows on a 16 processor Encore Multimax computer. He applied Conjugate Gradient and Chebyshev type iterative methods. Khan and Atta [7] calculated the two-dimensional laminar flow around an airfoil using an unstructured grid. They executed the calculations on a Connection Machine (CM2). Braaten [2] developed a parallel algorithm for the prediction of laminar two-dimensional flows on an Intel iPSC parallel computer. The algorithm was applied on a 32 node scalar and a 8 node vector machine. Thompson, Cowell and Leaf [14] describe the parallelization of an adaptive multigrid code for the calculation of laminar two-dimensional flow problems. Experience was gained on a Sequent Symmentry machine with 20 processors. Schreck and Peric [12] calculated two-dimensional laminar flows using a parallel multigrid algorithm. They used the domain decomposition method and applied different methods for the coupling of the subdomains. Two-dimensional problems were calculated by Farhat et al. [3] on a Connection Machine (CM2). They predicted the viscous flow around an airfoil using an unstructured mesh. This overview of topical literature shows, that most of the researchers were engaged in the calculation of two-dimensional laminar flows.

In contrast, the present study deals with the development and evaluation of algorithms for the calculation of two- and three-dimensional turbulent flows on parallel computing systems. In the context of our work on the development of the CFD-code EPOS (Elliptic Pakage on Shear Flows) for gas turbine combustors (cfe. Noll and Wittig [9], Noll [10]), extensive experience has been gained during the last year on CFD-applications with parallel computing systems. In this study a detailed performance analysis and a comparision of CPU-requirements and convergence rates of CFD-applications on a vector computer (SNI S600/20) and the Transputer Cluster GCel (PARSYTEC) are outlined. The vectorized and the parallelized versions of EPOS were applied to calculate isothermal two- and three-dimensional flows, representing typical flow configurations of gas turbine combustors. Furthermore the numerical results are compared with our experimental data.


## NUMERICAL METHOD

As frequently discussed, the set of governing equations describing turbulent flows consists of the continuity, the momentum and the two equations of the $k, \epsilon$-turbulence model. These equations can be written in a generalized form as:

$$div(\rho \vec{c} \phi) = div(\Gamma_\phi \cdot grad\phi) + S_\phi. \tag{1}$$

The finite-volume method is applied to discretize the partial differential equations. The diffusive fluxes and the source terms are discretized by the central-differencing scheme. In contrast to this, the convective fluxes are discretized by the 'Monotonized Linear Upwind' scheme (MLU) of second order accuracy developed by Noll [10]. The system of transport equations is solved by a block-iterative procedure. Only the spatial coupling with the neighbouring points of the same variable is considered implicitly, whereas the coupling with all the other variables is done via an outer iteration. The system of algebraic equations of one flow quantity for the whole computational domain is solved alternatively by the Strongly Implicit Procedure (SIP, cfe. Stone [13]) or the preconditioned Conjugate Gradient (ILU-CG) method (cfe. Noll and Wittig [9]). As it is well-known, special emphasis has to be put on the velocity-pressure coupling. Since a nonstaggered arrangement of the variables is used, the SIMPLEC algorithm (cfe. Van Doormal and Raithby [15]) in combination with the interpolation of the cell-face velocities proposed by Rhie and Chow [11] is applied.

## PARALLELIZATION PROCEDURE

Obviously, the domain decomposition method seems to be best suited for the parallelization of a CFD-code. The whole computational domain is subdivided into several subdomains (cfe. Fig. 1) and each processor of the parallel computer manages one subdomain. The division into subdomains can be performed in three space directions, according to the available number of processors.

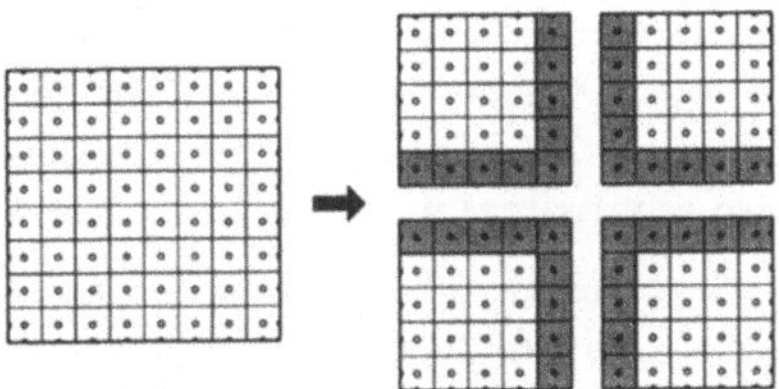

**Fig. 1:** Domain decomposition.

The solution strategy applying the domain decomposition method is as follows:

(1) Provide initial values for all variables in each subdomain.

(2) Solve all the algebraic systems in each subdomain.

(3) Exchange the data at the interior boundaries.

(4) Check for convergence, if the convergence criterion is not fullfilled go back to Step (2).

The Steps (2), (3) and (4) form one outer iteration. In contrast to this, the steps performed to solve one system of algebraic equations are called inner iterations. The inner and outer iterations are performed until a specified convergence criterion is fullfilled.

The data transfer between the subdomains at the interior boundaries is of prime importance for the efficiency of the parallel process. Since the subdomains are spatially decoupled by the domain decomposition, the data transfer is required to recover the coupling of the subdomains in order to retain good convergence characteristics. Otherwise the convergence rate of the parallel process would decrease. A sufficient coupling of the subdomains causes an additional overhead in communication and may therefore lead to a bad efficiency. In this context several relations are used to analyse the efficiency of the parallel process. The total efficiency is defined as

$$\epsilon_{tot} = \frac{T_1}{NP \cdot T_{NP}} \cdot 100 \quad [\%].$$ (2)

$T_i$ is the total CPU-time required for the calculation of a flow problem and $NP$ is the number processors used. In Eq. (2) the CPU-time $T_1$ of the best sequential algorithm should be used. In this study $\epsilon_{tot}$ is always computed using the CPU-time $T_1$ of the ILU-CG algorithm of Noll and Wittig. The total efficiency can be devided into two parts, which are estimates for the numerical efficiency

$$\epsilon_{num} = \frac{n_1}{n_{NP}} \cdot 100 \quad [\%]$$ (3)

and the parallel efficiency

$$\epsilon_{par} = \frac{\bar{t}_1}{NP \cdot \bar{t}_{NP}} \cdot 100 \quad [\%].$$ (4)

$n_i$ denotes the number of outer iterations and $\bar{t}_i$ is the average CPU-time required for one outer iteration. For the calculation of $\epsilon_{par}$ and $\epsilon_{num}$ the same holds as for the total efficiency concerning the reference values $n_1$ and $\bar{t}_1$. The Eqs. (3) and (4) are good estimates, if the number of inner iterations and the computational effort for the sequential and the parallel algorithm are approximately equal.

Three different types of data transfer have been evaluated and investigated in this study (cfe. Kurreck and Wittig [8]). The first type is the explicit coupling method within each outer iteration (EOC). Applying this approach, relatively few time for the data transfer is required and merely a weak coupling of the subdomains is achieved. The algorithm for solving the system of algebraic equations is unchanged in this case, i.e. both aforementined algorithms, SIP and ILU-CG, can be used. In this study the ILU-CG algorithm was used in combination with the EOC coupling method. The EIC method performs the data transfer within the SIP algorithm within each inner iteration. The third method (IIC) uses a parallelized version of the sequential ILU-CG algorithm proposed by Noll and Wittig [9]. The preconditioning is done locally in each subdomain and therefore in parallel. The CG step is performed globally using the data of all the subdomains.

Independent of the coupling procedure applied, pressure source terms and coefficients have to be exchanged to perform a correct pressure correction algorithm. Additional data for a second order coupling of the subdomains are also exchanged.

## Two-dimensional turbulent flow in a model combustor

The turbulent isothermal flow in our model combustor was calculated using the parallelized version of EPOS. A detailed discussion of the flow in the model combustor was published by Wittig et al. [16]. The symmetric two-dimensional flow field calculated is dipicted in Fig. 2. At the lefthand side of the channel the air enters with a mean velocity of $60\,\frac{m}{s}$ through two nozzles of $8\,mm$ in height. The rectangular channel with a cross sectional area of $100\,mm * 300\,mm$ has a length of $400\,mm$. Due to the sudden enlargement at the inlet of the channel two large recirculation zones are generated.

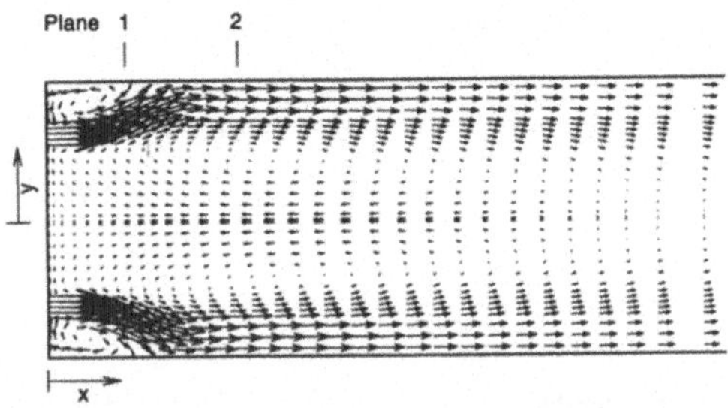


**Fig. 2:** Velocity field (model combustor).

The computational domain was discretized with $37 \times 33$ grid points. The calculations were performed on the GCel Transputer Cluster using 2 to 32 processors with different spatial distributions of subdomains in the $x-$ and $y-$ coordinate direction.

In Fig. 3 the total efficiency is plotted for all the coupling methods and processor configurations. The total efficiency decreases for all the coupling methods if the number of processors is increased. The best total efficiency is reached using the IIC method. For

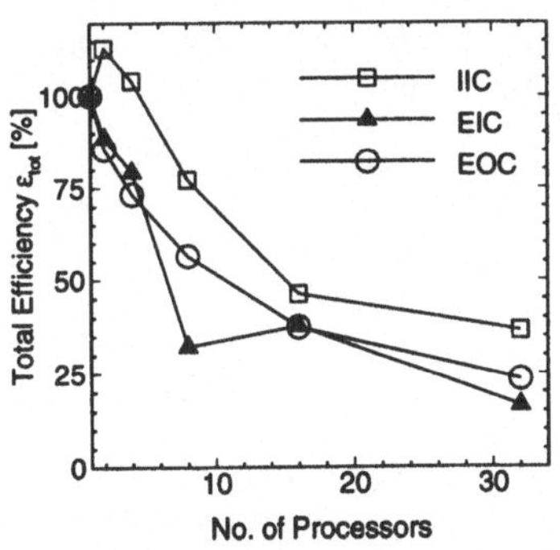


**Fig. 3:** Total efficiency.

the IIC method and 2 and 4 processors the total efficiency is larger than $100\,\%$. This is caused by a numerical efficiency of approximately $100\,\%$ and a parallel efficiency larger than $100\,\%$ (cfe. Fig. 4). The parallel efficiency itself is influenced by the number of inner iterations which may vary between one and fourty. The parallel efficiencies of the IIC and the EOC method are larger than 100 %, because for the parallel calculations the number of inner iterations is smaller than the number of inner iterations of the sequential cases. The EIC method gives parallel efficiencies larger than 100 %, too. In contrast to the IIC and the EOC method this is due to the fact that the numerical effort to performe one inner iteration using the EIC method is much smaller compared to the other methods. The IIC and the EOC methods are based on the ILU-CG algorithm, whereas the EIC method uses the SIP algorithm. The SIP algorithm requires less arithmetic operations than the preconditioned ILU-CG procedure.

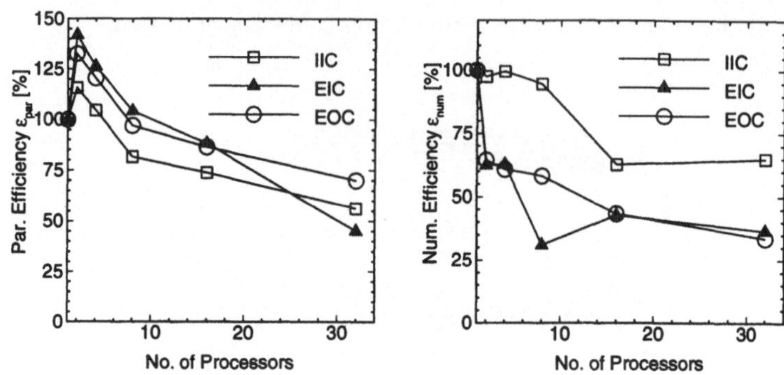

Fig. 4: Parallel and numerical efficiency.

Considering the numerical efficiency (cfe. Fig. 4) the IIC method again gives the best results. For this testcase the IIC method is more stable than the others. The EOC method is very unstable and sometimes failes due to divergence. This unstable behaviour is caused by the weak coupling of the EOC method.

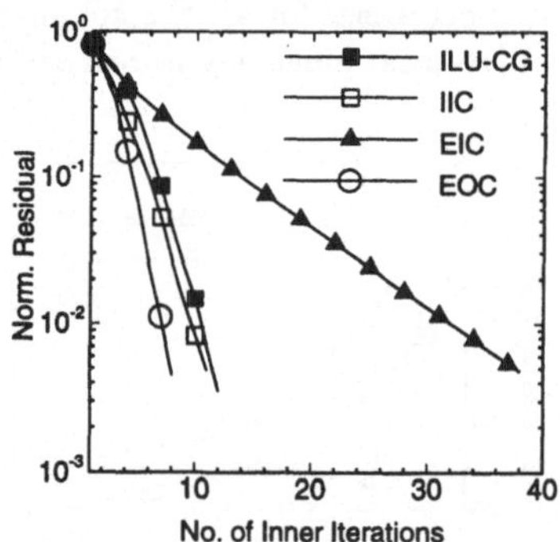

Fig. 5: Reduction of normalized residual norm of the system of equations for p' (first outer iteration).

To evaluate the performance of each of the algorithms and coupling methods, the normalized residual norm versus the number of inner iterations for the determination of the pressure correction in the first outer iteration is shown in Fig. 5. As a testcase the flow in the model combustor using one (ILU-CG) and eight processors was considered. The sequential ILU-CG, the EOC and the IIC algorithm give nearly the same convergence rates, whereas the convergence rate of the EIC method is approximately four times slower. For this application the EOC method reduces the residual of subsystems of the pressure correction most rapidly, but if the overall iterative process is considered the total efficiency of the IIC method is superior.

A comparison of the calculated and measured u-velocity component at two locations within the model combustor is given in Fig. 6. In general, good agreement between data calculated by the sequential and the parallel algorithms and the experimental data can be observed. The maximum values of the velocity are slightly underpredicted using 32 processors.

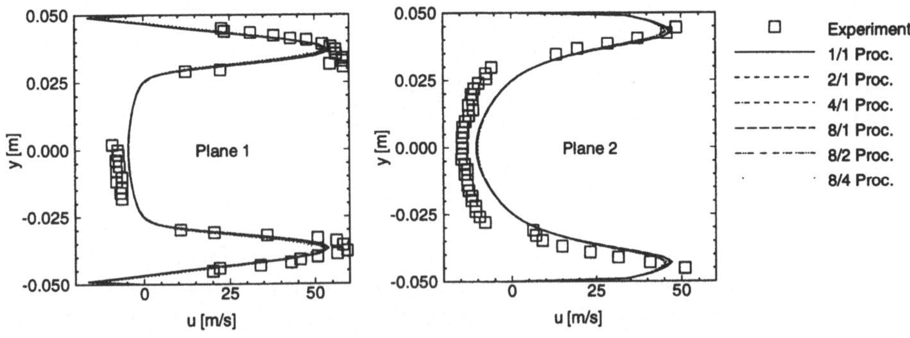


**Fig. 6:** Velociy profiles model combustor – IIC coupling method
(Plane 1: $x = 30\,mm$, Plane 2: $x = 70\,mm$)
Experiment: Wittig et al. [16].

## Three-dimensional flow in an experimental combustion chamber

The isothermal flow in our experimental combustion chamber shown in Fig. 7 is calculated with the described parallel algorithms. A detailed examination of the flow in the combustor can be found in Jeckel et al. [4] and Jeckel and Wittig [5]. Since jet-stabilization is applied, the air enters the flame tube through four jets $60\,mm$ downstream from the nozzle. The flame tube diameter is $80\,mm$, the jet diameter is $8\,mm$ and the jet velocity is $38.5\,\frac{m}{s}$. Because of symmetry properties only the eighth part in circumferential direction of the flame tube is discretized. Two different grids were used. The first grid consists of $14 \times 23 \times 59$ grid points (G1) in circumferential, radial and axial direction. In contrast, the number of grid points in radial and axial direction is increased for the second grid to 41 and 114 (G2).

The flow field in the experimental combustor was calculated on the GCel Transputer Cluster with different numbers of processors in the range of 56 to 448. The computational domain was subdivided in three space directions. Since both mesh sizes are too large for the calculation of the problem on one transputer only, the reference time $T_1$ was determined by various test calculations on a SUN-workstation. Therefore, it is assumed that the CPU-time for the calculation of the same problem is 5.6 times higher on one transputer compared to the SUN-workstation. The estimated CPU-time for the calculation of the flow on one transputer using grid G1 is 112000 seconds, whereas the CPU-time using grid G2 is 855507 seconds. Both CPU-times represent 2000 outer iterations. It should be

163

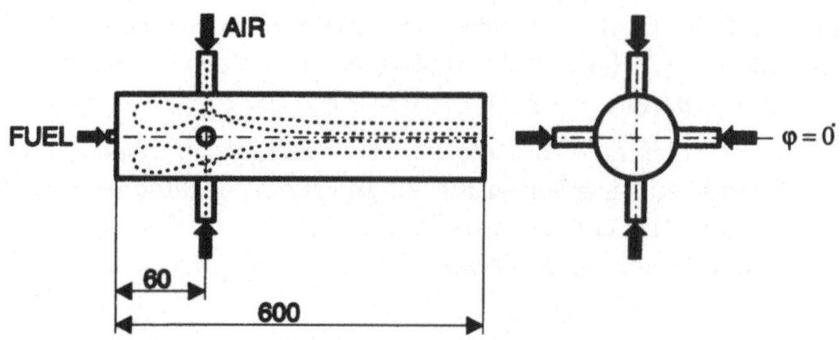



**Fig. 7:** Experimental combustion chamber.

pointed out, that all the efficiencies depend strongly on the aforementioned performance factor between the SUN-workstation and the transputer.

In Fig. 8, the parallel efficiency is shown for the EIC and IIC coupling methods. The efficiencies of both methods are nearly identical. For the coarse grid efficiencies in the range of 25 to 30 % are achieved. In contrast, the efficiency for grid G2 is approximately 45 % for 448 processors. Due to the finer grid, the computational effort for each processor is approximately four times higher for each outer iteration. However, the number of data at the interior boundaries is only two times higher compared to the coarse grid. Therefore, a much better efficiency is achieved using grid G2.

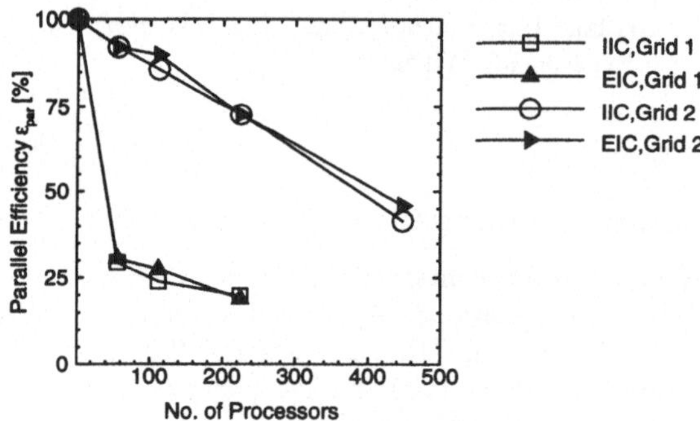



**Fig. 8:** Parallel efficiency.

The CPU-times required for the calculation of the flow on the SNI S600/20 vector computer of the University of Karlsruhe are 232 seconds (G1) and 701 seconds (G2). Both CPU-times represent 2000 outer iterations. The CPU-time is 11.42 times (G1, 224 processors, IIC) and 5.94 times (G2, 448 processors, EIC) higher on the parallel system compared to the vector computer.

Besides the efficiency analysis a comparison of the calculated and the measured data has been made. In Fig. 9 the axial velocity profiles at the plane $\varphi = 0°$ and two axial locations $z = 22\,mm$ and $z = 98\,mm$ are plotted. The velocites predicted are in good agreement with the measured data. The results presented here, give evidence that the solutions calculated by the domain decomposition methods provides the same values as the sequential algorithm. At the axis of symmetry and the walls some differences can be observed, which may be caused by the second order coupling procedure of the subdomains. Further investigations are necessary to analyse the reasons for this behaviour.

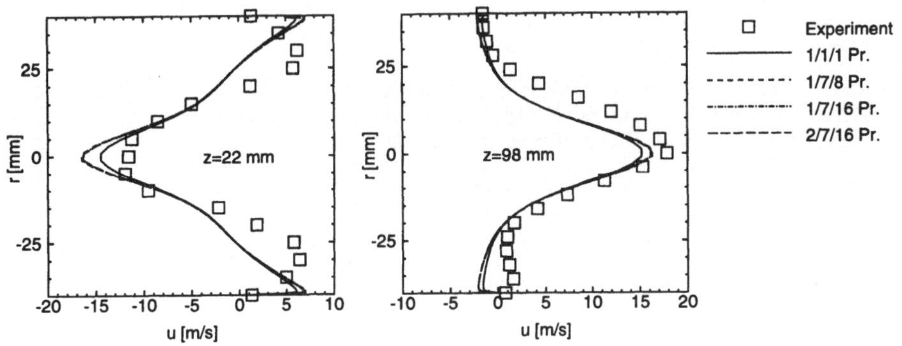


**Fig. 9**: Velocity profiles combustion chamber (IIC method, G1)
Experiment: Jeckel et al. [4].

## CONCLUSIONS

A parallelized finite-volume scheme for the calculation of three-dimensional turbulent flow problems has been outlined. In applying the domain decomposition method, three different methods for the coupling of the subdomains have been presented. Both implicit methods EIC and IIC are more stable than the explicit coupling method EOC. The algorithms presented in this study give good efficiencies for a wide range of processor configurations. The prediction of turbulent flow problems, representing typical flow configurations of gas turbine combustors, show, that the algorithms are highly accurate. In the near future reacting three-dimensional flows will be predicted using the IIC and the EIC methods.

## REFERENCES


[1] Agarwal R.K.: Development of a Navier-Stokes Code on A Connection Machine. Proceedings of the Fourth Conf. on Hypercubes, Conf.: Monterey, CA, USA 6-8 March 1989, 1989. pp. 917- 924, vol.2 of 2 vol.

[2] Braaten M.E.: Development of a Parallel Computational Fluid Dynamics Algorithm on a Hypercube Computer. Int. J. for Numerical Methods in Fluids, 1991. Vol. 12, pp. 947-963.

[3] Farhat C., Fezoui L., Lanteri S.: Two-dimensional Viscous Flow Computations on the Connection Machine: Unstructured Meshes, Upwind Schemes and Massively Parallel Computations. Computer Methods in Appl. Mech. and Eng., 1993. Vol. 102, pp. 61-88.

[4] Jeckel R., Noll B., Wittig S.: Three Dimensional Time-resolved Velocity Measurements in a Gas Turbine Model Combustor. 6$^{th}$ int. Symposium on the Application of Laser Techniques to Fluid Mechanics, Lissabon, 1992.

[5] Jeckel R., Wittig S.: Time-resolved Measurements in a Three Dimensional Model Combustor. AGARD – PEP 81$^{st}$ Symposium on "Fuels and Combustion Technology for Advanced Aircraft Engines", 1993. Colleferro, Rom, Italien, (10. –14. Mai 1993).

[6] Keyes D.E.: Domain Decomposition Methods for the Parallel Computation of Reacting Flows. Computer Physics Communications, 1989. Vol. 53, pp. 181-200.

[7] Khan M.M., Atta E.H.: Prediction of Laminar Flows using Tetrahedral Meshes and Massively-Parallel Computers. ASME, Recent Advances & Applications in Computational Fluid Dynamics, 1990. Dallas, Texas, Nov. 25-30.

[8] Kurreck M., Wittig S.: Numerical Simulation of Combustor Flows on Parallel Computers - Potential, Limitations and Practical Experience. Submitted for Publication, 39th IGTI Conference, The Hague, The Netherlands, June 13-16, 1994.

[9] Noll B., Wittig S.: Generalized Conjugate Gradient Method for the Efficient Solution of Three-Dimensional Fluid Flow Problems. Numerical Heat Transfer, 1991. Part B, Vol.20, pp. 207-221.

[10] Noll B.: Evaluation of a Bounded High-Resolution Scheme for Combustor Flow Computations. AIAA-Journal, 1992. Vol. 30, No. 1, pp. 64-69.

[11] Rhie C.M., Chow W.L.: Numerical Study of the Turbulent Flow Past an Airfoil with Trailing Edge Separation. AIAA-Journal, 1983. Vol. 21, No. 11, pp. 1525-1532.

[12] Schreck E., Peric M.: Computation of Fluid Flow with a Parallel Multigrid Solver. Int. J. for Numerical Meth. in Fluids, 1993. Vol. 16, pp. 303-327.

[13] Stone H.L.: Iterative Solution of Implicit Approximations of Multidimensional Partial Differential Equations. SIAM J. Num. Anal., 1968. Vol. 5, No. 3, pp. 530-558.

[14] Thompson C.P., Wayne R.C., Leaf G.K.: On the Parallelization of an Adaptive Multigrid Algorithm for a Class of Flow Problems. Parallel Computing 18, 1992. pp. 449-466.

[15] Van Doormal J.P., Raithby G.D.: Enhancement of the SIMPLE method for predicting incompressible fluid flows. Numerical Heat Transfer, 1984. Vol. 7, pp. 147-163.

[16] Wittig S., Klausmann W., Noll B.: Turbulence Effects on the Droplet Distribution behind Airblast Atomizers. AGARD CP-422, 1987.

# A HIGH–RESOLUTION FLUX SPLITTING SCHEME FOR THE SOLUTION OF THE COMPRESSIBLE NAVIER–STOKES EQUATIONS ON TRIANGULAR GRIDS


P.R.M. Lyra[†], K. Morgan[†] and J. Peraire[‡]

[†]Department of Civil Engineering, University College,
Swansea SA2 8PP, United Kingdom.

[‡]Department of Aeronautics and Astronautics,
MIT, Cambridge, MA02139, USA.


## ABSTRACT


The AUSM flux vector splitting scheme is employed as the basis for the construction of a high-resolution MUSCL type algorithm for the simulation of high speed compressible flows. A high-order conservative shock–capturing scheme is achieved for multidimensional simulation by the adoption of a Galerkin finite element formulation, implemented with a side–based representation for the grid. The resulting scheme is extended for the solution of the full Navier Stokes equations by computing the viscous terms via a mixed formulation.


## INTRODUCTION

Special attention has been devoted recently to the development of upwind based schemes for the solution of the compressible Euler and Navier–Stokes equations. Upwind schemes are generally regarded as being parameter free and this can be extremely important when challenging hypersonic viscous flow computations are addressed. Several upwind schemes have been proposed over the last decade [16], of which the most popular are the flux vector splitting schemes of Steger and Warming [18] and Van Leer [22] and the flux difference splitting schemes of Roe [15] and Osher [11]. Each of these methods has its associated advantages and drawbacks, representing in general a compromise between efficiency and accuracy.

Shock–capturing schemes operate by the selective addition of numerical viscosity which can severely affect the accuracy of the results when viscous problems are simulated e.g. certain upwind schemes introduce an excessive amount of artificial viscosity when handling the high gradients which are present in viscous shear layers. When viscous computations are attempted, better results have generally been achieved by the use of flux difference splitting, but the differentiability properties and the simplicity of schemes based upon flux vector splitting ideas motivate the recent search for improved methods of this class. Liou and Steffen [3] proposed a very simple and efficient splitting called AUSM (Advection Upwind Split Method) which satisfies the entropy condition, has the positivity property and is free from the anomalies [14] which plague other methods. The AUSM method has been shown to produce results that compare well with those of simulations employing flux difference splitting. Jameson [1] has recently proposed the CUSP (Convective Upwind and Split Pressure) scheme, which involves a construction of the artificial dissipation which is approximately equivalent to that found in AUSM.

The main objective of the present work is to describe a method of implementing the AUSM flux splitting within an unstructured triangular mesh algorithm for two dimensional flow simulations. The interest in the unstructured mesh approach is driven by the geometrical flexibility offered by the use of a discretization based upon triangles. To achieve the implementation, a side–based data structure [12] is employed for the grid and this allows for a direct incorporation of 1–D methodologies in multi–dimensions [7, 9]. This approach also results in considerable savings in both CPU time and memory requirements when a 3–D extension is attempted [10, 13].

The practical algorithm is obtained by achieving a higher order spatial discretization by the use of the MUSCL [21] concept. The extension of the approach for viscous flow computations follows by computing the viscous fluxes at the nodes via a variational recovery procedure.

<h2 align="center">GOVERNING EQUATIONS</h2>

The equations which govern the unsteady laminar flow of a compressible viscous fluid can be written, in the absence of external source terms, in the conservation form

$$\frac{\partial U}{\partial t} + \frac{\partial F^j}{\partial x_j} = \frac{\partial G^j}{\partial x_j} \qquad \text{for} \quad j = 1, 2 \tag{1}$$

where the summation convention is adopted for the index $j$ and $x_1, x_2$ are cartesian coordinates. In equation (1), $U$ is the vector of the conservative variables, while the vectors $F^j$ and $G^j$ denote the inviscid and viscous fluxes in the direction $x_j$, respectively i.e. in a non–dimensionalised form [10]

$$U = \begin{bmatrix} \rho \\ \rho u_1 \\ \rho u_2 \\ \rho \epsilon \end{bmatrix} \quad F^j = \begin{bmatrix} \rho u_j \\ \rho u_1 u_j + p\delta_{1j} \\ \rho u_2 u_j + p\delta_{2j} \\ \rho H u_j \end{bmatrix} \quad G^j = \frac{1}{Re_\infty} \begin{bmatrix} 0 \\ \tau_{1j} \\ \tau_{2j} \\ u_i \tau_{ij} + \frac{\mu}{P_r}\frac{\partial T}{\partial x_j} \end{bmatrix}. \tag{2}$$

In these equations, $\rho$, $p$, $\epsilon$ and $H$ denote the dimensionless density, pressure, total specific energy and total specific enthalpy of the fluid, respectively. The parameter $\delta_{ij}$ is the Kronecker delta and $u_j$ denotes the dimensionless velocity in direction $x_j$. In addition, $Re_\infty$ is the free stream Reynolds number and $Pr$ is the local Prandlt number, which is assumed to be constant. For the laminar flows, which are of interest here, the components of the viscous stress tensor are defined by

$$\tau_{ij} = \mu \left( \frac{\partial u_i}{\partial x_j} + \frac{\partial u_j}{\partial x_i} \right) - \frac{2}{3}\mu \frac{\partial u_k}{\partial x_k}\delta_{ij} \tag{3}$$

where the Stokes relation has been assumed. The viscosity coefficient $\mu$ varies with temperature according to Sutherland's empirical relation. The equation set is closed by the ideal gas assumption, which is expressed by the perfect gas state equations in the non–dimensional form

$$p = (\gamma - 1)\rho(\epsilon - 0.5u_k u_k) \qquad\qquad T = \frac{\gamma p}{(\gamma - 1)\rho} \tag{4}$$

where $\gamma = C_p/C_v$, and $C_p$ and $C_v$ are the specific heats of the fluid at constant pressure and at constant volume respectively.

The solution of this equations set is sought over a closed spatial domain $\Omega$, with boundary surface $\Gamma$. To render the initial/boundary value problem well posed, suitable boundary conditions and an initial condition must be specified. Here, we assume that

$$\begin{aligned}
\boldsymbol{F}^n &= n^j \boldsymbol{F}^j = \bar{\boldsymbol{F}}^n \\
\boldsymbol{G}^n &= n^j \boldsymbol{G}^j = \bar{\boldsymbol{G}}^n
\end{aligned} \qquad \text{on } \Gamma \text{ for all } t > t_m \qquad (5)$$

and

$$\boldsymbol{U}(\boldsymbol{x}, t_m) = \boldsymbol{U}_0(\boldsymbol{x}) \qquad \text{for all } \boldsymbol{x} \text{ in } \Omega \text{ at time } t = t_m \qquad (6)$$

where $n^j$ denotes the component, in direction $x_j$, of the unit outward normal vector to $\Gamma$. $\bar{\boldsymbol{F}}^n$ and $\bar{\boldsymbol{G}}^n$ are the normal inviscid and viscous fluxes at the boundary, which depend on the solution, and $\boldsymbol{U}_0$ is a known function.

## UNSTRUCTURED GRID FLUX VECTOR SPLITTING FOR INVISCID FLOWS

We consider initially the solution of the 2–D compressible Euler equations, which are obtained from the original system (1) by removing the viscous terms. The inviscid flux vector will be split into convective and pressure contributions, and the spatial discretization achieved via the Galerkin finite element method. The resulting ordinary differential equation system is discretized explicitly in time and a multi–dimensional central difference type scheme is obtained. By replacing the actual flux by the AUSM consistent numerical flux, a stable first–order scheme results, with high resolution following from the application of the limited variable extrapolation (MUSCL) approach.

### The Central–Difference Scheme

Following the procedure presented by Peraire et al [12], the Galerkin finite element approximation for the Euler equations, using linear triangular elements, can be expressed, at a typical internal node $I$, in the discrete form

$$\left[ M \frac{dU}{dt} \right]_I = - \sum_{s=1}^{m_I} \frac{C_{II_S}^j}{2} (\boldsymbol{F}_I^j + \boldsymbol{F}_{I_S}^j) \qquad (7)$$

where $m_I$ is the number of sides of the mesh connected to node $I$. In this expression, $C_{II_S}^j$ denotes the weight that must be applied to the average value of the flux in the $x_j$ direction on the side $S$, which joins nodes $I$ and $I_S$, to obtain the contribution made by the side to node $I$. The definition of these weighting coefficients can be found in reference [12]. The associated data structure consists of the list of the nodes $I$ and $I_S$ for each side of the mesh.

The inviscid flux vectors $\boldsymbol{F}^j$ are split into the convective flux $\boldsymbol{F}_{(C)}^j$ and pressure flux $\boldsymbol{F}_{(P)}^j$ according to

$$
F^j = F^j_{(C)} + F^j_{(P)} = u_j \overbrace{\begin{bmatrix} \rho \\ \rho u_1 \\ \rho u_2 \\ \rho H \end{bmatrix}}^{\Theta} + p \overbrace{\begin{bmatrix} 0 \\ \delta_{1j} \\ \delta_{2j} \\ 0 \end{bmatrix}}^{\Delta^j} . \tag{8}
$$

With this splitting, the Euler equations are expressed as

$$
\frac{\partial U}{\partial t} + \frac{\partial F^j_{(C)}}{\partial x_j} + \frac{\partial F^j_{(P)}}{\partial x_j} = 0 \qquad \text{for} \quad j = 1, 2 \tag{9}
$$

and for a typical interior node $I$, we obtain the alternative side–based discrete form

$$
\left[ M \frac{dU}{dt} \right]_I = - \sum_{s=1}^{m_I} \frac{C^j_{IIs}}{2} \left\{ [(u_j)_I \Theta_I + (u_j)_{Is} \Theta_{Is}] + [p_I + p_{Is}] \Delta^j \right\} . \tag{10}
$$

For notational convenience, we define the vector $C_{IIs}$ and the quantities $\mathcal{L}_{IIs}$ and $\mathcal{S}^j_{IIs}$ according to

$$
C_{IIs} = (C^1_{IIs}, C^2_{IIs}) \qquad \mathcal{L}_{IIs} = |C_{IIs}| \qquad \mathcal{S}^j_{IIs} = \frac{C^j_{IIs}}{|C_{IIs}|} . \tag{11}
$$

The convective velocity in the direction of $C_{IIs}$ is computed as

$$
V_I = \mathcal{S}^j_{IIs} (u_j)_I \tag{12}
$$

so that we can write equation (10) as

$$
\left[ M \frac{dU}{dt} \right]_I = - \sum_{s=1}^{m_I} \mathcal{L}_{IIs} F_{IIs} = - \sum_{s=1}^{m_I} \mathcal{L}_{IIs} \frac{1}{2} \left\{ [V_I \Theta_I + V_{Is} \Theta_{Is}] + [p_I + p_{Is}] \mathcal{S}^j_{IIs} \Delta^j \right\} . \tag{13}
$$

This expression will need to be suitably modified for nodes which lie on the computational boundary.

## Time Discretisation

Using a simple forward difference approximation in time, equation (13) can be discretised to give the time stepping procedure

$$
U^{m+1}_I = U^m_I + \Delta t \, [M_L]^{-1}_I R_I \tag{14}
$$

where $U^m_I$ denotes the solution at node $I$ at time $t_m$, $\Delta t$ represents the time step and $R_I$ is the right hand side of equation (13). The consistent finite element mass matrix $M$ has been replaced by the standard lumped (diagonal) mass matrix $M_L$, as only steady state computations are of interest here. This produces a truly explicit scheme. As the correct modeling of the transient development of the flow is not of interest, local time stepping is employed to accelerate the convergence rate towards steady–state.

## First–Order Upwind Scheme

It is apparent that the use of the actual flux $\boldsymbol{F}_{II_S}$ in equation (13) represents a central difference type scheme. To obtain a practical algorithm, a high order diffusion must be added in the form of a consistent numerical flux. A classical way of achieving this requirement is by the introduction of upwinding into the discretization process.

In the AUSM approach of Liou and Steffen [3], the convective term, which is treated in equation (13) as a simple average of fluxes, is now replaced by a weighted average

$$V_I \Theta_I + V_{I_S} \Theta_{I_S} \simeq V_{II_S}(\Theta_I + \Theta_{I_S}) \tag{15}$$

where $V_{II_S}$ can be considered as a suitably defined interface velocity, given by

$$V_{II_S} = f(V_I, V_{I_S}) = (V_I)^+ + (V_{I_S})^-. \tag{16}$$

The velocity splitting $(V_I)^\pm$ can be defined in various ways [2]. Following Liou and Steffen [3], we employ the splitting

$$(V_k)^\pm = \begin{cases} \dfrac{[(V_k) \pm |(V_k)|]}{2} & \text{if} \quad |(V_k)| \geq a_k \\[3mm] \pm \dfrac{[(V_k) \pm a_k]^2}{4a_k} & \text{otherwise} \end{cases} \tag{17}$$

where $a_k$ is the speed of sound at node $k$.

For the pressure term, the simple average of equation (13) is replaced by

$$p_I + p_{I_S} \simeq 2p_{II_S} \tag{18}$$

where $p_{II_S}$ represents an interface pressure defined according to

$$p_{II_S} = f(p_I, p_{I_S}) = p_I^+ + p_{I_S}^- . \tag{19}$$

Different methods of splitting the pressure, as a function of the Mach number $M_k$, now become possible [2]. Here, the pressure is split according to

$$p_k^\pm = \begin{cases} \dfrac{p_k[M_k \pm |M_k|]}{2M_k} & \text{if} \quad |M_k| \geq 1.0 \\[3mm] \dfrac{p_k[1 \pm M_k]}{2} & \text{otherwise} \end{cases} \tag{20}$$

so that a simple first order expansion is used in the subsonic range. Substitution of the relations (15) and (18) into equation (13), and the addition of a scalar diffusion

$(|V_{II_S}|[\Theta_{I_S} - \Theta_I])$, results in a generalized AUSM numerical flux for a 2–D triangular grid given by

$$\boldsymbol{\mathcal{F}}_{II_S} = \frac{1}{2}\left\{V_{II_S}[\Theta_I + \Theta_{I_S}] - |V_{II_S}|[\Theta_{I_S} - \Theta_I] + 2p_{II_S}\mathcal{S}^j_{II_S}\boldsymbol{\Delta}^j\right\}. \tag{21}$$

Instead of using an interface convective velocity $V_{II_S}$, an interface Mach number $M_{II_S}$ can be adopted, defined according to

$$M_{II_S} = (V_I)^+/a_I + (V_{I_S})^-/a_{I_S} . \tag{22}$$

In this case the numerical flux of equation (21) is replaced by

$$\boldsymbol{\mathcal{F}}_{II_S} = \frac{1}{2}\left\{M_{II_S}[a_I\Theta_I + a_{I_S}\Theta_{I_S}] - |M_{II_S}|[a_{I_S}\Theta_{I_S} - a_I\Theta_I] + 2p_{II_S}\mathcal{S}^j_{II_S}\boldsymbol{\Delta}^j\right\}. \tag{23}$$

It is our experience that these two different flux formulations are numerically equivalent for the inviscid flow computations which we have tested. For viscous computations, the split–velocity produces oscillations in the solution in the vicinity of viscous walls. For these problems, therefore, the use of the split–Mach version is essential if meaningful results are to be obtained.

## Second–Order MUSCL Extension

The upwind numerical flux described in the previous section leads to a first–order accurate solution in space. To obtain a high–order upwind approximation, an extension of the MUSCL concept of Van Leer [21] to unstructured grids is used. Standard MUSCL schemes compute the state variables at interfaces by an extrapolation between neighbouring cell values on structured meshes. Here we can achieve an equivalent effect by either the computation of the local gradients of the variables or by the computation of the variables at nodes $I_L$ and $I_R$, which are located in the mesh as described in figure 1. The second option was adopted due to its robustness and accuracy, as demonstrated in previous studies [5, 9]. Full details of the procedure are described in [6, 7]. Interface values $U^L_{II_S}$, $U^R_{II_S}$ are determined, using 1–D concepts, from

$$\begin{aligned}
U^L_{II_S} &= U_I + \frac{1}{4}[(1 - \varepsilon)(U_I - U_{I_L}) + (1 + \varepsilon)(U_{I_S} - U_I)] \\
U^R_{II_S} &= U_{I_S} - \frac{1}{4}[(1 + \varepsilon)(U_{I_S} - U_I) + (1 - \varepsilon)(U_{I_R} - U_{I_S})] .
\end{aligned} \tag{24}$$

This represents a combination of backward and forward extrapolation weighted by the choice of the free parameter $\varepsilon$. In particular, for $\varepsilon = -1$ this linear one–sided extrapolation results in the second–order fully upwind scheme, which is adopted in the present study. To prevent overshoots and undershoots in the numerical solution, slope limiters $\Phi$ are introduced such that the interface values are computed according to

$$\begin{aligned}
\hat{U}^L_{II_S} &= U_I + \frac{1}{2}\Phi^+_{I_L I}(U_I - U_{I_L}) \\
\hat{U}^R_{II_S} &= U_{I_S} - \frac{1}{2}\Phi^-_{I_S I_R}(U_{I_R} - U_{I_S}) .
\end{aligned} \tag{25}$$

For the computations performed in this study, the limiter functions designed by Thomas [19] were adopted. These functions are specially developed to ensure the positivity of the thermodynamic variables $(\rho, p)$ and are recommended for high Mach number computations.

The AUSM/MUSCL scheme replaces equation (23) by

$$\mathcal{F}_{IIS} = \frac{1}{2} \Big\{ M_{L/R}[a_L \Theta_L + a_R \Theta_R] - |M_{L/R}|[a_R \Theta_R - a_L \Theta_L] \\ + 2p_{L/R}\,[\mathcal{S}^1_{L/R}\Delta^1 + \mathcal{S}^2_{L/R}\Delta^2] \Big\} \tag{26}$$

where $a_L \Theta_L$, $a_R \Theta_R$ are evaluated with interface values, respectively. The interface Mach number $M_{L/R}$ and pressure $p_{L/R}$ are computed using equations (17, 22) and (19, 20), but also now using the interface values of the state variables.

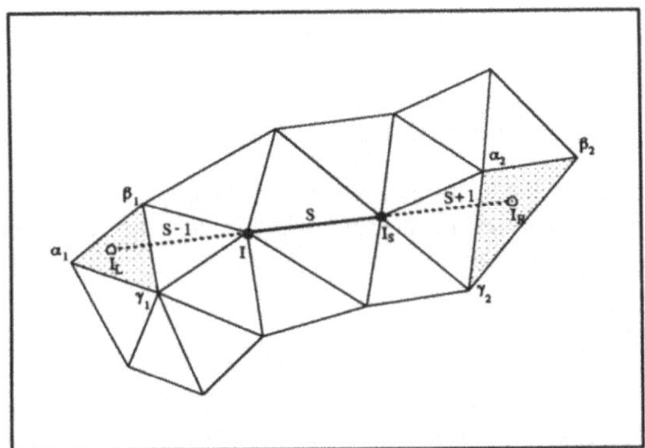

**Figure 1** : Location of Ghost Nodes

## EXTENSION TO VISCOUS FLOWS

The simulation of general viscous flows, involving strong viscous–inviscid interactions, requires the use of the full system of Navier–Stokes equations. In this case, the inviscid fluxes are treated in exactly the same manner as described previously. The viscous terms are discretized by a central difference scheme, and this is accomplished by the use of a finite element mixed formulation, which is consistent with the use of the side–based data structure for the mesh.

Applying the approach followed to derive equation (7), the values of the gradients of the variables at internal nodes $I$ can be obtained as

$$\left[ M_L \frac{\partial U}{\partial x_j} \right]_I = \sum_{s=1}^{m_I} \frac{C^j_{IIS}}{2}(U_I + U_{I_S}) \tag{27}$$

and the nodal values of the viscous fluxes can therefore be directly evaluated as

$$G^j_I = G^j \left( U_I, \left. \frac{\partial U}{\partial x_1} \right]_I, \left. \frac{\partial U}{\partial x_2} \right]_I \right). \tag{28}$$

The Navier–Stokes equations are then discretised in the same form as the Euler equations, leading to the expression

$$\left[M\frac{dU}{dt}\right]_I = -\sum_{s=1}^{m_I}\frac{C_{II_S}^j}{2}\left\{(F_I^j + F_{I_S}^j) + (G_I^j + G_{I_S}^j)\right\}. \tag{29}$$

The use of a standard finite element approach to compute the viscous flux terms results in an expression which is difficult [12] to evaluate in assembled form via the side data structure. An alternative approach can be devised [4], but its storage overhead requirements are prohibitive. The discretization of the viscous terms described above involves information from two layers of points surrounding the point under consideration. The standard finite element approach would use information from one layer only. Peraire et al [12] performed some numerical experiments to compare these procedures and the numerical predictions were found to be practically identical. Based on this preliminary study, and some recent comparisons performed by Manzari [8] for hypersonic regimes, the mixed formulation approach outlined above was adopted here.

## NUMERICAL APPLICATIONS

Simple geometries are involved in the applications which are considered in this paper. This allows for comparison with other numerical predictions, and also means that structured triangular meshes can be adopted.

### Supersonic Flow Past a Flat Plate

This example tests the performance of the basic algorithm on a standard test problem involving the development of a boundary layer on a flat plate of unit length.

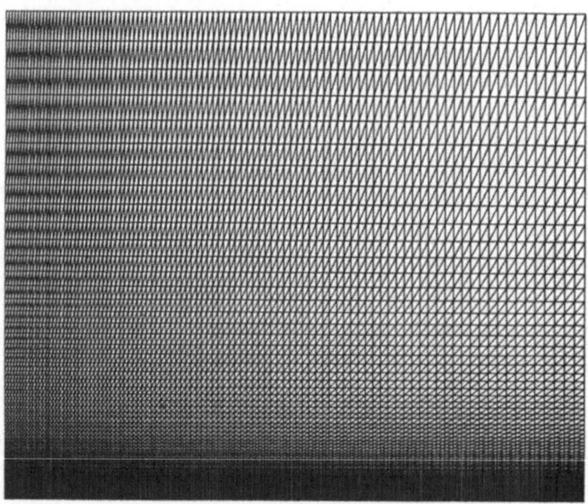

**Figure 2** : Flate Plate Mesh.

A leading edge shock is generated at the nose of the plate. The free stream flow conditions correspond to a Mach number of 4.0 and a temperature at infinity of $392.4^0 R$. The Reynolds number per unit length is $4,000,000$ and the local Prandlt number is 0.75. The computational mesh is displayed in figure 2 and consists of 20,000 elements and 10,201 nodes. This mesh uses highly stretched elements. In the streamwise direction, the spacing distribution increases as we move away from the leading edge (aspect ratio 1/50) to the trailing edge (aspect ratio 1/175.25). In the normal direction, we also have an increase in mesh spacing as we move away from the flat plate, as seen in the figure. The initial conditions are chosen to simulate a flat plate that is suddenly exposed to the free stream. No slip, adiabatic wall conditions are imposed on the plate. At the upstream boundary, all flow variables are prescribed at free stream values. At others boundaries, the viscous flux $\bar{G}^n$ and inviscid flux $\bar{F}^n$ are made equal to the fluxes $G^n$ and $F^n$ computed from the local numerical solution.

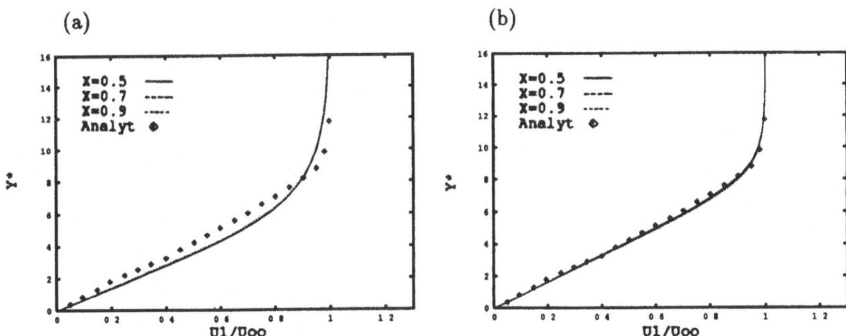



**Figure 3** : Velocity Profile.

Figure 3(a) shows the comparison between the van Driest similarity solution [20] and the velocity profiles computed using the first–order AUSM numerical flux, at different chord lengths downstream of the leading edge, scaled by introducing the dimensionless length

$$Y^{\star} = \frac{Y}{X}\sqrt{Re_X} \; . \tag{30}$$

A similar comparison for the results of the high–order MUSCL extension can be seen in figure 3(b). The velocity profiles are in good agreement with the analytical solution. An idea of the higher resolution achieved with the proposed high–order scheme when compared with the first–order upwind solution can also be seen when both figures are compared.

**Hypersonic Flow Over a Compression Corner**

A second application involves hypersonic flow over a compression corner of $24^0$ angle. This problem has been extensively studied in the literature and represents a challenging application in the validation of viscous algorithms. A schematic description of the flow behavior is shown in figure 4.

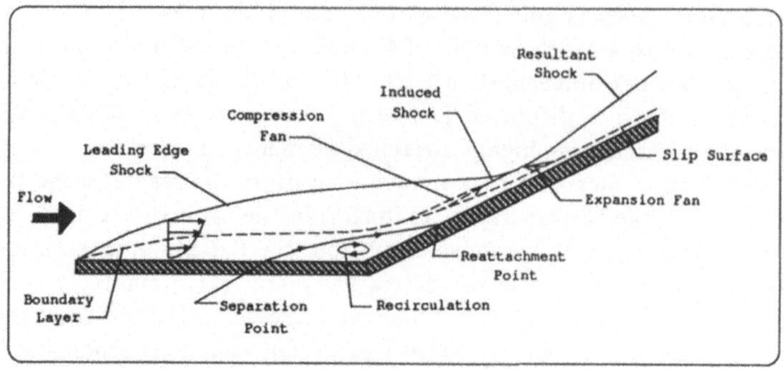

**Figure 4** : Hypersonic Flow over a Compression Corner

This test case includes most of the difficulties encountered in boundary layer–shock wave interaction simulation and is an important problem which is of interest to the designer of the propulsion system of hypersonic flight vehicles.

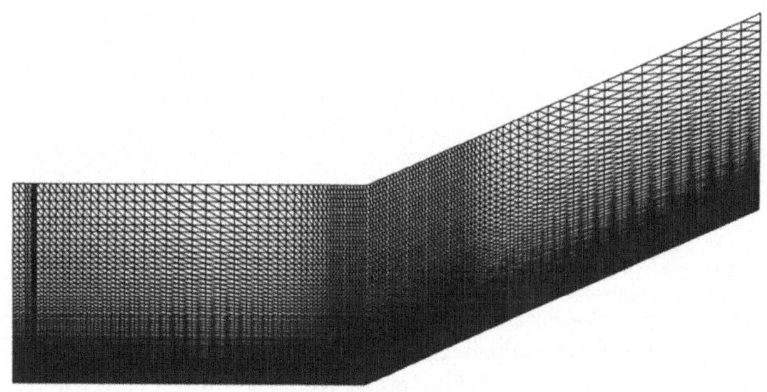

**Figure 5** : Mesh.

The free stream Mach number is 14.1, and the Reynolds number, based upon a flat plate length of 1.44 ft, is $103,680$. The temperature of the free stream is $160^0 R$ and the constant value of 0.72 is assumed for the Prandlt number. The wall temperature $T_w$ is $535^0 R$. The Reynolds number is low enough to ensure that the flow remains completely laminar and the free stream temperature is low enough so that there are no significant real–gas effects [17]. Computations were made using the grid shown in figure 5. The domain of computation is extended in front of the leading edge. The no slip isothermal condition has been imposed at the wall. At the upstream boundary, all flow variables are set to free stream values. At other boundaries, the viscous flux $\bar{G}^n$ and inviscid flux $\bar{F}^n$ are made equal to the fluxes $G^n$ and $F^n$ computed from the local numerical solution.

Figure 6 shows the density, pressure coefficient and Mach number contours for the MUSCL scheme. Qualitative aspects of the flow field, such as separation zone and shock–

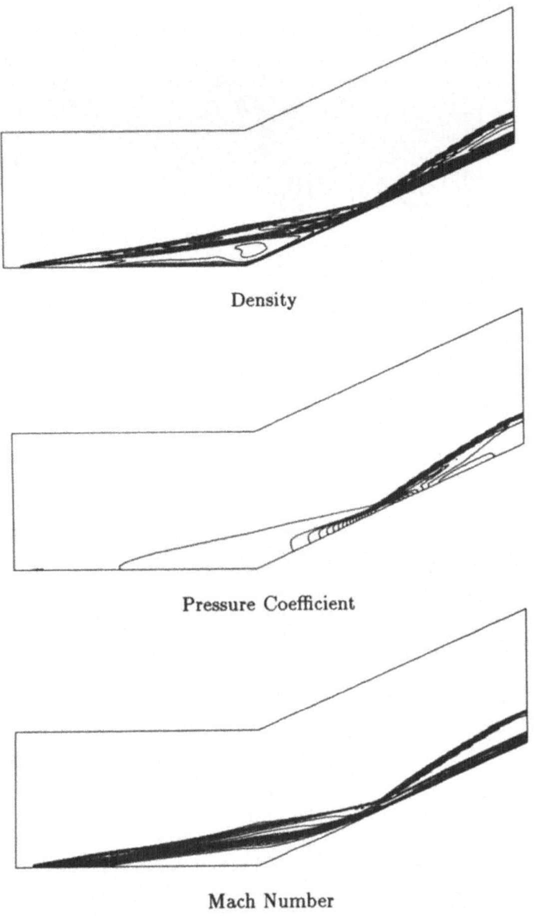

Density

Pressure Coefficient

Mach Number

**Figure 6** : Computed contours.

boundary layer interaction, are well represented. The velocity vectors are presented in figure 7 where most of the features shown schematically in figure 4 are apparent.

We compare computed wall profiles with the results produced with the code CFL3D by Rudy et al [17]. Coefficients of pressure, heat transfer and skin friction, defined according to

$$C_p = p/(\frac{\rho_\infty U_\infty U_\infty}{2}) \qquad C_h = \frac{\partial T}{\partial n}/[\rho_\infty U_\infty (H_\infty - H_w)] \qquad C_f = T_w/(\rho_\infty U_\infty^2/2) \quad (31)$$

are compared. The variation of $\log(10^2 C_p/2)$ along the wall is shown in Figure 8(a) for and the solution is in very good agreement with CFL3D results. Slight differences occur in the

177

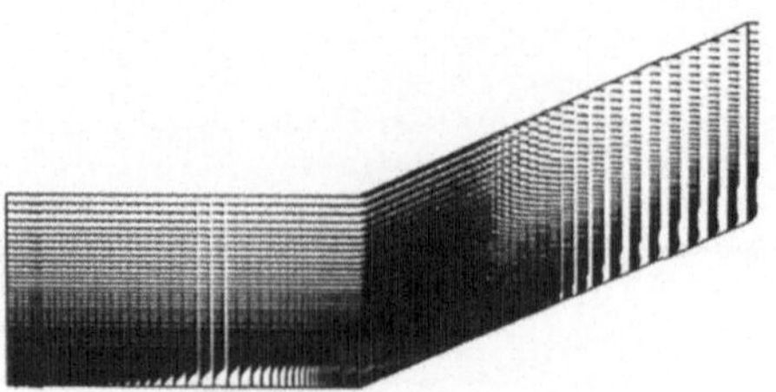

**Figure 7** : Velocity Vectors

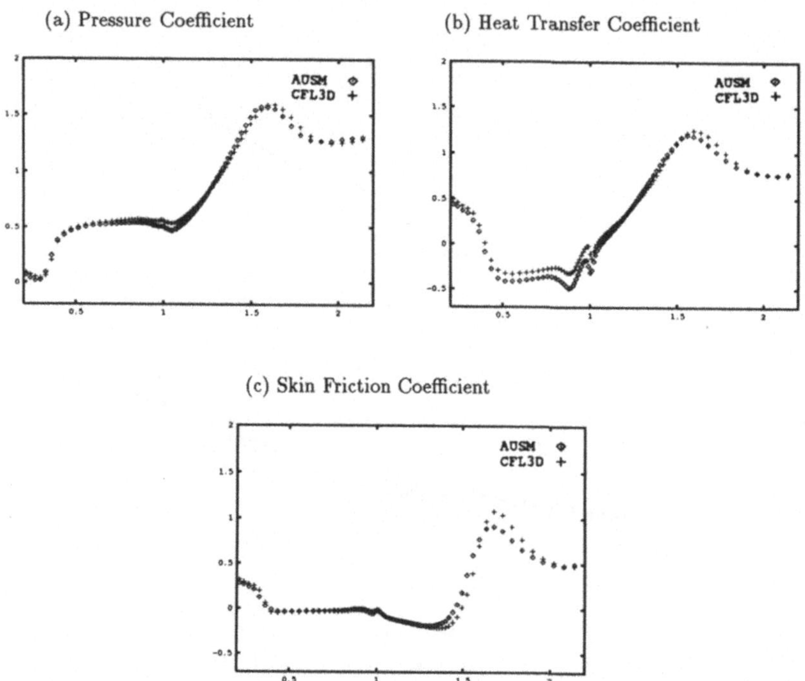

(a) Pressure Coefficient

(b) Heat Transfer Coefficient

(c) Skin Friction Coefficient

**Figure 8** : Comparison of Computed Surface Distributions.

separation zone and this has also been experienced by the authors in a previous work [5], where a symmetric TVD algorithm and a MUSCL scheme, both with a Roe numerical flux, were used. The variation of $\log(10^3 C_h)$ along the wall is compared in figure 8(b). The results differ some what from those produced by CFL3D in the separation zone. Figure 8(c) shows the variation of $10^2 C_f/2$ along the wall. In this case, the overall agreement is good with a discrepancy in the region downstream of the reattachment point. Again, a similar behavior has been observed by the authors previously [5]. It should be mentioned that the adoption of different limiters has been found to have a big influence on the magnitude of the discrepancies which are apparent in figure 8.

## CONCLUSIONS

The construction of a high–resolution non–oscillatory algorithm, based on the AUSM flux splitting, for the solution of inviscid and viscous high–speed flows on general triangular meshes has been described. The proposed algorithm is directly extendable for 3–D analysis on tetrahedral meshes. The choice of a split–Mach version was found to be crucial when Navier–Stokes equations are considered. It was also observed that the higher–order AUSM scheme gives sharp shock resolution. The numerical results obtained for the applications analysed, the flexibility of the suggested schemes and also the favorable properties of the AUSM formulation make the present scheme promising for viscous hypersonic complex blunt–body flow computations.

## ACKNOWLEDGEMENTS

It is a pleasure to acknowledge many stimulating discussions during the course of the present work with Dr. O. Hassan. The first author would like to acknowledge the support received from CNPq ( Brazilian Research Council ). The other authors acknowledge the partial support provided by the Aerothermal Loads Branch of the NASA Langley Research Center under research grants NAGW-1809 and NAGW-3290.

## REFERENCES

[1] A. JAMESON. Artificial Diffusion, Upwind Biasing, Limiters and their Effect on Accuracy and Multigrid Convergence in Transonic and Hypersonic Flows. Paper 93-3359, AIAA, 1993.

[2] M.-S. LIOU and C.J. STEFFEN. High-Order Polynomial Expansions (HOPE) for Flux-Vector Splitting. Paper 104452, NASA, 1991.

[3] M.-S LIOU and C.J. STEFFEN. A New Flux Splitting Scheme. Paper 104404, NASA, 1991.

[4] H. LUO, J.D. BAUM, R. LÖHNER, and J. CABELLO. Adaptive Edge-Based Finite Element Schemes for the Euler and Navier-Stokes Equations on Unstructured Grids. Paper 93-0336, AIAA, 1993.

[5] P.R.M. LYRA, M.T. MANZARI, and K. MORGAN. Side-Based Unstructured Grid Algorithms for Compressible Viscous Flow Computations. *Int. J. Num. Meth. in Engng.*, 1994. submitted.

[6] P.R.M. LYRA, K. MORGAN, J. PERAIRE, and J. PEIRÓ. TVD Algorithms for the Solution of the Compressible Euler Equations on Unstructured Meshes. Paper, University College of Swansea, 1993. also submitted for Int. J. Num. Meth. Fluids.

[7] P.R.M. LYRA, K. MORGAN, J. PERAIRE, and J. PEIRÓ. Unstructured Grid FEM/TVD Algorithm for Systems of Hyperbolic Conservation Laws. In *8th International Conference on Numerical Methods in Laminar and Turbulent Flow*, Swansea/U.K., 1993.

[8] M.T. MANZARI. Privite Communication, January 1994.

[9] M.T. MANZARI, P.R.M. LYRA, K. MORGAN, and J. PERAIRE. An Unstructured Grid FEM/MUSCL Algorithm for the Compressible Euler Equations. In *VIII International Conference on Finite Elements in Fluids: New Trends and Applications*, Barcelona/Spain, 1993.

[10] K. MORGAN, J. PERAIRE, and J. PEIRÓ. Unstructured Grid Methods for Compressible Flows. In *Special Course on Unstructured Grid Methods for Advection Dominated Flows*, pages 5.1–5.39, France, 1992. AGARD/VKI.

[11] S. OSHER and F. SOLOMON. Upwind Difference Schemes for Hyperbolic Systems of Conservation Laws. *Mathematics of Computation*, 38:339–374, 1982.

[12] J. PERAIRE, K. MORGAN, M. VAHDATI, and J. PEIRÓ. The Construction and Behavior of Some Unstructured Grid Algorithms for Compressible Flows. In *ICFD Conference on Numerical Methods for Fluid Dynamics*. Oxford Univeristy Press, 1992.

[13] J. PERAIRE, J. PEIRÓ, and K. MORGAN. A 3-D Finite Element Multigrid Solver for the Euler Equations. Paper 92-0449, AIAA, 1992.

[14] J.J. QUIRK. A Contribution to the Great Riemann Solver Debate. Report 92-64, ICASE, 1992.

[15] P.L. ROE. Approximate Riemann Solvers, Parameter Vectors and Difference Schemes. *J. Comp. Phys.*, 43:357–372, 1981.

[16] P.L. ROE. A Survey of Upwind Differencing Techniques. *Lecture Notes in Physics*, 323:69–78, 1989.

[17] D.H. RUDY, J.L. THOMAS, A.KUMAR, P.A. GNOFFO, and S.R. CHAKRAVARTHY. A Validation Study of Four Navier-Stokes Codes for High-Speed Flows. Paper 89-1838, AIAA, 1989.

[18] J.L. STEGER and R.F. WARMING. Flux Vector Splitting of the Inviscid Gasdynamic Equations with Application to Finite-Difference Methods. *J. Comp. Phys.*, 40:263–293, 1981.

[19] J.L. THOMAS. An Implicit Multigrid Scheme for Hypersonic Strong-Iteration Flowfields. In *Proceedings of the Fifth Copper Mountain Conference on Multigrid Methods*, 1991.

[20] E.R. VAN DRIEST. Investigation of Laminar Boundary Layer in Compressible Fluids Using the Crocco Method. Technical Note 2597, NASA, 1952.

[21] B. VAN LEER. Towards the Ultimate Conservative Difference Scheme. V. A Second Order Sequel to Godunov's Method. *J. Comp. Phys.*, 32:101–136, 1979.

[22] B. VAN LEER. Flux-Vector Splitting for the Euler Equations. *Lecture Notes on Physiscs*, 170:507–512, 1982.

[23] B. VAN LEER, J.L. THOMAS, P.L. ROE, and R.W. NEWSOME. A Comparison of Numerical Flux Formulas for Euler and Navier-Stokes Equations. Paper 87-1104, AIAA, 1987.

# THE NUMERICAL APPROXIMATION OF A FREE GAS-VACUUM SURFACE

C.-D. Munz, R. Schneider, O. Gerlinger
Kernforschungszentrum Karlsruhe
Institut für Neutronenphysik und Reaktortechnik
Postfach 3640
76021 Karlsruhe

## Summary

We present a simple approach of the tracking of a gas-vacuum interface within a fixed Eulerian grid in conjunction with a scheme in conservation form. The tracking algorithm gives an estimation of the movement of the interface. This information is used to modify the numerical flux at the vacuum boundary to prevent the numerical smearing and is used to identify the vacuum region. The numerical method is applied to grid zones containing gas only.

## 1. Introduction

The approximation of gas flow with a gas-vacuum boundary may give rise to severe difficulties in the numerical schemes. If the conservation laws of compressible fluid flow are formulated in an Eulerian frame of reference, which is fixed in space, the computational domain contains the vacuum region and a material vacuum boundary. As the conservation equations are based on the continuum assumption, they are no longer valid in the vacuum and a numerical approximation based on these equations will fail. In this paper we propose a tracking method for a material vacuum interface by means of which the difficulties mentioned above are avoided. The propagation of the gas-vacuum boundary is followed, the information about its actual location is used to identify the vacuum region and to determine the numerical fluxes between the grid zones at the vacuum interface. The numerical method for the gas flow is never applied within a vacuum region. The basic ideas of this method are described in one space dimension. Due to the invariance of the Euler equations according to a rotation, these considerations are also valid in the multidimensional case in normal direction of the gas-vacuum boundary.

Our investigations are motivated by the numerical simulation of an anode plasma within a vacuum diode which are based on a fluid model. Due to the electromagnetic fields the transition of plasma to vacuum remains relatively sharp so that the gas-vacuum boundary can be approximated by that continuum-vacuum transition. In other problems a region of dilute gas may occur where the mathematical modelling as continuum fails. In this case a gas kinetic description based on the Boltzman equations has to be introduced and an asymptotic analysis should be appropriate to analyze the transition. Other fields of application for our method may be super- or hypersonic fluid flow where regions of very low densities occur which may be treated as a vacuum region or jet engines for satellite steering.

## 2. The Problem, the Equations, the Numerical Approximation

The one-dimensional Euler equations, written in the conservation form, read as

$$u_t + f(u)_x = 0 \tag{1}$$

with the vector of the conserved variables and the fluxes

$$u = \begin{pmatrix} \rho \\ \rho v \\ e \end{pmatrix}, \quad f(u) = \begin{pmatrix} \rho v \\ \rho v^2 + p \\ v(e + p) \end{pmatrix}, \tag{2}$$

respectively. Here, $\rho$ is the density, $\rho v$ is the momentum per unit volume, $v$ is the fluid velocity and $e$ is the total energy per unit volume. The pressure $p$ is related to the conserved variables via the equation of state $p = p(\rho, \epsilon)$, where we assume for the sake of simplicity that of a perfect gas

$$p = (\gamma - 1)\rho\epsilon \tag{3}$$

with the adiabatic exponent $\gamma$; $\epsilon$ is the specific internal energy

$$\epsilon = \frac{e}{\rho} - \frac{1}{2}v^2 \quad . \tag{4}$$

We will give some remarks about a refusion of the methods proposed to a real equation of state within the last section.

We consider a numerical method for the Euler equations (1) which can be written in the conservation form

$$u_i^{n+1} = u_i^n - \frac{\Delta t}{\Delta x_i} \left( g_{i+1/2}^n - g_{i-1/2}^n \right) \quad . \tag{5}$$

Here $\Delta t$ denotes the time step $t_{n+1} - t_n$, $u_i^n$ is an approximation of the mean value of the conservative variables in the grid zone $[x_{i-1/2}, x_{i+1/2}]$ of length $\Delta x_i$, $g_{i+1/2}^n$

is the approximation of the flux $f(u)$ at the right grid zone boundary $x_{i+1/2}$ during the time step $\Delta t$. Numerical methods for the Euler equations are usually written in this conservation form, because it reproduces the exact integral conservation and guarantees the right propagation rates of shock waves.

As discussed in the introduction the Euler equations (1) are no longer valid within the vacuum and, hence, (5) cannot be a consistent approximation. A common procedure to treat a vacuum region within the computational domain is to replace the vacuum by a dilute gas. It is assumed that the gas flow into this rarefied gas is quite similar to the flow into real vacuum. But it should be not too dilute , because then any approximation error may lead to negative values of the internal energy (see [3]). This is due to the fact that the dominant energy mode becomes kinetic near the vacuum. If we use the scheme in conservation form (5), then at the new time level $t_{n+1}$ the values

$$\rho_i^{n+1} \,, \; (\rho v)_i^{n+1} \; \text{and} \; e_i^{n+1} \tag{6}$$

are calculated. The values of the primitive variables are then extracted from these one by computing the internal energy, which is obtained by subtracting the kinetic energy from the total energy. These are values which are almost identical. The kinetic energy is obtained by dividing the momentum and the density which both tend to zero. Hence, small rounding errors may be accentuated and produce unphysical negative values of pressure. If they are simply replaced by small positive ones to continue the calculation, the conservation laws are violated and nonlinear instabilities may be generated.

## 3. Tracking of Gas-Vacuum Interfaces

The idea in this paper is to track within each time step the gas-vacuum boundary in a first step. This will give an estimation of the real movement of the gas-vacuum boundary. Here we will propose two different ways in the section 3.1 and 3.2, respectively. In a second step we use this information twice: ($i$) We do not apply the numerical method in grid zones containing pure vacuum; ($ii$) We modify the numerical fluxes of the grid zones containing the gas-vacuum interface in such a way that the interface is kept sharp. This is described in section 3.3.

### 3.1 Vacuum Riemann Problem Approach

In [4] Godunov proposed a scheme in conservation form where the main building block is the solution of a Riemann problem (RP) which is used to determine the numerical fluxes. The RP is an initial-value problem of the Euler equations with piecewise constant initial data: A left constant state $u_l$ for $x < 0$ and a right state $u_r$ for $x > 0$. A detailed formulation of the RP and its solution for a perfect gas is reviewed, e.g., by Chang and Hsiao [1]. The general solution of the RP is given by a fixed point problem, and consists of four constant states $u_l$, $u_1$, $u_2$ and $u_r$, separated by three elementary waves, where the left and the right are shocks or rarefactions. The intermediate states $u_1$, $u_2$ are separated by a contact discontinuity, where the density is discontinuous, while the pressure and velocity are continuous.

From now on we consider the situation where the right state $u_r$ coincides with the vacuum. This means, that the conserved variables forming the right state are zero, and both pressure $p_r$ and sound velocity $c_r$ also vanish. We note that the specification of a vacuum fluid velocity has no physical meaning. In the following, we call this problem for the Euler equations (1) with the initial values specified by

$$u(x,0) = \begin{cases} \left(\rho_l, (\rho\, v)_l, e_l\right) & \text{for} \quad x < 0 \\ (0,0,0) & \text{for} \quad x > 0 \end{cases} \tag{7}$$

the "vacuum Riemann problem (VRP)". In reality it is no initial value problem, because vacuum is no solution of the Euler equations. Hence, the VRP is a free boundary problem. As it is found and dicussed by Halter [5] (see also [1]), the solution of the VRP may be obtained as a limit of the solution of the usual Riemann problem. The general structure of the solution of the VRP can be summarized as follows. The contact discontinuity, travelling with the local fluid velocity , sets up in general the interface between different materials. Hence, the right elementary wave of the usual RP disappears. Because the pressure has to be constant across the contact (namely zero), the contact discontinuity also disappears in the sense that it coincides with the right boundary of the left elementary wave connecting $u_l$ with $u_1$. Consequently, the left state is directly connected to the vacuum by only one elementary wave, see Fig. 1. Since the Rankine-Hugoniot condition cannot be satisfied in this case, the elementary wave has to be a rarefaction wave. For a more complete discussion as well as for the explicit solution of (1), (7) we refer to [1], [5] and [11].

We describe next how to use this VRP to get insight into the time evolution of a gas-vacuum boundary. Suppose that the location of plasma vacuum boundary $(x_V^n)$ is situated in the i-th grid zone at the time level $t = t_n$ (see Fig.2). For the sake of simplicity we assume in the following that the computational region $[a, b]$ is divided into that of gas $[a, x_V^n]$ and that of vacuum $(x_V^n, b]$, where the gas expands to. Refer to the exact solution of the VRP, the location of the plasma vacuum boundary at the time level $t = t_{n+1}$ is obtained from

$$x_V^{n+1} = x_V^n + (t_{n+1} - t_n)\, v_V^n \quad \text{with} \quad v_V^n = v_l + \frac{2}{\gamma - 1}\, c_l. \tag{8}$$

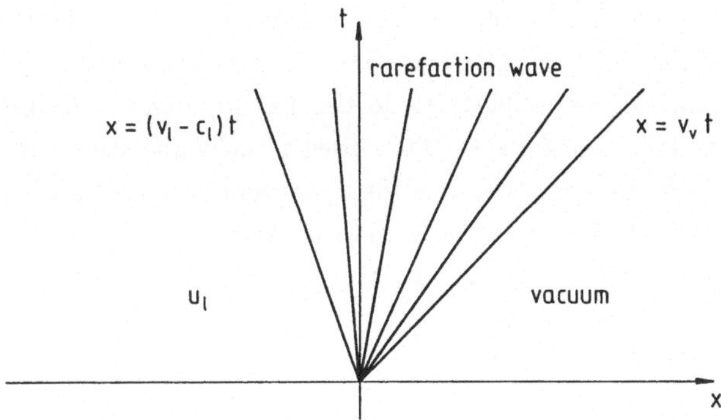


Fig. 1: (x,t)-diagram of a solution of the vacuum Riemann problem

The left state $u_l$ required in (8) may be obtained by the formulae

$$u_l = \theta\, u_i^n + (1 - \theta)\, u_{i-1}^n, \qquad \theta = \frac{1}{\Delta x_i}\left(x_V^n - x_{i-1/2}\right), \tag{9}$$

where the knowledge of the location of the plasma vacuum boundary at the time level $t_n$ is taken into account, explicitly. The calculation of the left state $u_l$ as proposed by (9), can be understood in the following way. Using only $u_i^n$ as the left state, waves generated at $x_{i-1/2}$ will interact with the VRP and can change its solution. To guarantee that these waves do not reach the gas-vacuum interface a constant left state integrated and averaged over the interval of length $\Delta x_i$ must be introduced. Because the i-th grid interval $\left[x_{i-1/2}, x_{i+1/2}\right]$ contains gas as well as vacuum, we ought to use the information about the gas-vacuum location and redistribute the integral value $u_i^n$ over the interval $\left[x_{i-1/2}, x_V^n\right]$ :

$$< u_i^n > = \frac{1}{\theta}\, u_i^n = \left(\bar{\rho}_i^n, (\bar{\rho}\bar{v})_i^n, \bar{e}_i^n\right)^T. \tag{10}$$

In the numerical calculations we found that it is favorable to calculate the right hand side of (10) not in the conserved but in the primitive variables $\rho$, $v$ and $p$. Thereupon, the vector $< w_i^n >$ in the primitive variables, corresponding to (10) reads as

$$< w_i^n > = \left(\bar{\rho}_i^n, v_i^n, \bar{p}_i^n\right)^T, \tag{11}$$

where we have to emphasize that the velocity is not obtained by its redistributed integral value. Performing the calculation in the primitive variables ensures that $p_l$ is a convex average of $\bar{p}_i^n$ and $p_{i-1}^n$ and remains positive. Otherwise, the pressure $p_l$ has to be recalculated from the average of the conservative variables, which may lead in low density regions to the numerical difficulties mentioned already above. The integration (9) may be avoided if the interaction of the solution of the RP at $x_{i-1/2}$ and the solution of the VRP at $x_V^n$ is taken into account. This can only be done in an approximative way. One possibility to do this has been shown by LeVeque and Shyne in [8] for tracking shock waves. They used a technique based on large time step wave propagation methods. Because the expansion into vacuum is continuous contrary to shock waves or material interfaces we assume that the simpler integration (9) is accurate enough within our context.

### 3.2 $G$-Function Approach

An alternative approach for tracking the interface between gas and vacuum is to introduce a so-called level set function $G$. This method has been proposed by Mulder et al. [9] for the tracking of material interfaces and is excellently reviewed and extended in the work of Davis [2]. The basic idea within this approach is to define a scalar function $G = G(x,t)$, with the properties that this level set function is transported with the same velocity as the material and has a root at the gas-vacuum interface. Being more precisely, the temporal evolution of the $G$-function is given by a simple transport equation

$$G_t + v\, G_x = 0 \quad , \tag{12}$$

where $v$ denotes the fluid velocity. The initial form of the function $G(x, t = 0)$ is chosen in such a way that ($i$) $G$ is smooth, ($ii$) $G$ has a zero at the initial gas-vacuum location: $G(x_V^0) = 0$, ($iii$) $G$ is monotonic in the neighborhood of $x_V^0$.

In each time step the level set equation is solved numerically before the calculation of the gas flow. This is performed by a non-conservative extension of the MUSCL scheme of van Leer, including the flux calculation by an upwind scheme, similar to the methods proposed in [2] and [9]. The location of the interface at time $t_{n+1}$ are extracted from the discrete values $G_i^{n+1}$. The cells adjacent to the gas-vacuum interface can be found out by a simple search through all cells which stops when the condition

$$G_i^{n+1} G_{i+1}^{n+1} < 0$$

is achieved. From this we have the information that the gas-vacuum boundary lies within grid zone $i$ or $i + 1$ and we have next to decide whether it is $i$ or $i + 1$. If the values of the level set function are initially chosen skew symmetrical with respect to the gas-vacuum boundary then a simple algorithm may give this information: it is this grid zone with the smaller absolute value of $G$.

The main problem in this approach is the definition of an appropiate velocity $v$ in $G$ within the whole computational domain. Good results have been obtained by using the velocity (12) within each grid zone and a constant continuation into vacuum. To avoid the calculation $\frac{p}{\rho}$ for small values an isentropic approximation of the sound velocity is introduced.

## 3.3 Modified Flux Calculation

By the algorithms proposed in section 3.1 as well as 3.2 we obtain an approximation of the gas-vacuum boundary at time $t_{n+1}$ in a first step. In the second step, we use this information to identify the vacuum region and to calculate the numerical flux near the gas-vacuum boundary. Two cases are to distinguish (see Fig. 2).

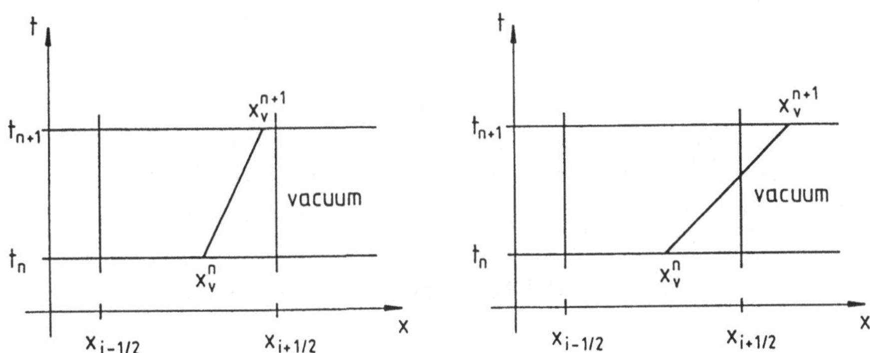


Fig. 2: Tracking of the gas-vacuum boundary within the Eulerian grid: (a) case 1; (b) case 2

<u>Case 1:</u> According to Fig. 2a, the grid zone interface at $x = x_{i+1/2}$ belongs to vacuum during the whole time step $\Delta t$. Hence, the numerical flux there must be zero:

$$\text{If} \quad x_V^{n+1} \leq x_{i+1/2} \qquad \text{then} \quad g_{i+1/2} = 0. \tag{13}$$

For all $j > i + 1$ the scheme in conservation form is not applied. The information on the real movement of the gas-vacuum boundary obtained by the tracking methods and the conclusion (13) keeps the approximation of the gas-vacuum interface sharp in the sense that no values of physical quantities are introduced into vacuum as numerical artifacts.

<u>Case 2:</u> If $x_V^{n+1} > x_{i+1/2}$, then the gas-vacuum boundary moves across the grid zone interface during the time step, evidently from Fig. 2b. In this case the flux will be positive. If the scheme in conservation form is positively conservative for fluid flow of low density (see [3]) then the usual flux calculation can be applied. Otherwise an

extension of Godunov's idea to the gas-vacuum boundary may be used: The VRP
is solved at the grid zone boundary $x_{i+1/2}$ with $u_i^n$ as left state, the numerical flux
is then defined to be the physical flux of this VRP. If the VRP-tracking method is
applied, another good numerical flux calculation is obtained by using the fluxes of the
exact VRP, taken at $\delta x^n = x_{i+1/2} - x_V^n$ and averaged over the time step, afterwards.
Incorporating further informations of the tracking, we assume that the gas-vacuum
interface intersect the vertical line $x = x_{i+1/2}$ at $t_n^* = t_n + \frac{\delta x^n}{v_V^n}$ (see Fig. 2). Then, if
$v_l - c_l$ is positive, the intersection of the left rarefaction fan boundary with $x = x_{i+1/2}$
happens at $t_{n+1}^* = \min\left(t_{n+1}, t_n + \frac{\delta x^n}{v_l - c_l}\right)$. Otherwise, if $v_l - c_l < 0$, $t_{n+1}^* = t_{n+1}$. This
can be summarized in the compact form

$$g_{i+1/2} = \frac{1}{t_{n+1} - t_n}\left\{\int_{t_n^*}^{t_{n+1}^*} f\left[u_0(\delta x^n, t)\right] dt + \left(t_{n+1} - t_{n+1}^*\right) f(u_l),\right\} \qquad (14)$$

where $u_0(\delta x^n, t)$ denotes the exact solution of the VRP. In principle, the integrals (14)
can be determined analytically. But, in order to reduce the numerical effort, it seems
to be favorable to approximate the integrals numerically with methods, accurate up
to second order.

## 4. Numerical Results and Conclusions

As a test problem we consider a simple gas-vacuum expansion wave which is generated
by the VRP initial data

$$(\rho, v, p) = \begin{cases} (1.0, 0.0, 1.0) & \text{for} \quad x < 0 \\ vacuum & \text{for} \quad x > 0 \end{cases}.$$

The numerical calculations are performed on a grid with 100 grid zones and the
computational region is $[-0.3, 0.7]$. Fig.3 shows numerical results of the Godunov
scheme combined with the tracking algorithm based on the VRP Riemann problem at
time $t = 0.1$ in comparison with the exact solution ($\circ\circ\circ$ numerical results, —— exact
solution). Fig.4 shows results obtained with the $G$-function tracking method. Beside
the density, the pressure and the velocity as function of $x$ we plotted the location of
the gas-vacuum boundary as a function of time. The figures indicate that this gas-
vacuum boundary is captured very well. The movement is slightly underestimated.
The different tracking algorithms lead to almost the same results. The numerical
dumping at the rarefaction which is introduced by the first-order accurate Godunov
scheme will be considerably reduced using a second-order extension.

In the two-dimensional case, the situation is much more complicated, since the
gas-vacuum interface is now a curve within the computational domain. Our one-
dimensional considerations are valid into the normal direction of the gas-vacuum

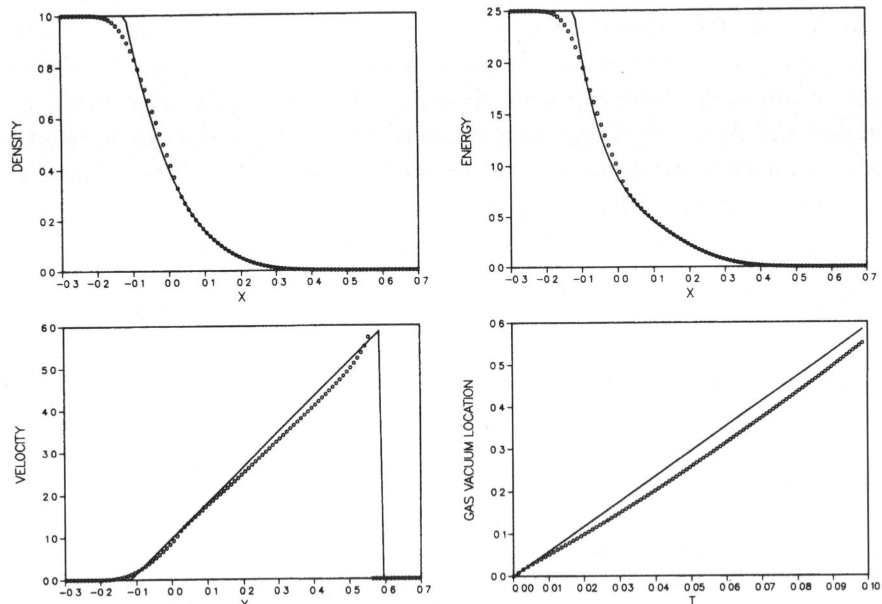

Fig.3: Numerical results at $t = 0.1$ using the VRP tracking algorithm

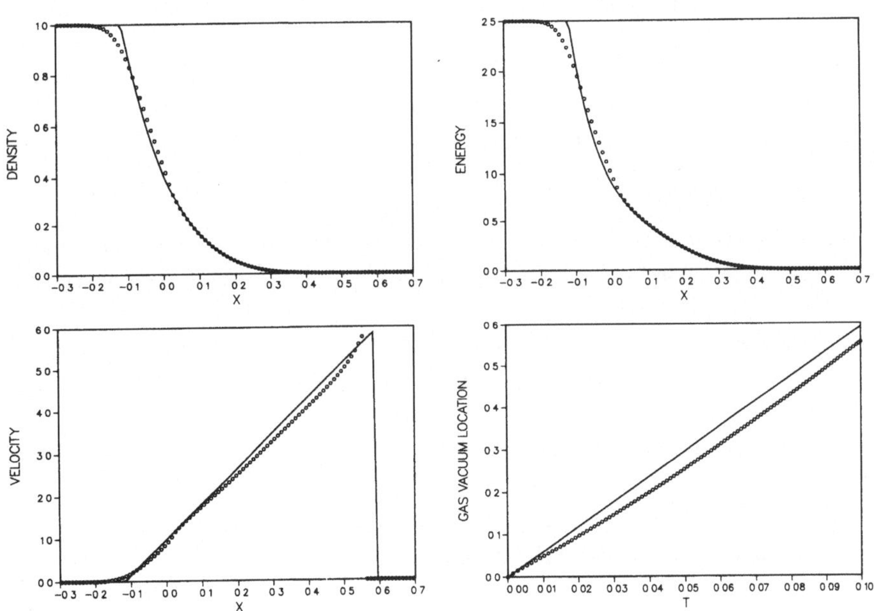

Fig.4: Numerical results at $t = 0.1$ using the $G$-function tracking

interface and they have to be combined with a two-dimensional tracking method as reviewed in [7]. The simplest extension to two dimensions appears to be the $G$-funcion approach. In this case a two-dimensional transport equation has to be solved numerically. Again the change of the sign of $G$ indicates the position of the interface. A simple dimensional splitting technique has been used in [2] for material interfaces and produces good results.

# References


1. Chang, T. and Hsiao, L.: The Riemann Problem and Interaction of Waves in Gas Dynamics; Longman, UK, Essex, 1989.

2. Davis, S.F.: An interface tracking method for hyperbolic systems of conservation laws; Applied Numerical Mathematics **10**, 447-472 (1992).

3. Einfeldt, B., Munz, C.-D., Roe, P.L. and Sjögreen, B.: On Godunov-type methods near low densities; J. Comput. Phys. **92**, 273-295 (1991).

4. Godunov, S.K.: Finite difference method for numerical computation of discontinous solutions of the equations of fluid dynamics; Math. Sbornik **47**, 271-306 (1959) *(in russian)*.

5. Halter, E.: A fast solver for Riemann Problems; Math. Meth. in the Appl. Sci. **7**, 101-107 (1985).

6. Harten, A., Lax, P.D., and van Leer, B.: On upstream differencing and Godunov-type schemes for hyperbolic conservation laws; SIAM Rev. **25**, 35-61 (1983).

7. Hyman, J.M.: Numerical methods for tracking interfaces, in: A.R. Bishop, L.J. Campbell and P.J. Channell, eds. *Fronts, Interfaces and Patterns* (Elsevier, New York, 1984).

8. LeVeque, R.J. and Shyne, K.M.: Shock tracking based on high resolution wave propagation methods; University of Washington, Seattle, Technical Report no. 92-3, January 1992.

9. Mulder,W., Osher,S. and Sethian,J.A.: Computing interface motion in compressible gas dynamics; CAM Rept. 90-03, University of California, Los Angeles, CA (1990).

10. Munz, C.-D.: On Godunov-type schemes for Lagrangian gas dynamics; SIAM J. Numer. Anal. **31**, (1994), in press.

11. Munz, C.-D.: A tracking method for gas flow into vacuum; Math. Meth. in the Appl. Sci. **16**, (1994), in press.

# A CHIMERA GRID SCHEME FOR THE CALCULATION OF THE FLOW AROUND PARTICLES IN DIFFICULT GEOMETRIES


Nirschl, H., Dwyer, H.A.[*], Denk, V.

Technische Universität München
Lehrstuhl für Fluidmechanik und Prozeßautomation
85350 Freising
Germany

[*]University of California Davis
Department of Mechanical and Aeronautical Engineering
Davis, CA 95616
U.S.A.


## SUMMARY


In this investigation a Chimera grid scheme has been developed for the calculation of particle flows. The Chimera scheme is an overset grid scheme approach, where each configuration of a complex geometry is grided separately and then overset onto a main grid. In general the method can handle efficiently difficult problems in computational fluid dynamics like particle wall interaction and the flows around multiple particle systems. The scheme has been introduced into a two as well as three dimensional and incompressible Navier-Stokes computer code. The system of equation was solved by a finite volume formulation including the thermal energy equation. It was tested for some simple problems like the two dimensinal flow through a heat exchanger configuration and the well known problem of the flow over a spherical particle. In the results some calculations concerning particle wall interaction will be presented.


## INTRODUCTION

The transport of particles and drops is of interest in a lot of engineering applications. The majority of previous studies have been for a single particle in an uniform or shear flow with a finite velocity at the axis of the particle. An extensive review of theoretical and experimental work on bubbles, drops and particles is given by *Clift, Grace and Weber* /1/. In most practical flows the particle is influenced either by other particles or by walls. Especially for low Reynolds numbers the flow is dominated by viscous effects. This usually implies a strong influence of the particle on the flow field far away from its surface. Therefore a wall near a particle can have a significant effect on the local distribution of the forces on the particle, as well as on the global variables such as drag, lift, torque and heat transfer. It is a primary purpose of this investigation to extend our ability to treat multiple particle flows and/or particle wall flows.

In this study we will describe a Chimera grid method for two as well as three dimensional flow calculations. The Chimera grid method allows a favorable treatment of multiple body configurations and it gives the possibility to handle difficult problems like particle-wall interactions. Each part of a Chimera configuration is grided separately and then overset onto a main grid. In our cases we use for the main grids simple rectangular grids stretched over the entire computational domain. A minor grid is generated around the particle to resolve the details of the flow close to the surface. In general the minor grids are used to resolve features of the geometry or flow that are not sufficient resolved by the major grid. The fundamental ideas of the Chimera grid scheme approach are described in References /2/, /3/ and /4/.

Another great advantage of the approach is obvious considering the calculation of unsteady flow problems. When the particle has to move because of external forces it is not necessary to generate a new grid for the whole configuration. With Chimera each particle will keep his own mesh and only the relative position of the meshes to each other will change. All the geometric quantities of a mesh, like volumes, areas and coordinate transformations will still be the same. The generation of a grid around the whole configuration is not only very difficult, esspecially in 3D, but also very time consuming.

## PROBLEM STATEMENT AND NUMERICAL APPROACH

The basic geometry will be a particle or particles held fixed in a large rectangular computational domain. We assume that the particle is spherical with a diameter D and rigid with no slip at the surface. The fluid is incompressible with a constant density $\rho$ and dynamic viscosity $\mu$. In this paper we want to consider two different problems in detail. The first example is a simple shear flow over a particle where the shear is generated by two in opposite direction moving walls. The two walls are moving with the same speed and the particle is located in the middle of the gap (see figure 1). The influence of particle rotation dependend on the Reynolds number and the gap width has been included. The second example will be the flow around a particle close to a wall (see figure 2). This problem is of interest in particle loaden flows through tubes, in boundary layers or in filtration processes, where the efficiency of the filtration apparatus can depend on the forces on the particle.

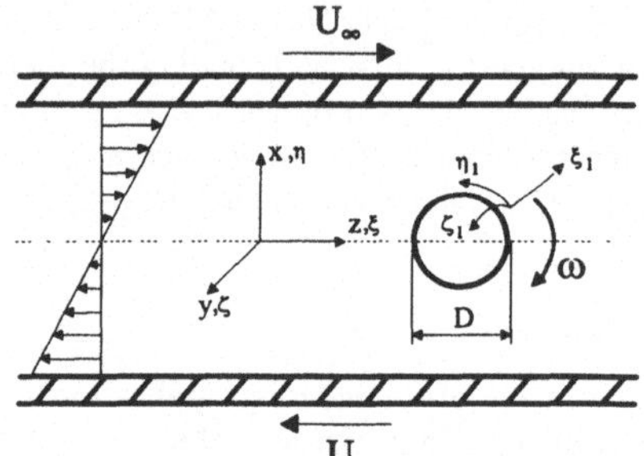

Figure 1:  Schematic description of the simple shear flow around a particle

Figure 2:  Schematic description of the flow around a particle close to a wall

The equations describing the flow and the heat transfer are the continuity equation, the Navier-Stokes and the thermal energy equation in integral and dimenionless form. The equations are formulated in the Cartesian coordinate system x,y,z where the geometries of the particles or the rectangular meshes are described with the body fitted coordinates $\xi$, $\eta$ and $\zeta$. The dimensionless form of the governing equations are the following

$$\iint_S \bar{v} \cdot \bar{n} dA = 0 \ , \tag{1}$$

$$\frac{\partial}{\partial t} \iiint_V \bar{v} dV + \iiint_V \bar{v} \cdot \bar{\nabla} \bar{v} dV = -\iint_S p\bar{n} dA + \frac{1}{Re} \iint_S \bar{n} \cdot \bar{\bar{\tau}} \, dA \ , \tag{2}$$

$$\frac{\partial}{\partial t} \iiint_V T dV + \iiint_V \bar{v} \cdot \bar{\nabla} T dV = \frac{1}{Re\,Pr} \iint_S (\bar{\nabla} T) \cdot \bar{n} dA \ , \tag{3}$$

192

where the velocity vector is $\vec{v} = u\hat{i} + v\hat{j} + w\hat{k}$, corresponding to the Cartesian coordinate system, p the dynamic flow pressure, $\overline{\overline{\tau}}$ the viscous stress tensor, and T temperature. The Prandtl number is $Pr = \upsilon/\alpha$, with $\nu$ as the kinematic viscosity and $\alpha$ as the thermal diffusivity. For the simple shear flow problem the Reynolds number is defined $Re = \gamma D^2/\upsilon$, with $\gamma$ as the shear rate. For the second case the Reynolds number $Re = U_\infty D/\upsilon$ is calculated with the undisturbed fluid velocity along the particle axis. The dimensionless distance of the wall to the particle midpoint h is related to the diameter D of the particle.

The system of equations was solved by a finite volume formulation. The solution method was successive line relaxation with replacement. As the first step in the procedure we calculate the three velocity components and the temperature by using a predictor corrector scheme. The pressure correction algorithm was calculated by solving the continuity equation based on a Poisson equation. The solution algorithm for the equations is described in detail in Reference /5/. In general it can be said that the stability and convergence properties of the numerical method with the multiple Chimera meshes have been similar to a single mesh calculation. However, there has been an approximate thirty to fifty percent increase in the number of iterations needed for convergence, and this is due to the transfer of information between the meshes. In the present investigation we have used second order central differences everywhere in the flow, and an artificial viscosity has not been introduced into the numerical method. However at higher Reynolds numbers it may be necessary to introduce smoothing techniques if grid resolution requirements become excessive.

## THE CHIMERA GRID SCHEME APPROACH

Normally, for relatively simple problems like the uniform flow around a single particle, the system of equation is solved on a single structured mesh with the implicit algorithm as described before. The geometry is evaluated with one mesh by using curvilinear coordinates. The Chimera overset grid approach essentially allows each component of the configuration to be grided separately and then overset onto a main grid. Usually there is a main grid stretched over the entire computational domain, in our case a simple rectangular grid. A minor grid is generated around the particle to resolve the flow around it and the regions of high gradients. In general the minor grid can have any shape and the mesh boundaries do not have to join the major grid in a special way. The minor grid can be located at any position relative to the main grid. It is also possible to introduce a lot of particles into the computational domain, each particle with its own grid. The grids can also interset or touch each other. For unsteady flow calculations the particles will keep their grids, only the relative position to the main grid will change. The details of the scheme are described in Reference /6/.

The Chimera scheme can be divided into the following topics:

1. Finding the holes and fringe points,

2. Determination of the location of a fringe point in the corresponding mesh,

3. Interpolation of the solution between the meshes.


1. The holes and the fringe points

The points on the major mesh where the rigid body is located must be excluded from the solution. These excluded points in the mesh are defined as holes. Figure 3 shows a section of a Chimera grid configuration of a spherical particle in a rectangular major computational domain. The blanked squares are the holes in the major mesh, but it is also possible that there exist

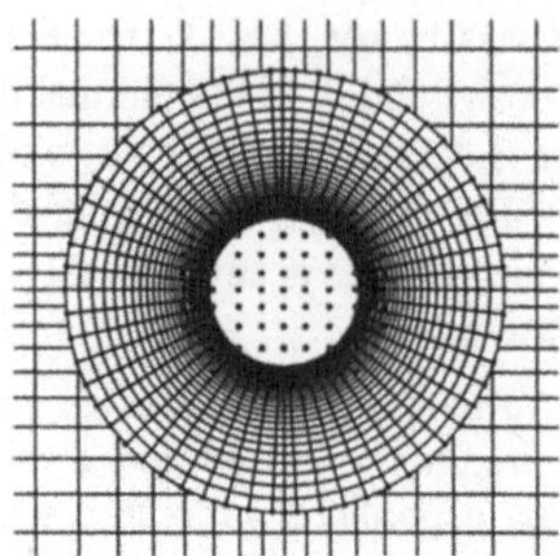

Figure 3: Typical Chimera grid configuration

holes in the minor mesh, too. This can occur when a part of the minor grid is located outside the computational domain.

The flow variables between major and minor meshes are exchanged at points known as fringe points. These points are marked in the figure with solid squares. The fringe points are actually not solved for in the mesh where they are used. They serve as the boundary points where information is exchanged between the two grids. So in the solution algorithm the fringe points are handled as a Dirichlet boundary condition. The solution variables of the fringe points are determined by interpolation in the cell where the fringe points are located. Finally the holes must be entirely enclosed by the fringe points.

The identification of the holes can be done by using simple vector algebra. With a scalar product between the vector to the considered point from any point of the surface and the area normal we can determine on which side of the closed surface the point is located. For many geometries it is possible to develop simpler methods to find the holes in the mesh, and we have employed special methods to preprocess and eliminate grid points. However, it is usually necessary to apply the above mentioned method to make an exact determination if a point is a hole.

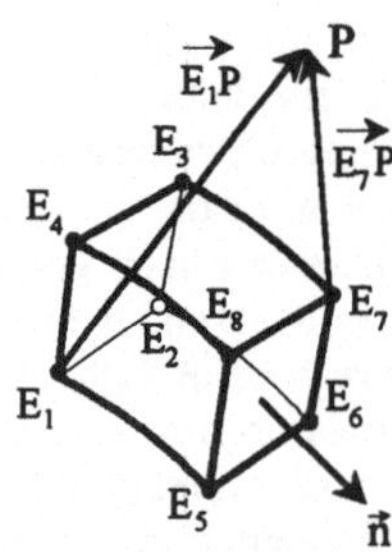

Figure 4: The search algorithm

## 2. The search algorithm

The solution variables at the fringe points are determined by trilinear interpolation from the corresponding major grid. To obtain these values it is necessary to determine the correct three dimensional cell or volume for the interpolation. The search algorithm uses vector algebra to locate the proper points, and it is shown in figure 4. On each area of a six sided control volume defined by points 1 to 8 the six normal vectors are calculated. Two vectors are then formed with the area normal $\bar{n}$ and the vector that is in the corner of the area under consideration. The sign of the calculated scalar product shows on which side of the area the fringe points is located. For the fringe points to be located inside the cell it is necessary for all six scalar products to be negative. The search algorithm was tested in two and three dimensional meshes, and we have always been able to find the fringe points in the major mesh.

## 3. The interpolation scheme

Finally it is necessary to interpolate the variables at a fringe point from the variables of the corresponding cell in the other mesh. This is an important aspect to relate solution variables between the various meshes. Typically, a grid point from one Chimera mesh $(x',y',z')$ is surrounded by eight neighbors of another overlapping Chimera mesh $(x,y,z)$. Linear interpolation of a variable, f, using the eight surrounding grid points would yield

$$f(x',y',z') = a_1 + a_2 x' + a_3 y' + a_4 z' + a_5 x'y' + a_6 y'z' + a_7 z'x' + a_8 x'y'z' \qquad (4)$$

where the coefficients $a_i$ must be determined with eight simultaneous equations from the surrounding eight known values of f. Although this procedure is not difficult, it can be computationally expensive. A much more efficient method is trilinear interpolation in the logical space of curvilinear coordinates, which takes advantage of the fact that this space is cubical.

The principal difficulty in the use of this form of trilinear interpolation is that the logical coordinates of the point (x', y', z') are not known *a priori* in the coordinate system x, y, z. The location of the point in logical space $(\xi', \eta', \zeta')$ can be found from the following nonlinear equations

$$x' = a_1 + a_2\xi' + a_3\eta' + a_4\zeta' + a_5\xi'\eta' + a_6\eta'\zeta' + a_7\zeta'\xi' + a_8\xi'\eta'\zeta'$$
$$y' = b_1 + b_2\xi' + b_3\eta' + b_4\zeta' + b_5\xi'\eta' + b_6\eta'\zeta' + b_7\zeta'\xi' + b_8\xi'\eta'\zeta' \qquad (5)$$
$$z' = c_1 + c_2\xi' + c_3\eta' + c_4\zeta' + c_5\xi'\eta' + c_6\eta'\zeta' + c_7\zeta'\xi' + c_8\xi'\eta'\zeta'$$

where again the coefficients $a_i$, $b_i$, $c_i$ are determined from the eight surrounding grid locations. In our work these three nonlinear equations were linearized with Newton's method and solved iteratively for the logical space locations $(\xi', \eta', \zeta')$. For further details see Reference /6/.

A summary of the implementation of the scheme into the computer code is the following:

1.  Calculation of a main mesh and one or more minor meshes,

2.  Determination of the holes in the main mesh, and if necessary in the minor meshes too,

3.  Determination of the fringe points in the main mesh; in the minor mesh all points at the outer boundary are fringe points; if a minor mesh intersects a wall we have also fringe points inside the minor mesh,

4.  Calculation of the curvilinear coordinates $(\zeta', \eta', \xi')$ of the fringe point in the other mesh by Newton's method; these values are stored in an array in the computer code,

5.  Iteration of the Navier Stokes equation in the minor mesh or meshes,

6.  Interpolation of the variables at the fringe points of the main mesh from the solution of the minor mesh,

7.  Iteration of Navier Stokes equation in the major mesh with the fringe points as boundary points near the holes; the holes are skipped from the solution,

8.  Interpolation of the outer boundary of the minor mesh or meshes and fringe points from the solution of the major mesh,

9.  Repeat step five to eight until the solution is converged.


## TEST CALCULATIONS

Figure 5 shows the Chimera grids for the calculation of the flow through a heat exchanger configuration in two dimensions. The geometry of each tube was described by a spherical grid overset onto the rectangular major grid. The holes are marked with the blanked, the fringe points with the filled squares. We have used 31/31 grid points for the minor meshes and 61/81 points for the major mesh. Figure 6 gives a comparison of the pressure loss $\lambda$ in the configuration versa the Reynolds number Re between the calculated values and a purely viscous approximation from Reference /7/. The approximation does not include any inertia effects which is the reason for the differences for Re > 100.

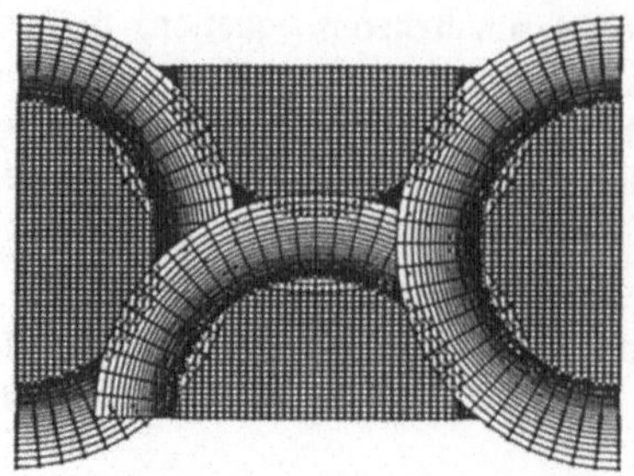

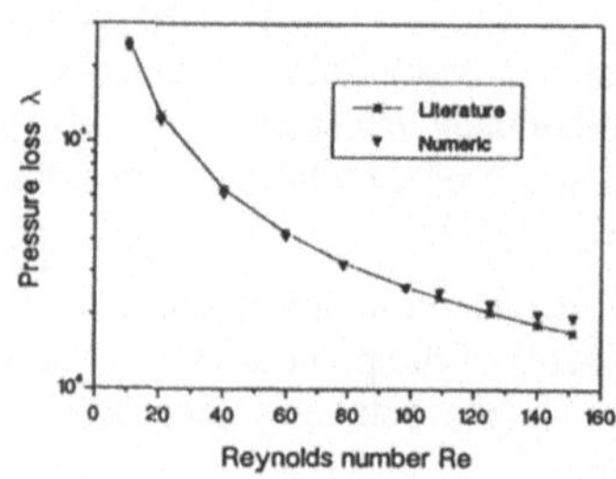


Figure 5:    Chimera grids for the heat exchanger calculation

Figure 6:    Pressure loss coefficient versus Reynolds number

As another test problem we have chosen the well known problem of the axissymmetric flow over a spherical particle in an unbounded fluid. Although it is expensive to do this calculation in three dimensions it is a good test case for the Chimera scheme. The flow was calculated in the Reynolds number region $0.1 < Re < 100$, since in this region the values of drag and heat transfer are well documented in the literature (see Ref. /1/). For the main grid we have taken a simple rectangular one. Around the particle we haven chosen a three dimensional spherical grid, and this minor grid was located at the midpoint of the major grid.

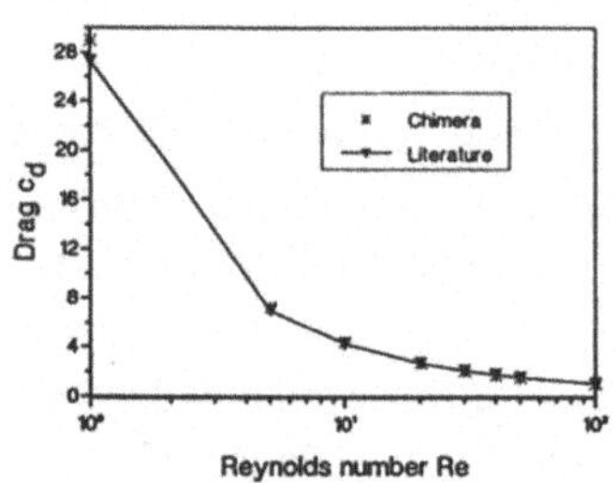


Figure 7:    Comparison of the drag coefficient $c_d$ over the Reynolds number

Figure 7 shows a comparison of the uniform flow over a spherical particle of the drag coefficient $c_d$ versus the Reynolds number, Re. The agreement with previous results for the drag coefficient is quite good over the entire Reynolds number range.

## RESULTS AND DISCUSSION

When particles are freely suspended they will rotate with the flow. To simulate this condition we consider a particle in a shear flow, where the shear has been generated by two, in opposite direction and same speed, moving walls (see figure 1). The particle is located in the middle of the gap and there will be no net drag or lift on the particle. The major and minor meshes are shown in figure 8, and it can immediately be seen that the minor grid intersects the walls on

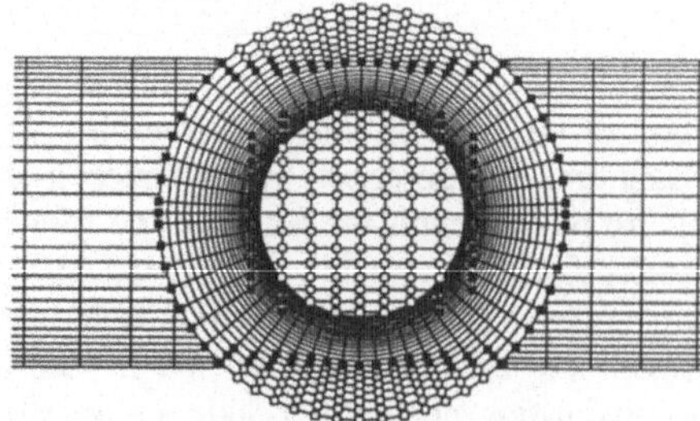

Figure 8:  Chimera grid configuration

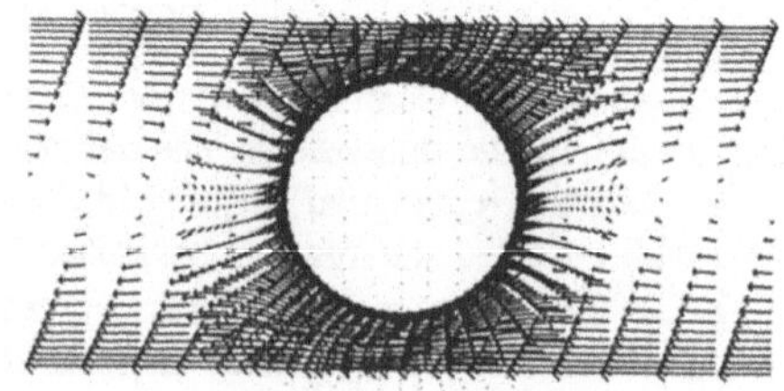

Figure 9:  Velocity vectors

both sides. This geometry generates additional holes and fringe points in the mesh. The velocity field for a flow Reynolds number of 10 is shown in figure 9. In this simulation the particle is rotating at an angular velocity of one half the shear gradient, and this flow is close to the condition where the net torque on the particle is zero. Presently, detailed calculations are being carried out to determine the exact angular velocity for the zero torque condition.

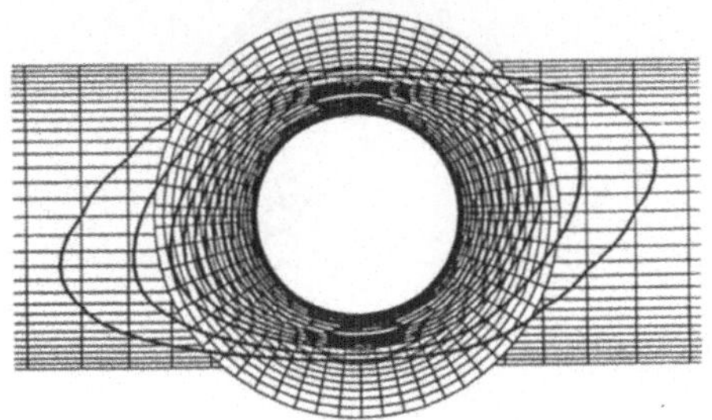

Figure 10:  Temperature contours, Prandtl=1

The temperature contours shown in figure 10 reflect the symmetry of the flow, and the strong influence of the wall. High heat transfer rates are obtained in the regions where the walls are close to the surface of the particle. It is also interesting to note that the transition between minor and major meshes do not cause significant changes in the temperature gradients. The simulation has shown that the influence of the wall increases the Nusselt number on the particle surface of around 100%.

The next example deales with the flow around a particle in a shear flow which is close to a rigid wall. The Reynoldsnumber for this flow is based directly on the particle diameter and the undisturbed fluid velocity along the particle axis in the flow upstream. The upstream flow condition is a linear shear flow on a wall and the mesh geometry is given in figure 11. Both the major and minor meshes have been stretched to reflect the flow fields, and it can be seen that the minor mesh extends outside the major mesh geometry. In this case we have holes or invalid flow points in the main and in the minor mesh. All minor grid points lying outside of the major domain are treated the same way as holes in the major mesh. Additionally, as in the previous discussed case we must define fringe points in the minor mesh close to the wall in order that information from the minor mesh can be transferred to the major mesh.

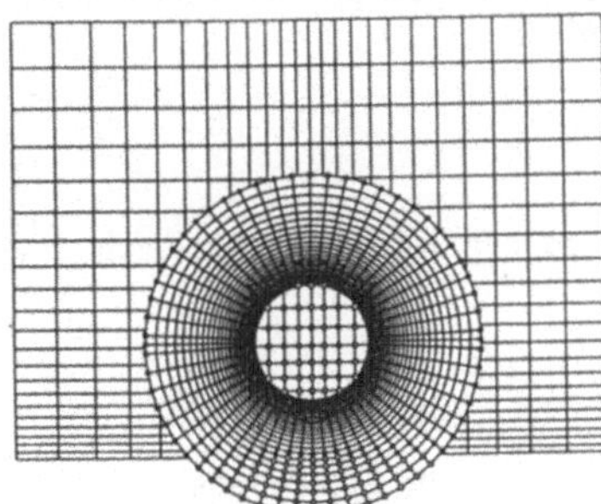

Figure 11: Chimera grid configuration

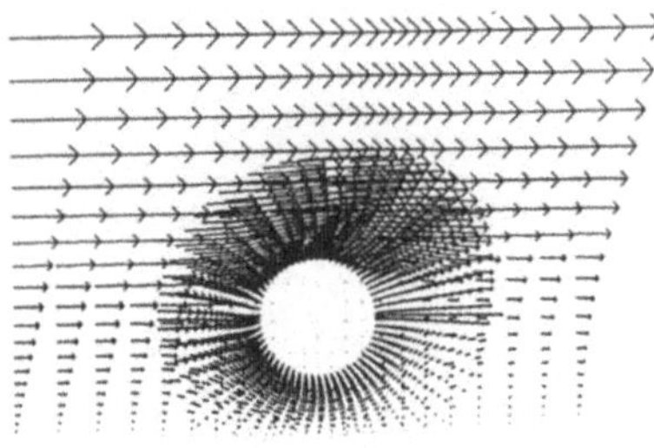

Figure 12: Velocity vectors

The solution for  the velocity field at a Reynolds number of ten is presented in figure 12 where the shear flow enters the computational domain from the left. The symmetry plane flow field between the particle and the wall has been resolved well, and the flow blockage between the wall and the particle can be clearly seen. The temperature contours shown in figure 13 show a large difference between the high and low velocity sides of the particle. On the high velocity side the contours are very close to the surface and the heat transfer is high, while on the low velocity side the contours are not dense, and they reflect the blocked nature of the flow, which yields low heat transfer. The pressure distribution over the sphere, figure 14, is unusual in that the stagnation point has been moved toward the high velocity side. The lowest pressure region on the surface occurs on the high velocity side.

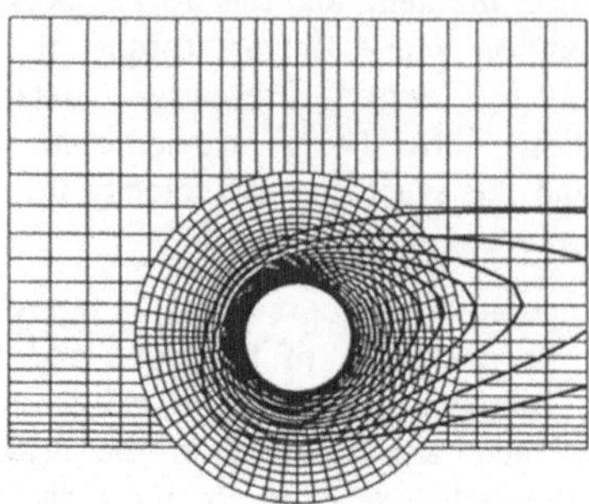

Figure 13: Temperature contours

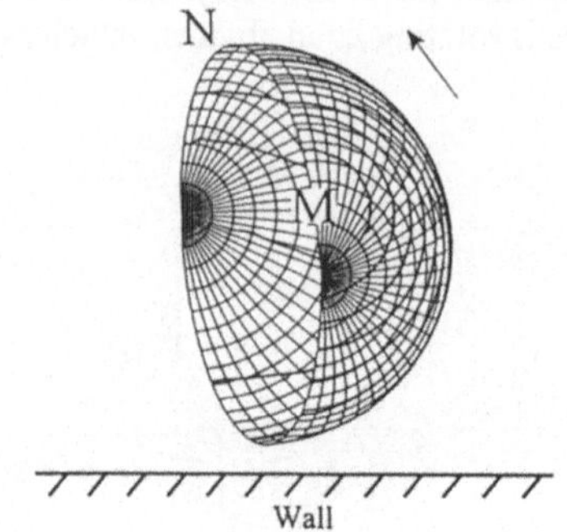


Figure 14: Pressure distribution at the surface (M=Maximum, N=Minimum)

## CONCLUSIONS

1. A three dimensional Chimera grid scheme has been introduced into a full Navier Stokes code for incompressible flows. The scheme works well in two and three dimensions with an implicit numerical method.

2. The paper presents results for relatively difficult fluid flows around individual particles in shear flows, particle wall interactions and a multiple particle configuration. The results show clearly that the Chimera technique has a future in the calculation of multi particle flows in difficult and complex configurations.

3. For the intermediate Reynolds numbers used trilinear interpolation yielded high quality solutions. These solutions were accurate at mesh boundaries and the were no practical problems with the conservation of the physical fluxes. For all test problems the quality of the Chimera solution were similar to single mesh solutions.

## REFERENCES


/1/ Clift, R., Grace, J.R., Weber, M.E.: "Bubbles, drops and particles", Academic Press (1978).

/2/ Benek, J.A., Steger, J.L., Dougherty, F.C. and Buning, P.G.: "Chimera: A grid-embedding technique", AEDC-TR-85-64, Arnold Air Force Station, TN (1986).

/3/ Dougherty, F.C.: "Development of a Chimera grid scheme with apllications to unsteady problems", PhD thesis, Stanford University (1985).

/4/ Buning, P.G., Chiu, I.T., Obayashi, S., Rizk, Y.M. and Steger, J.L.: "Numerical simulation of the integrated space shuttle vehicle in ascent", AIAA Atmospheric Flight Mechanics Conference, Minneapolis, Minnesota, AIAA Paper-88-4359-CP (1988).

/5/ Dwyer, H.A.: "Calculation of droplet dynamics in high temperature environments", Prog. Energy Combust. Sci., Vol. 15, 131.158 (1989).

/6/ Nirschl, H., Dwyer, H.A., Denk, V.: "A Chimera grid scheme for the calculation of particle flows", submitted to Journal of Computational Physics (1993).

/7/ VDI-Wärmeatlas, 6. Edition (1991).

# A new multi-domain algorithm for the spectral solution of the incompressible Navier-Stokes equations


A. PINELLI and A. VACCA

Von Karman Institute for Fluid Dynamics Belgium

Chauseé de Waterloo, 72 B-1640 Rhode-St-Geneśe - Belgium



## Abstract

The two dimensional incompressible Navier-Stokes equations in primitive variables have been solved by a new spectral multi-domain method using a semi-implicit fractional step scheme. Each scalar problem obtained after spatial collocation is solved using the Projection Decomposition Method (P.D.M.) [1] that is a new numerical tool that combines domain decomposition technique with the Galerkin method.


## 1  Introduction

The main features of the present procedure are somehow related to the ones of the spectral element method and differ from the multi-domain spectral methods for elliptic equations. All the mentioned techniques rely on use of orthogonal polynomials expansion having a support locally defined on each element. The present method (as in the case of the spectral elements) uses a weak formulation with trial functions that are $C^0$ across the element boundary, with flux continuity at the element interface satisfied as part of the convergence process. The multi-domain spectral method does not use a weak formulation and as a results the function and derivative continuity conditions must be separately imposed at the interface [2]. The main difference between the present approach and the one on which the spectral element method is based, lies in the particular choice of the test and trial functions at the interfaces. The latter ensures good conditioning properties for the trace problems associated to the treatment of the interfaces. Moreover the corresponding algebraic systems are symmetric. These characteristics allow for an effective use of conjugate gradient technique.
The special choice of the basis functions at the interfaces implies special care for the weak projection step.
All these topics will be illustrated in the present paper together with some test cases that have been solved to validate the algorithm. The preliminary results that have been achieved indicate that the present procedure might become a viable alternative to the spectral element technique for accurate flow prediction.

## 2   Equations and time-splitting scheme

We consider here the incompressible Navier-Stokes equations in primitive variables formulation $(\vec{u}, p)$ with the non-linear terms expressed in a skew-symmetric form to minimize the aliasing effects:

$$\frac{\partial \vec{u}}{\partial t} = -\nabla p \; + \; \nu \, \triangle \, \vec{u} - \frac{1}{2} \left( \vec{u} \cdot \nabla \vec{u} + \nabla \cdot (\vec{u}\vec{u}) \right) \tag{1}$$

$$\nabla \cdot \vec{u} \; = \; 0 \tag{2}$$

where $\vec{u}$ is the velocity field, $p$ the pressure and $\nu = \frac{\mu}{\rho}$ the kinematic viscosity ($\rho$ is constant). In the incompressible Navier-Stokes equations the velocity and the pressure are coupled together by the incompressibility condition which makes the equations difficult to solve. Classical procedures to overcome this drawback are provided by time splitting schemes [3] [4]. The basic idea is to decouple the pressure and the velocity computation at each time step. The terms associated with the spatial derivatives appearing in the given equations might be computed at old, new or some intermediate time step. Implicit treatment of the viscous terms allows one to overcome the most severe time step restriction met when dealing with spectral methods [5]:

$$\triangle t \sim Re \frac{1}{N^4} \quad Re = Reynolds \; number; \; N \; number \; of \; nodes \,. \tag{3}$$

For the present work we selected the "pressure correction scheme" developed by Van Kan [6] (here given for non-slip conditions):

$$\frac{\vec{\hat{u}} - \vec{u^n}}{\triangle t} - \frac{\nu}{2} \triangle \left( \vec{\hat{u}} + \vec{u^n} \right) + \nabla p^n \tag{4}$$

$$= -\frac{3}{2} \left( \vec{u^n} \cdot \nabla \right) \vec{u^n} + \frac{1}{2} \left( \vec{u}^{n-1} \cdot \nabla \right) u^{n-1}$$

$$\vec{\hat{u}}_{\partial \Omega} = \vec{u} \left( (n+1) \triangle t \right) \tag{5}$$

$$\frac{\vec{u}^{n+1} - \vec{\hat{u}}}{\triangle t} + \frac{1}{2} \nabla \left( p^{n+1} - p^n \right) = 0 \tag{6}$$

$$\nabla \cdot \vec{u}^{n+1} = 0 \,. \tag{7}$$

In the first step, we solve an intermediate velocity field $\vec{\hat{u}}$ which is not physical. In fact, $\vec{\hat{u}}$ does not satisfy the incompressibility condition. Then in the second step we project $\hat{u}$ onto the divergence free space to get an adequate velocity approximation $\vec{u}^{n+1}$.

The scheme with the given boundary conditions is nothing else than a second order Crank-Nicolson Adams-Bashforth scheme with an $\mathcal{O}\left(\triangle t^2\right)$ deviation in the tangent direction to the boundary

$$\vec{u}^{n+1} \cdot \tau|_{\partial \Omega} = \vec{u} \left( (n+1) \triangle t \right) \cdot \tau - \triangle t \nabla \left( p^{n+1} - p^n \right) \cdot \tau \,. \tag{8}$$

By applying the divergence operator to (6), we find that the latter is equivalent to

$$\Delta \left(p^{n+1} - p^n\right) = \frac{2}{\Delta t}\nabla \cdot \vec{u} \tag{9}$$

$$\frac{\partial p^{n+1}}{\partial \mathbf{n}}|_{\partial\Omega} = 0 \tag{10}$$

$$\vec{u}^{n+1} = \vec{\tilde{u}} - \frac{\Delta t}{2}\nabla \left(p^{n+1} - p^n\right). \tag{11}$$

At each time step, we have to solve a cascade of scalar boundary value problems: two Helmholtz equations for the predicted value of velocity and a Poisson one for the pressure. Having treated the diffusive part implicitly, the only stability restriction on the time step is given by the Courant (CFL) condition

$$\Delta t \sim \frac{1}{U}\frac{1}{N^2}; \quad U = max \ velocity \tag{12}$$

that is less severe, at least for low Reynolds number, than the one related with the viscous terms (3). In the next section we will focus our attention on the solution of each scalar boundary value problem arising after the given time discretization.

## 3   Space Discretization

We consider here the following problem as representative of one of the elliptic scalar problems mentioned in the previous section:

$$-\Delta^2 u + \alpha u = f \quad on \ \Omega \tag{13}$$

$$u = 0 \quad on \ \partial\Omega. \tag{14}$$

The equivalent weak formulation being: find $u$ in $H_0^1(\Omega, \partial\Omega)$ such that

$$l(u,v) = f(u,v) \quad \forall \, v \in H_0^1(\Omega, \partial\Omega). \tag{15}$$

Where $H_0^1(\Omega, \partial\Omega)$ is the subspace of $H^1(\Omega)$ with $u$ identically vanishing on $\partial\Omega$:

$$H_0^1(\Omega, \partial\Omega) = \{u \in H^1(\Omega): \ u \, |_{\partial\Omega} = 0\}.$$

$l(u,v)$ is the following bilinear form:

$$l(u,v) = \int_\Omega (\nabla u \cdot \nabla v + \alpha uv)d\Omega \tag{16}$$

and $\alpha \geq 0$ is either identically equal to zero (i.e., for the Poisson problem related with the pressure) or is equal to $2/\Delta t\nu$ (i.e., for one of the momentum equations). We consider a non-overlapping partitioning of the domain $\Omega$ as a union of $N_s$ elemental rectangles $\Omega_i$ $i = 1, \cdots, N_s$. First we rewrite problem (15) as the system:

$$\begin{cases} u = \tilde{u} + w \quad \tilde{u} \in H_0^1(\Omega_i, \partial\Omega_i) \ w \in H_0^1(\Omega, \partial\Omega) \ \forall \, i = 1, \cdots, N_s \\ l(\tilde{u}, z) = (f, z) \quad \forall z \in H_0^1(\Omega_i, \partial\Omega_i) \ \forall \, i = 1, \cdots, N_s \\ l(w, z) = (f, z) - l(\tilde{u}, z) \quad \forall z \in H_0^1(\Omega, \partial\Omega). \end{cases} \tag{17}$$

Meaning that we consider a set of $N_s$ uncoupled Dirichlet problems with test functions identically vanishing at the sub-domains interfaces and an interface problem that might be re-written as $S\phi = h$ where the $S$ operator is defined as:

$$(S\phi, \psi) = l(E_\sigma \phi, E_\sigma \psi) = (h, \phi) = (f, E_\sigma \phi) - l(\tilde{u}, E_\sigma \phi) \qquad (18)$$

where $E_\sigma$ is an "harmonic" extension operator that maps $\phi \in H_0^{1/2}(\Gamma, \partial\Omega)$ on the interfaces $\Gamma$ to $w \in H_0^1(\Omega, \partial\Omega)$. Here with $H_0^{1/2}(\Gamma, \partial\Omega)$ we mean the following space of traces:

$$H_0^{1/2}(\Gamma, \partial\Omega) = \{\phi \in L^2(\Gamma) : \exists u \in H_0^1(\Omega, \partial\Omega) : u \mid_\Gamma = \phi\}.$$

Thus the final algorithm reads as follows:

- Solve $N_s$ equations (17) in $\tilde{u}$ by using as trial functions Legendre polynomials and as test functions Lagrange ones constructed on the Gauss-Lobatto nodes (i.e. polynomials that are identically equal to zero in all the nodes except in the $i^{th}, j^{th}$ node where the $i^{th} j^{th}$ polynomial is identically equal to 1) and identically vanishing on both the physical boundaries and on the interfaces of each subdomain.

- Solve problem (18) to obtain $\phi = w \mid_\Gamma$.

- Solve the $N_s$ uncoupled harmonic problems with Dirichlet value $\phi$ at the subdomains interfaces with the same test and trial functions as for (17) to get $w$.

- Obtain the solution $u = \tilde{u} + w$.

The first point and the last points are simply achieved by direct inversion of the $N_s$ matrices that represent the discrete counterpart of the problem:

$$\sum_{h,k}\{[\alpha \tilde{u}_{hk} - \Delta^2 \tilde{u}_{hk} - f_{hk}]\omega_h \omega_k Le_{ij}(x_h, y_k)\} = 0 \qquad (19)$$

where $\omega_k$ is an appropriate Gauss-Lobatto weight and where $Le_{ij}(x_h, y_k) = 1$ only if $i = j = k = h$ are the Lagrange polynomials built on the Gauss-Lobatto nodes. The interfaces problem is solved iteratively, by applying a Galerkin method with special test and trial functions that are dense in $H_0^{1/2}$ and that guarantee matrix $S$ to be symmetric and with a conditioning number independent of the number of subdomains that have been used [1].

Clearly the algorithm is readily applicable to Neumann problems as well, by just considering the flux of the normal derivatives in equation (19).
Some care has to be taken when considering the projection step:

$$\frac{\vec{u}^{n+1} - \vec{\tilde{u}}}{\Delta t} + \frac{1}{2}\nabla\left(p^{n+1} - p^n\right) = 0 . \qquad (20)$$

In fact, a strong formulation would not provide a continuous velocity field at the interfaces because the jumps of the normal pressure derivatives at the subdomain

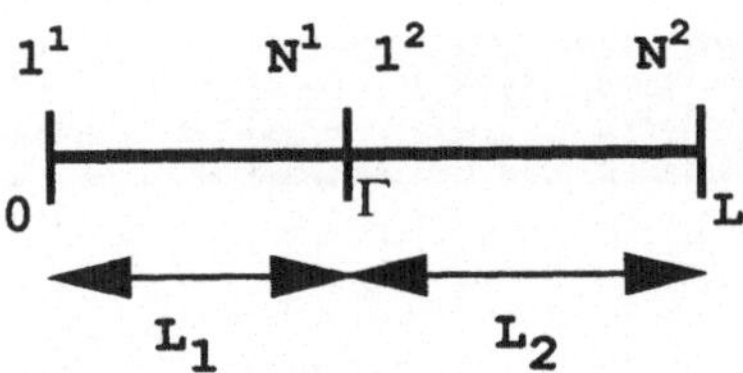

Figure 1: 1D case

interfaces are continuous only in a weak sense. Strictly speaking one should proceed to a weak update consistent with the test and trial functions that have been used in the solution procedure for both the predicted velocity components and for the pressure. Nevertheless, it is possible to prove that for problems with sufficiently smooth solutions all the different discrete inner products that have been considered so far, converge quickly (i.e.; "spectrally") to their continuous counterparts. For such a reason we kept on using Legendre polynomials as trial functions and the previously defined Lagrange ones as test functions. Such a position leads to the following update to be applied to the subdomains interfaces, here given for an equivalent 1-D problem:

$$\int_0^L u^{n+1} Le_i dx = \int_0^L \hat{u} Le_i d\dot{x} + \int_0^L \frac{dp}{dx} dx \ . \tag{21}$$

Using the usual Gauss-Lobatto quadrature and exploiting the way the Lagrange basis ($Le_i$) has been built (in this case $Le_i(x_j) = \delta_{ij}$) we can express the discrete updating in the case of 2 subdomains as:

$$\begin{aligned}
&for \ the \ internal \ nodes &&u_j^{n+1}\omega_j = \hat{u}\omega_j + \frac{dp}{dx}\big|_{x_j} \omega_j \\
&for \ the \ interface \ nodes &&u_\Gamma^{n+1}(\omega_1^{dom\ 2} + \omega_N^{dom\ 1}) = \\
&&&\hat{u}_\Gamma(\omega_1^{dom\ 2} + \omega_N^{dom\ 1}) + (\frac{dp}{dx}{}^1_N\omega_N^1 + \frac{dp}{dx}{}_2\omega_1^2) \ .
\end{aligned} \tag{22}$$

Having indicated with the super-scripts 1 and 2 the subdomains and with the sub-scripts 1 and $N$ the respective Gauss-Lobatto nodes at the interface $\Gamma$ (figure 1).

## 4  Validation

Laminar and turbulent flow in a pipe or channel expansion is a complex flow situation often used as a test for the numerical and experimental techniques. We choose the problem of the flow in a asymmetric channel expansion to demonstrate the viability of the present multi-domain technique.

The channel configuration is the same that has been proposed in the "Analysis of Laminar Flow over a Backward Facing Step" GAMM Workshop (1984)

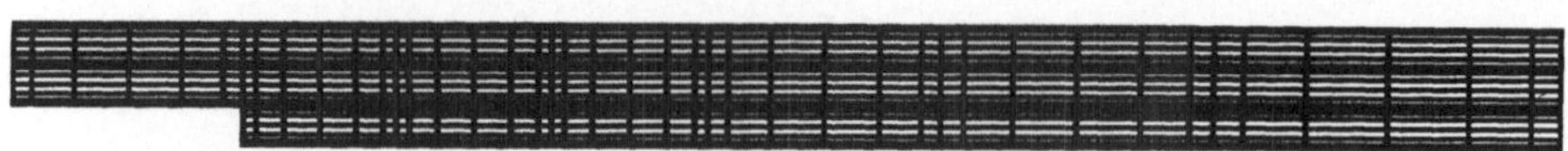

Figure 2: Geometry and grid configuration

[7], and is given in figure 2 to present the selected spatial discr.tization (20 elements, each one with order 7 Legendre polynomial expansion). It is assumed that the channel length previous to the step is sufficiently long to allow the imposition of the parabolic profile at the inlet. The Navier-Stokes equations are non-dimensionalized with respect to the step height $h^\star = H - h$, and the maximum velocity at the inlet $\left(Re = \frac{U_{max}h^\star}{\nu}\right)$.

For the computation we restricted ourselves to 2 laminar, moderate Reynolds numbers (i.e., 50 and 150). There are several criteria on which comparisons can be made with previous numerical work and experiments: for the present work we selected and compared with the ones proposed in the mentioned work-shop.

In the following tables we compare our predicted recirculation lengths versus both the experimental ones [8] and the ones predicted by other authors [9].

Table 1: Reattachmet lengths

| $x_{rea}$ | Present | Experim. | Glowinski et al |
|---|---|---|---|
| $Re = 50$ | 2.959 | 3. | 2.5 |
| $Re = 150$ | 6.451 | 6. | 5.75 |

In figure 3 the stream-lines contours at the 2 considered Reynolds numbers are shown together with the predicted pressure fields, and finally in table 2 we compare the predicted maximum and minimum stream-wise velocity component locations after the step with the experimental values (x refers to the distance from the step).

Table 2: Location of min. and max. values of u

| x | Present MIN | Present MAX | Exp. MIN | Exp. MAX |
|---|---|---|---|---|
| $Re = 50$ at $x = 1.6\,h^\star$ | −0.048 | 0.910 | −0.040 | 0.898 |
| $Re = 50$ at $x = 4.\,h^\star$ | 0. | 0.783 | 0. | 0.772 |
| $Re = 50$ at $x = 8.\,h^\star$ | 0. | 0.695 | 0. | 0.688 |
| $Re = 150$ at $x = 1.6\,h^\star$ | −0.072 | 0.970 | −0.070 | 0.972 |
| $Re = 150$ at $x = 4.\,h^\star$ | −0.053 | 0.910 | −0.046 | 0.928 |
| $Re = 150$ at $x = 8.\,h^\star$ | 0. | 0.817 | 0. | 0.819 |

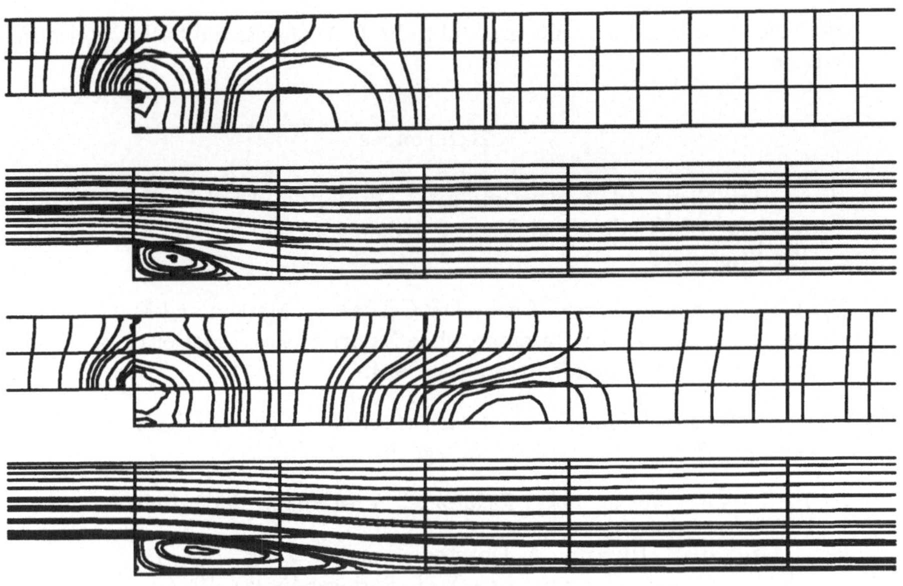

Figure 3: Top, Re=50; Bottom, Re=150 (pressure and stream-lines contours)

## 5 Conclusions

A new spectral multi-domain technique that exploits the features of fractional step methods for the incompressible Navier-Stokes equations has been presented. The results are encouraging and proved that accurate solutions might be achieved even with low order expansions. The multi-domain approach allows to recover spectral accuracy in the subdomains not involved with singularities where the solution is expected to be smooth enough. On the other side the presented iterative multi-domain algorithm proved [1] to be almost insensitive to the number of subdomains in which the original domain is partitioned. This last feature is indeed important and allows opportune and flexible domain decomposition strategies.

## Acknowledgments

The above text presents research results of the Belgian Incentive Program *Information Technology* - Computer Science of the Future, initiated by the Belgian State - Prime Minister's Service - Science Policy Office. The scientific responsibility is assumed by its authors. A large part of the present work has been carried out at Crs4 (Centro di Ricerca, Sviluppo e Studi Superiori in Sardegna) Italy.

Both the authors are specially grateful to Prof. A. Quarteroni for his advice and support.

## References


[1] V. Agoshkov and E. Ovchinnikov. Projection Domain Decomposition Method. CRS4 (Centro di Ricerca, Sviluppo e Studi Superiori in Sardegna) Pre-Print, 1993.

[2] A. Pinelli and A. Vacca. 'Chebyshev Collocation Method and Multi-Domain Decomposition for the Incompressible Navier-Stokes Equations'. *Int. J. for Num. Methods in Fluids*, To appear, 1994.

[3] A. Chorin and J. Marsen. *'A Mathematical Introduction to Fluid Mechanics'*. Springer: New York. Springer-Verlag, 1979.

[4] R. Temann. *'Navier-Stokes Equations'*. Amsterdam. North-Holland, 1977.

[5] C. Canuto, M. Hussaini, A. Quarteroni, and T. Zang. *'Spectral Methods In Fluid Dynamics'*. Springer: New York. Springer-Verlag, 1988.

[6] J. Van Kan. 'A Second Order Accurate Pressure-Correction Scheme for Viscous Incompressible Flow'. *J. Sci. Stat. Comp.*, 7:870–891, 1986.

[7] Various Authors. 'analysis of laminar flow over a backward facing step'. In K. Morgan, J. Periaux, and F. Thomasset, editors, *Notes in Numerical Fluid Mechanics, Volume 9*. Vieweg, 1984.

[8] J. Kueni and G. Binder. 'viscous flow over backward facing steps an experimental investigation'. In K. Morgan, J. Periaux, and F. Thomasset, editors, *Notes in Numerical Fluid Mechanics, Volume 9*. Vieweg, 1984.

[9] R. Glowinski, B. Mantel, J. Periaux, and O. Tissier. 'analysis of laminar viscous flow over a step by non-linear least squares and alternating direction methods'. In K. Morgan, J. Periaux, and F. Thomasset, editors, *Notes in Numerical Fluid Mechanics, Volume 9*. Vieweg, 1984.

# ON THE CONSTRUCTION OF ROBUST SMOOTHERS FOR INCOMPRESSIBLE FLOW PROBLEMS


Henrik Reichert and Gabriel Wittum

Institut für Computeranwendungen (ICA/Numerik), Universität Stuttgart

Pfaffenwaldring 27, D-70 569 Stuttgart


## Summary


We introduce and compare several smoothers for the stationary incompressible Navier-Stokes equations in primitive variables on unstructured and locally refined grids. Special emphasis is laid on the robustness of the linear multigrid solver in view of large convecting velocities and bad aspect ratios in the grid. We describe a new streamwise numbering algorithm for the unknowns with a special treatment of the cyclic dependencies due to vortices. Further we describe the implemented smoothers and show diagrams of the convergence rates (per grid level) versus the aspect ratios of the elements.


## 1  Introduction and the Notion of "Robustness"

The final goal of our recent work is a highly efficient solver for general CFD problems by using multigrid methods on locally refined and adapted unstructured meshes. The present paper is concerned with the first step on this way we have reached at. We discuss the following problem: Since we use a linear multigrid as inner solver we want it to be able to yield good convergence rates for all problems the discretization passes to it. Especially the cases of large convecting velocities (see section 2) and bad aspect ratios (see section 3) of the elements, which appear frequently in boundary layer fitted grids, should cause no severe problems for the smoother.

To achieve this we tried the following strategy: Take some variant of $\text{ILU}_\beta$ as smoother and possibly decouple the equations by a transforming approach [Wi1]. Then the algorithm should prove to be robust in view of bad aspect ratios.

For the convection dominated case we will choose a special streamwise numbering of the unknowns.

## 2  Robustness in View of Large Convecting Velocities

### 2.1  Description of the numbering algorithm

Let us switch off the diffusion for the time being. Due to the quasi Newton linearization and to the upwind scheme a given node depends only on its upwind neighbours. If we can find a global ordering of the unknowns in a way that the stiffness matrix has nonzero entries only in the lower triangle, then of course we will be able to solve the system of equations in one step even by a Gauß-Seidel method. Unfortunately in most of the relevant cases there are vortices in the flow and therefore cyclic dependencies. But nevertheless we will obtain good results if we introduce arbitrary cuts through the vortices by removing just enough of the "cyclic" nodes to get rid of the cylic dependencies. We start the numbering at the inlet going in layers downstream but taking only nodes depending on the already numbered ones (those nodes will form the beginning of our new list). In a similar way we go upstream from the outlet (those nodes will make up the end of our new list). Finally we are left with nodes with cyclic dependencies. We cut it, appending those nodes to the beginning of our list. Then steps one to three are repeated until every node is processed.

After this rough description we introduce the following algorithm that does the job (the basic ideas can be found in [BW]):

```
while (some nodes are not numbered)
{
    /* find FIRST set */
    do {
        Find all nodes having at most such UPWIND
            neighbour nodes that are already numbered
        Number them starting with the least number not used yet.
    } while (no further nodes are found).

    /* find LAST set */
    do {
        Find all nodes having at most such DOWNWIND
            neighbour nodes that are already numbered
        Number them starting with the greatest number not used yet.
    } while (no further nodes are found).

    /* find CUT set (only cyclic dependencies are left) */
    Cut one vortex transverse to the streamlines.
    Number the nodes on this cut starting with the least number
        not used yet.
}
```

---

**Algorithm 1:** Streamwise numbering

An example of the resulting sparsity pattern of the stiffness matrix could look like this:

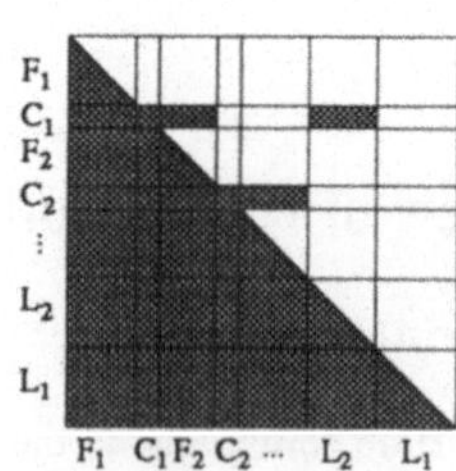

For the Backward Facing Step this could look like:

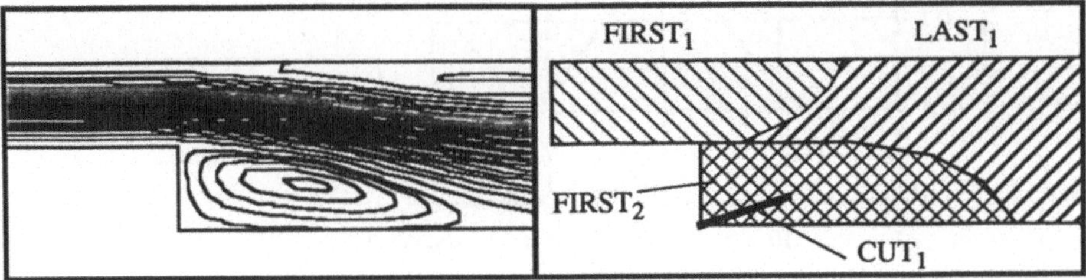


**Fig. 1:** Streamwise numbering of the Backward Facing Step.

For the Driven Cavity at a Reynolds number of 500 as a more complicated example the FIRST, CUT and LAST sets are shown in the following pictures:

**Fig. 2:** Left side: FIRST$_1$ (marked nodes): Dirichlet boundary nodes (LAST$_1$ is empty)
middle: CUT$_1$ (black nodes), FIRST$_2$ (medium gray nodes) and LAST$_2$ (gray nodes)
right side: CUT$_2$ (black), FIRST$_3$ (gray) (LAST$_3$ is empty).

## 2.2 Results

As test examples for the efficiency of this numbering strategy we chose a simple Pipe flow and the Backward Facing Step from above. We calculated a velocity field for Re=100 (called $\vec{u}_{\mathrm{old}}$) and treated the equations linearized in $\lambda\vec{u}_{\mathrm{old}}$ with our smoother:

$$-\Delta\vec{u} + Re\,(\lambda\vec{u}_{\mathrm{old}} \cdot \nabla)\,\vec{u} + \nabla p = 0$$

$$\nabla\vec{u} = 0 \qquad\qquad (2.2.1)$$

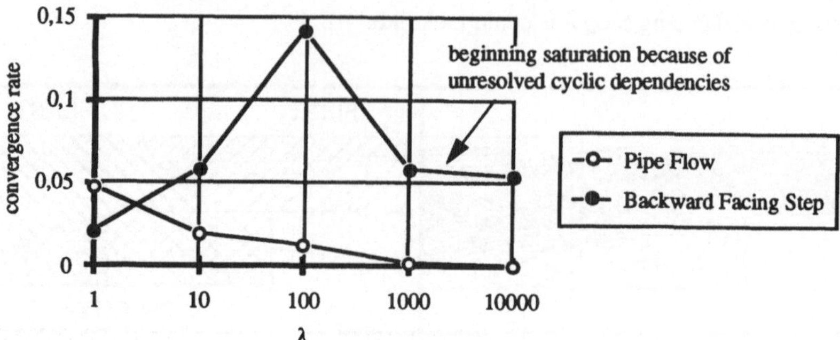


**Fig. 3:** Convergence rate over $\lambda$ for streamwise numbering.

# 3  Robustness in View of Large Aspect Ratios

For this section we want to introduce the following simplifications: a) we stick to the Stokes equation (pure diffusion) to avoid the mixing of various effects and b) we use a rectangular equidistant grid with meshsizes $h_x$, $h_y$ in $x$- and $y$-direction resp. for the calculations since we want to have only elements of one type with the same aspect ratio $s = h_x/h_y$.

## 3.1  Description of the implemented smoothers

For the classification and description of the implemented smoothers we need to introduce some notation:

We have to solve the linear system of equations

$$Kx = b \tag{3.1.1}$$

$$x = \begin{bmatrix} u \\ v \\ p \end{bmatrix} \qquad b = \begin{bmatrix} f_x \\ f_y \\ 0 \end{bmatrix} \tag{3.1.2}$$

with the stiffness matrix

$$K = \begin{bmatrix} -\Delta & 0 & \frac{\partial}{\partial x} \\ 0 & -\Delta & \frac{\partial}{\partial y} \\ \frac{\partial}{\partial x} & \frac{\partial}{\partial y} & -c_0\Delta \end{bmatrix} \tag{3.1.3}$$

where $c_0$ is the stability parameter introduced in a natural way by the discretization procedure. It behaves like $O(h^2)$ and is necessary to supress artificial pressure oscillations.

Once the ordering of the nodes in the grid is given there are two obvious ways to remove the remaining arbitraryness in the order of the $3N$ unknowns:

- $u_1, \ldots, u_N, v_1, \ldots, v_N, p_1, \ldots, p_N$
- $u_1, v_1, p_1, \ldots, u_N, v_N, p_N$

with their associated block structures, the first of which we will refer to as *equationwise* ($3\ N \times N$ blocks), the ladder as *nodewise* ($N\ 3 \times 3$ blocks).

Classification of implemented smoothers:

| variant / ordering | equationwise | nodewise |
|---|---|---|
| scalar | ILU | – |
| block | Gauß-Seidel<br>(inner solver ILU) | ILU, Gauß-Seidel<br>(inner solver exact) |

Additionally we implemented two (right) transforming smoothers.

The transforming iteration step reads:

$$x^{i+1} = x^i + \overline{K}M^{-1}(b - Kx^i) \tag{3.1.4}$$

where $M$ is a regular decomposition of

$$K\overline{K} = M - N \tag{3.1.5}$$

with some rest matrix $N$.

Now choose for $\overline{K}$

- **distributive ILU:**

$$
\overline{K} = \begin{bmatrix} 1 & 0 & \frac{\partial}{\partial x} \\ 0 & 1 & \frac{\partial}{\partial y} \\ 0 & 0 & \Delta \end{bmatrix} \Rightarrow K\overline{K} = \begin{bmatrix} -\Delta & 0 & \{\frac{\partial}{\partial x}\Delta - \Delta\frac{\partial}{\partial x}\} \\ 0 & -\Delta & \{\frac{\partial}{\partial y}\Delta - \Delta\frac{\partial}{\partial y}\} \\ \frac{\partial}{\partial x} & \frac{\partial}{\partial y} & \{\Delta - O(h^2)\} \end{bmatrix}
$$

if we neglect the commutators (in curly brackets) coupling the momentum equations to the pressure and if we then perform an ILU on the diagonal blocks we yield the TILU (transforming ILU) by Wittum [Wi4], which essentially is a DGS-like algorithm (Distributive Gauß-Seidel, introduced by Brandt/Dinar [BD]).

Or take

- **SIMPLE-ILU:**

$$\bar{K} = \begin{bmatrix} 1 & 0 & D^{-1}\frac{\partial}{\partial x} \\ 0 & 1 & D^{-1}\frac{\partial}{\partial y} \\ 0 & 0 & 1 \end{bmatrix} \Rightarrow K\bar{K} = \begin{bmatrix} -\Delta & 0 & 0 \\ 0 & -\Delta & 0 \\ \frac{\partial}{\partial x} & \frac{\partial}{\partial y} & \{-c_0\Delta - \nabla^T D^{-1}\nabla\} \end{bmatrix}$$

this holds exactly if we take $D$ to be the Laplacian. Now we leave the zero entries in the upper triangle but replace $D$ by some easily invertible approximation.

We took as a very crude approximation $D = \mathrm{diag}(\Delta)$. This leads to the well known SIMPLE-method by Patankar/Spalding [PS]. The resulting system we treated with ILU (instead of the original Gauß-Seidel).

As a much more elaborate version we treated the Schur complement with the Frequency-Filtering method by Wittum [Wi3]. There an approximate matrix is defined by its sparsity pattern and by requiring that it acts on a certain subspace exactly like the original matrix. We want to emphasize that our SIMPLE smoothers have nothing in common with the original SIMPLE but the very basic idea of choosing an approximation for $D$.

## 3.2 Results

To test the performance of our smoothers we measured the dependence of the convergence rate (mean value over 10 V-cycles, denoted as $\kappa_{10}$) on a rectangular grid for the Driven Cavity on the aspect ratio (the y-coordinates of the grid are scaled by powers of 2)

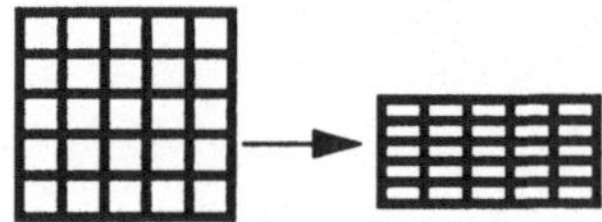

as well as the dependence of the mean convergence rate on the meshsize of the finest grid (h-independence expected).

We did the tests with various parameters as there are damping factors $\omega$ and the $\beta$-parameters of the $\mathrm{ILU}_\beta$ each chosen independently for the velocity and the pressure.

All calculations where done with V-type multigrid cycles and one pre- and two post-smoothing steps.

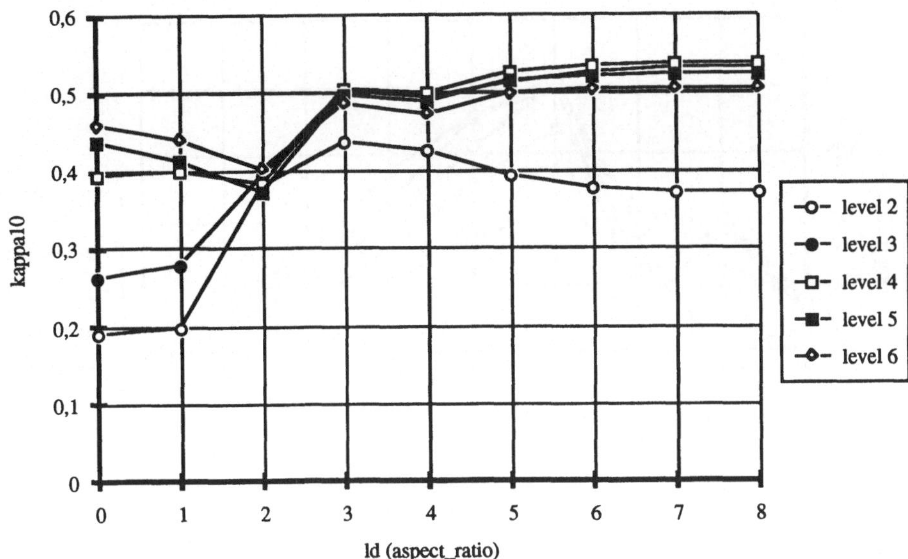

**Fig. 4:** Nodewise block Gauß-Seidel (inner solver exact), $\omega_u=1.4$, $\omega_p=0.5$.

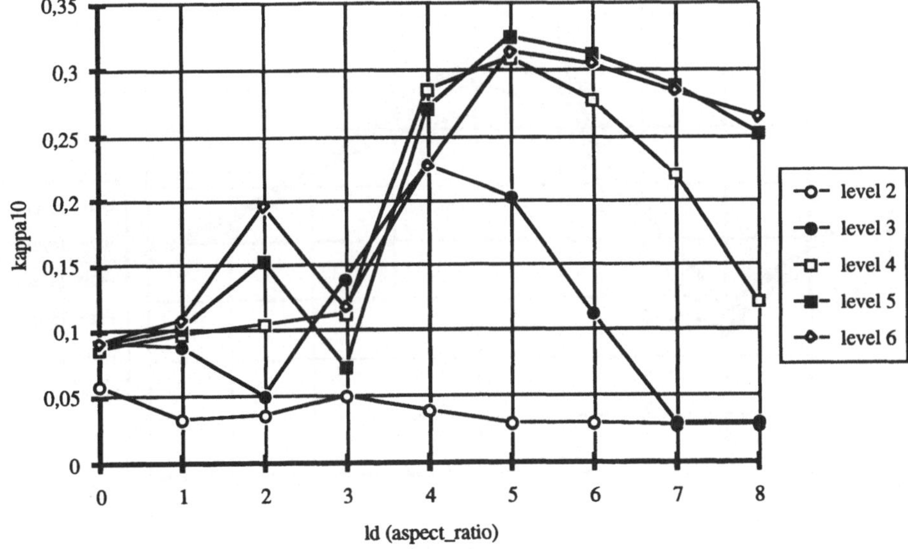

**Fig. 5:** Nodewise block ILU (inner solver exact), $\beta_u=0$, $\beta_p=0$, $\omega_u=1$, $\omega_p=1$.

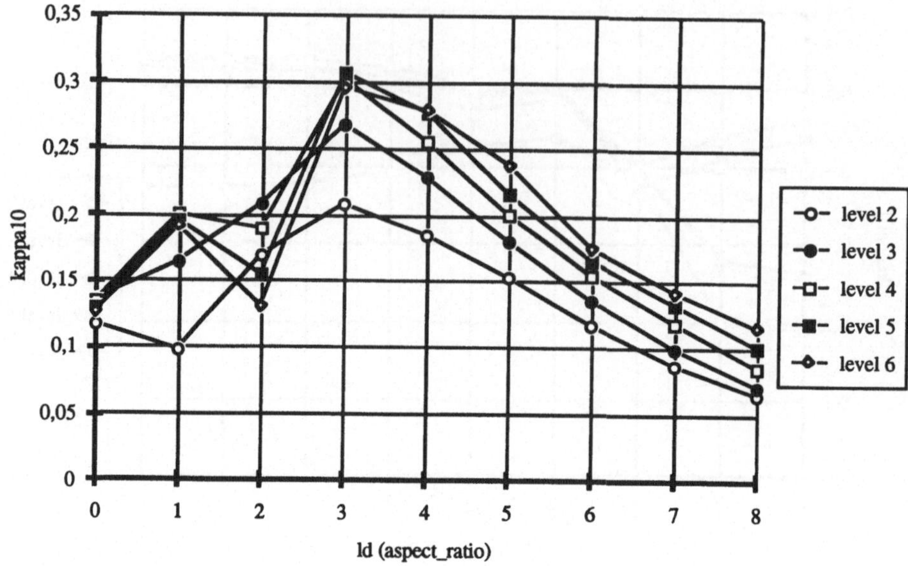

**Fig. 6:** Equationwise block Gauß-Seidel (inner solver ILU), $\beta_u=0$, $\beta_p=10$, $\omega_u=1$, $\omega_p=0.6$.

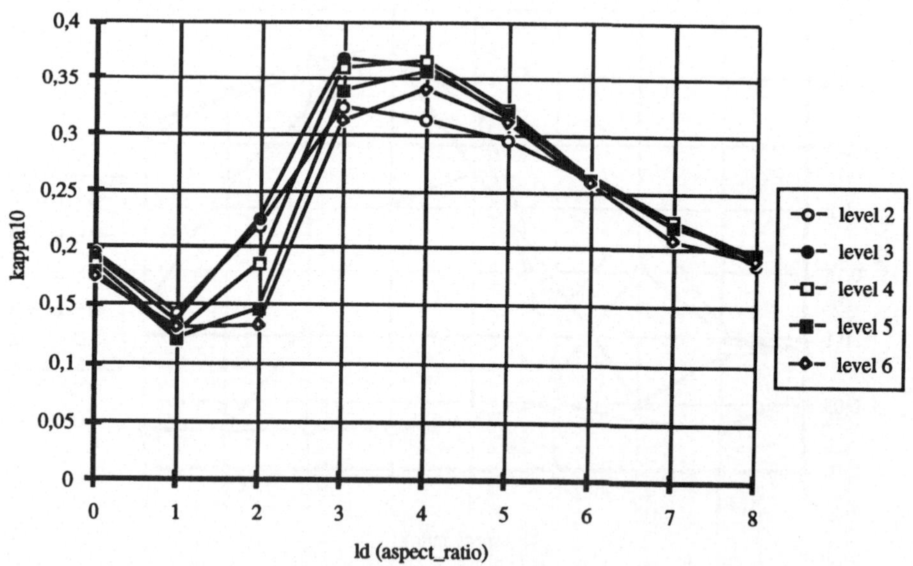

**Fig. 7:** SIMPLE-ILU, $\beta_u=0$, $\beta_p=10$, $\omega_u=1$, $\omega_p=1$.

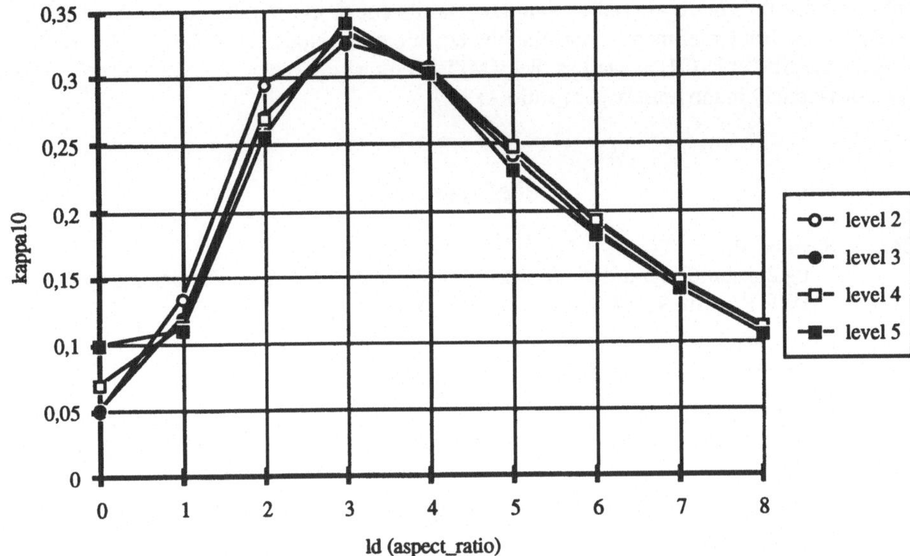


**Fig. 8:** SIMPLE-Frequency-Filter, $\omega_u=1$, $\omega_p=1$.

For the interpretation of the results one should have in mind that the finite volume discretization – similar to the finite element discretization – will not be singularly perturbed in the sense the finite difference discretizations are. There only the coupling to the two next neihgbours remains finite whereas the other ones vanish in the limit of the aspect ratio tending to $\infty$. That leads to an asymptotic stencil for minus the Laplacian of the type

$$\frac{1}{h_y^2}\begin{bmatrix}0 & -1 & 0\\ 0 & 2 & 0\\ 0 & -1 & 0\end{bmatrix}. \tag{3.2.1}$$

On the other hand our discretization yields (no factor of $1/h^2$ because of the volume integration)

$$\frac{h_x}{8h_y}\begin{bmatrix}-1 & -6 & -1\\ 2 & 12 & 2\\ -1 & -6 & -1\end{bmatrix}. \tag{3.2.2}$$

Note the "wrong" signs in the horizontal neighbours of the center which destroy the M-matrix property!

This behavior of the stencils explains why the Gauß-Seidel convergence rates are still bounded below 0.6 while in the scalar case with finite differences the convergence rates approach 1 for aspect_ratio $\to \infty$.

Unfortunately, the effects are rather complex, mainly due to the coupling of the three differential equations. But experiment shows that the equationwise block Gauß-Seidel with inner solver ILU, the SIMPLE-ILU as well as the SIMPLE-Frequency-Filter prove to tend to solve the equations exactly in the limit aspect_ratio $\to \infty$.

# References


[Ba]　*Bastian, P.:*
ug 2.0: Ein Programmbaukasten zur effizienten Lösung von Strömungsproblemen. IWR Preprint 92-14 (Univ. Heidelberg), 1992.

[BW]　*Bey, J., Wittum, G.:*
To appear.

[BD]　*Brandt, A., Dinar, N.:*
Multigrid solutions to elliptic flow problems. ICASE Report 79-15 (1979).

[Mo]　*Moffatt, K.H.:*
Viscous and resistive eddies near a sharp corner. Journal of Fluid Mech. 18 (1964) 1-18.

[PS]　*Patankar, S.V., Spalding, D.B.:*
A calculation procedure for heat and mass transfer in threedimensional parabolic flows. Int. J. Heat Mass Transfer 15 (1972), 1787-1806.

[SR]　*Schneider, G.E., Raw, M.J.:*
Control volume finite-element method for heat transfer and fluid-flow using colocated variables. Numer.Heat.Transf. 11 (1988) 363.

[Re]　*Reichert, H., Wittum, G.:*
Solving the Navier-Stokes-Equations on Unstructured Grids. NNFM, Vol. 39, p. 321-333, Vieweg, Braunschweig 1993.

[Wi1]　*Wittum, G.:*
On the Convergence of Multi-Grid Methods with Transforming Smoothers. Theory with Applications to the Navier-Stokes Equations. Numer. Math. 57 (1990) 15-38.

[Wi2]　*Wittum, G.:*
On the Robustness of ILU-Smoothing. SISSC 10 (1989) 699-717.

[Wi3]　*Wittum, G.:*
Filternde Zerlegungen: Ein Beitrag zur schnellen Lösung großer Gleichungssysteme. Habil., Univ. Heidelberg, 1990.

[Wi4]　*Wittum, G.:*
Distributive Iterationen für indefinite Systeme. Ph. D. Thesis, Univ. Kiel, 1986.

# A NONCONFORMING UNIFORMLY CONVERGENT FINITE ELEMENT METHOD IN TWO DIMENSIONS


H.-G. Roos    D. Adam    A. Felgenhauer

TU Dresden, Institute of Numerical Mathematics

D – 01062 Dresden, Germany


### Summary


We give an new analysis of a nonconforming Galerkin finite element method for solving linear singularly perturbed two-dimensional boundary value problems without turning points. The method is shown to be convergent, uniformly in the perturbation parameter, of order $h^{1/2}$ in an energy norm. The trial functions used are exponentials adapted to the differential operator.


## 1    Introduction

In this paper we consider the following singularly perturbed linear elliptic boundary value problem

$$
\begin{aligned}
Lu \equiv -\varepsilon\Delta u + b_1 u_x + b_2 u_y + cu = f \qquad & \text{on } \Omega = (0,1)\times(0,1) \\
u = 0 \qquad & \text{on } \partial\Omega
\end{aligned}
\tag{1}
$$

with $b_{1,2} = b_{1,2}(x,y)$, $c = c(x,y)$, $f = f(x,y)$, all assumed to be sufficiently smooth, $\varepsilon \ll 1$ and

$$
(\,b_1(x,y), b_2(x,y)\,) \geq (\beta_1, \beta_2) > (0,0) \quad , \quad c(x,y) \geq \gamma > 0 \;\; \text{on } \bar\Omega.
\tag{2a}
$$

Furthermore we assume

$$
c \geq \frac{1}{2} div\, b \qquad \text{on } \bar\Omega.
\tag{2b}
$$

To guarantee the existence of smooth classical solutions we require

$$
f(0,0) = f(0,1) = f(1,0) = f(1,1) = 0.
\tag{2c}
$$

This problem is a basic model of a steady-state convection-diffusion process. For small values of $\varepsilon$ the solution $u$ will in general vary rapidly in a layer region at $\{\,(x,y) \in \overline{\Omega} \mid x = 1$ or $y = 1\}$.

It is well known that classical numerical methods generate oscillations for small values of $\varepsilon$. Therefore, several upwind methods have proposed in the literature (see chapter 7 in [7]). However upwind methods yield accurate approximations only in smooth regions (away from the layer).

We are interested in *uniformly convergent* numerical methods. A method is called *uniformly convergent* with respect to a norm $\|.\|$ , if one can prove an inequality of the form

$$\|u - u_h\| \le C h^p,$$

where both $C$ and $p$ are assumed to be strictly positive and independent of $\varepsilon$ and the mesh width $h$. For the boundary value problem (1) it is natural to ask for an uniformly convergent numerical method in the energy norm

$$\|w\|_\varepsilon^2 \equiv \varepsilon |w|_1^2 + |w|_0^2 \qquad \text{for } w \in \mathrm{H}^1(\Omega).$$

The only result known in this direction is due to O'Riordan and Stynes. In [10] they investigated the case

$$\bar{b} \text{ is constant on } \bar{\Omega} \,,$$

while in [9] they were able to handle

$$b_1 = b_1(x) \text{ and } b_2 = b_2(y) \text{ on } \bar{\Omega} \,. \tag{3}$$

The aim of our work consists in avoiding the very restrictive assumption (3).


## 2   The Nonconforming Method

Let us introduce the bilinear form

$$a(u,v) := \varepsilon(\nabla u, \nabla v) + (b\nabla u + cu, v) \text{ for } u, v \in \mathrm{H}^1(\Omega).$$

Then, a standard *conforming* finite element method starts with a finite element space $V_h$, $V_h \subset \mathrm{H}_0^1(\Omega)$, and defines the approximation $u_h \in V_h$ to solve

$$a(u_h, v_h) = (f, v_h) \qquad \forall v_h \in V_h.$$

Due to

$$(b\nabla u, v) = -(b\nabla v, u) - (div\, b, uv)$$

it would be possible to modify the bilinear form, for instance to

$$a^\star(u,v) = \varepsilon(\nabla u, \nabla v) + \frac{1}{2}(b\nabla u, v) - \frac{1}{2}(b\nabla v, u) + \left((c - \frac{1}{2}\,div\,b)u, v\right).$$

However in a conforming method such a modification results only in some minor advantages, the discrete problems generated are equivalent.

The crucial point to achieve uniform convergence lies in the definition of the finite element space. Let us first define a grid

$$x_i = ih\ ,\ y_j = jh \text{ with } h = 1/N$$

and denote $\Omega_{ij} = (x_i, x_{i+1}) \times (y_j, y_{j+1})$. In the next step we define some splines on $\Omega_{ij}$ . We use piecewise projections $\bar{d}$ of continuous functions $d = d(x,y)$ defined by

$$d^{i,j} = \frac{1}{4}(d(x_i, y_j) + d(x_i, y_{j+1}) + d(x_{i+1}, y_j) + d(x_{i+1}, y_{j+1})) \text{ on } \Omega_{ij}$$
$$\text{and } \bar{d}\big|_{\Omega_{i,j}} = d^{i,j}$$

and introduce the basis functions $\varphi^i_j(x,y)$ to be tensor products of ordinary one-dimensional L-splines in $x$- and $y$-direction , i.e. we assume a representation $\varphi^i_j(x,y) = \varphi^i_{(kl)}(x)\varphi^{(kl)}_j(y)$ on every subdomain $\Omega_{kl}$. Such splines are defined for instance by

$$-\varepsilon\varphi^i_{xx}(x) + \bar{b}_1\varphi^i_x(x) = 0 \text{ on each subintervall and } \varphi^i(x_j) = \delta_{ij}.$$

Due to the fact, that we have to localize the $\varphi^i_j(x,y)$ to a special subdomain later, we use a wildcard (.) and get

$$\varphi^i_{(.)}(x) = \begin{cases} \varphi^{i,l}_{(.)}(x) = 1 - \dfrac{1 - \exp(-b_1^{i-1,\cdot}(x_i - x)/\varepsilon)}{1 - \exp(-b_1^{i-1,\cdot}h/\varepsilon)} & x \in [x_{i-1}, x_i] \\[3mm] \varphi^{i,r}_{(.)}(x) = \dfrac{1 - \exp(-b_1^{i,\cdot}(x_{i+1} - x)/\varepsilon)}{1 - \exp(-b_1^{i,\cdot}h/\varepsilon)} & x \in [x_i, x_{i+1}] \\[3mm] 0 & \text{elsewhere .} \end{cases}$$

The $\varphi^{(.)}_j(y)$ are defined analogously. Finally we get

$$\varphi^i_j(x,y) = \varphi^i_{(.)}(x)\varphi^{(.)}_j(y)$$

and for instance

$$\varphi^i_j(x,y)\big|_{\Omega_{i-1,j-1}} = \varphi^i_{(j-1)}(x)\varphi^{(i-1)}_j(y).$$

Due to $b_1^{i,j-1} \neq b_1^{i,j}$ in general, we especially obtain

$$\varphi^i_j(x,y_j^+) = \varphi^i_{(j)}(x) \neq \varphi^i_{(j-1)}(x) = \varphi^i_j(x,y_j^-).$$

*Thus our splines are not continuous.* Every element $\psi$ of our finite element space $S_h$ admits the representation

$$\psi = \sum_{i,j=1}^{N-1} v_{i,j}\varphi_j^i(x,y).$$

Now we define our nonconforming finite element method as follows. We use the bilinear form

$$a_h(w,v) := \sum_{i,j=0}^{N-1} \int_{\Omega_{ij}} \varepsilon \nabla w \cdot \nabla v + \frac{1}{2}(\bar{b}\cdot\nabla w)v - \frac{1}{2}(\bar{b}\cdot\nabla v)w$$
$$+ (\bar{c} - \frac{1}{2}\overline{div\,b})\,wv\,dx\,dy \qquad \text{for } w,v \in S$$

where $S = \{v$ with $v|_{\Omega_{ij}} \in \mathrm{H}^1$ for all $\Omega_{ij}\}$ and require $u_h \in S_h$ to solve

$$a_h(u_h,\psi) = (f,\psi)_h \qquad \forall\psi \in S_h$$

with

$$(f,\psi)_h = \sum_{i,j=0}^{N-1} \int_{\Omega_{ij}} f\psi\,dx\,dy\ .$$

**Remark 1** Exactly in the separable case (3) our splines are continuous. Then, our method is conforming and almost identical with the method of O'Riordan and Stynes.

$\square$

Adapted to our bilinear form we define the norm

$$\|v\|_\varepsilon^2 := \sum_{i,j} \int_{\Omega_{ij}} \varepsilon(\nabla v)^2 + v^2\,dx\,dy$$

for functions $v$ of $S$ . From the definition of the bilinear form it follows immediately (with $m = 1$)

**Lemma 1** *There exists a constant $m > 0$ independently of $\varepsilon$ such that*

$$a_h(v,v) \ge m\|v\|_\varepsilon^2 \qquad \forall v \in S.$$

Therefore the discrete problem admits a unique solution. It is well known that in every finite element analysis the interpolation error plays a crucial role. Let us suppose that $w$ is continuous and define $w^I \in S_h$ by

$$w^I(x,y) = \sum_{i,j=1}^{N-1} w(x_i,y_j)\varphi_j^i(x,y).$$

The aim of this paper is to estimate the discretization error $\|u - u_h\|_\varepsilon$ . We finally get the following

**Theorem 1 (discretization error)** *Assuming our problem (1) satisfies (2). Let furthermore*

$$F = \begin{pmatrix} (b_1)_x + c & (b_2)_x \\ (b_1)_y & (b_2)_y + c \end{pmatrix} \tag{L}$$

*be a strictly diagonal-dominated L-matrix and let*

$$(b_2)_x = (b_1)_y. \tag{I}$$

*Then we have*

$$\|u - u_h\|_e = O(h^{1/2}).$$

**Remark 2** We expect that the conditions (L) and (I), esspecially the integrability condition $(b_2)_x = (b_1)_y$, can be weekened. But so far we still need them to prove the a-priori estimates of the exact solution for getting the desired order of the interpolation error.

□

The proof of Theorem 1 follows the standard way to estimate the error in nonconforming finite element methods. First we split the discretization error into two parts

$$\|u - u_h\|_e \leq \|u - u^I\|_e + \|u^I - u_h\|_e.$$

By means of Lemma 1 we get

$$\begin{aligned}\|u^I - u_h\|_e^2 &\leq a_h(u^I - u_h, u^I - u_h) \\ &= a_h(u^I - u, u^I - u_h) + a_h(u - u_h, u^I - u_h).\end{aligned}$$

Thus

$$\|u^I - u_h\|_e \leq \sup_{\psi \in S_h} \frac{|a_h(u^I - u, \psi)|}{\|\psi\|_e} + \sup_{\psi \in S_h} \frac{|a_h(u - u_h, \psi)|}{\|\psi\|_e}. \tag{4}$$

While in section 3 we shall estimate the interpolation error, we bound the two expressions in (4) – the approximation error and the consistency error – in section 4 and 5.

## 3   The Interpolation Error

In order to estimate the interpolation error we need some information about the behaviour of the derivatives of our exact solution $u$. The key of getting the desired a-priori estimates for the first partial derivatives of the exact solution on the whole domain lies in the following maximum principle for systems of differential equations (see [5] for details):

**Lemma 2 (Maximum principle for systems)**
*Let*

$$L^* \underline{v} \le L^* \underline{w} \quad ; \quad L_k^* = L_k v_k + f_k(v_1, v_2) , \ k = 1, 2$$

*where $L_k$ are elliptic operators and*

$$f(v_1, v_2) = \begin{pmatrix} f_1(v_1, v_2) \\ f_2(v_1, v_2) \end{pmatrix} = \begin{pmatrix} f_{11} & f_{12} \\ f_{21} & f_{22} \end{pmatrix} \begin{pmatrix} v_1 \\ v_2 \end{pmatrix} =: F\underline{v}$$

*where $F$ denotes an L-Matrix (i.e. $f_{ii} > 0, f_{ij} \le 0$ ), satisfying*

$$f_{11} \ge \frac{1}{2} \mid f_{12} + f_{21} \mid \quad , \quad f_{22} \ge \frac{1}{2} \mid f_{12} + f_{21} \mid .$$

*Furthermore let $\underline{v} \le \underline{w}$ on $\partial\Omega$ . Then we get*

$$\underline{v} \le \underline{w} \qquad on \ \bar{\Omega}.$$

Having this maximum principle one gets

**Lemma 3** *Let $c \ge c_0$ with $c_0$ sufficiently large, $(b_2)_x \le 0$ , $(b_1)_y \le 0$. Then*

$$|u_x(x,y)| \le C^*(1 + \varepsilon^{-1} \exp(-\alpha_1(1-x)/\varepsilon)) \qquad for \ 0 < y < 1 \quad , \tag{5}$$
$$|u_y(x,y)| \le C^*(1 + \varepsilon^{-1} \exp(-\alpha_2(1-y)/\varepsilon)) \qquad for \ 0 < x < 1 \quad , \tag{6}$$

*where $\alpha_1 = \frac{1}{2} \min b_1$ , $\alpha_2 = \frac{1}{2} \min b_2$ .*

Replacing $v$ by the gradient of $u$ and using the following majorizing functions

$$w_1(x,y) = C_1(1 + \varepsilon^{-1} \exp(-\alpha_1(1-x)/\varepsilon)) + D \exp(-\alpha_2(1-y)/\varepsilon)) \quad ,$$
$$w_2(x,y) = C_2(1 + \varepsilon^{-1} \exp(-\alpha_2(1-y)/\varepsilon)) + E \exp(-\alpha_1(1-x)/\varepsilon))$$

which can be estimated against the desired upper bounds this Lemma is an easy consequence of the maximum principle cited above.

Essentially applying the usual maximum principle for elliptic operators (see, e.g. [6]) yields the following second estimate

**Lemma 4** *Let the a-priori estimates (5) and (6) be satisfied and furthermore let $(b_2)_x = (b_1)_y$. Then*

$$|-\varepsilon u_{xx} + b_1 u_x| \le C \qquad on \ \bar{\Omega},$$
$$|-\varepsilon u_{yy} + b_2 u_y| \le C \qquad on \ \bar{\Omega}.$$

Assuming these a-priori estimates of the exact solution in general, we now can prove the needed order of the interpolation error.

**Theorem 2 (interpolation error)** *Let*

$$|u_x(x,y)| \le C(1 + \varepsilon^{-1}\exp(-\alpha_1(1-x)/\varepsilon)) \quad ; \quad |-\varepsilon u_{xx} + b_1 u_x| \le C,$$
$$|u_y(x,y)| \le C(1 + \varepsilon^{-1}\exp(-\alpha_2(1-y)/\varepsilon)) \quad ; \quad |-\varepsilon u_{yy} + b_2 u_y| \le C.$$

*Then*

$$\|u - u^I\|_\varepsilon = O(h^{1/2}).$$

The proof of this Theorem mainly uses the splitting

$$C\|u - u^I\|_\varepsilon^2 \le a_h(u - u^I, u - u^I)$$
$$= (a_h - a)(u - u^I, u - u^I) + a(u - u^I, u - u^I)$$

and furthermore some tricky estimates partly based on the maximum principle for elliptic operators.

For all the details of this section see [2].

# 4 The Consistency Error

To handle this error we introduce a continuous approximation $v_h$ of our arbitrary (discontinuous) test function $\psi$. This approximation is characterized as follows:

- The values in the grid vertices are the same.

- $\bar{L}[v_h(x,y)] := -\varepsilon(v_h)_{xx} - \varepsilon(v_h)_{yy} + \bar{b}_1(v_h)_x + \bar{b}_2(v_h)_y = 0 \quad \forall \Omega_{ij}$ .

- $v_h$ is continuous on the entire domain $\Omega$.
  ( More detailed: $v_h$ satisfies one-dimensional versions of the operator $\bar{L}$ mentioned above on each grid edge as boundary conditions. A small modification in the approximation of the convection coefficients finally guarantees global continuouity. )

This allows us a new explicit representation of the consistency error:

$$a_h(u_h - u, \psi) = (f, \psi - v_h)_h + (a - a_h)(u, v_h) + a_h(u, v_h - \psi) \tag{7}$$

where

$$(f, w_h)_h := \sum_{i,j=0}^{N-1} \int_{\Omega_{ij}} f w_h \, dx \, dy.$$

After proving some further estimates for our nonconforming spline $\psi$ (for instance relations between different norms of this spline) and estimates for the difference of $\psi$ and its continuous counterpart $v_h$ (for all the details see [3]) we have the three terms of (7) in hand (see [3], Lemma 16 – 18) and end with the final estimate

**Theorem 3 (consistency error)** *Let*

$$|u_x| \le C(1 + \varepsilon^{-1} \exp(-\beta_1(1-x)/\varepsilon)) \tag{8a}$$

$$|u_y| \le C(1 + \varepsilon^{-1} \exp(-\beta_2(1-y)/\varepsilon)). \tag{8b}$$

*Then we have*

$$|\, a_h(u - u_h, \psi)\,| \le C\|\psi\|_\varepsilon\, h^{1/2}.$$

**Remark 3** In the previous section we proved that the exact solution $u$ satisfies (8) if we assume the condition (L) of Theorem 1.

$\square$

# 5 The Approximation Error

Recalling our discretized bilinear form we have to estimate the following explicit expression

$$a_h(u - u^I, \psi) = \sum_{i,j=0}^{N-1} \int_{\Omega_{ij}} \left\{ \varepsilon[\nabla(u - u^I) \cdot \nabla\psi] + \left( c_i - \frac{1}{2}(\overline{div\,b}) \right)(u - u^I)\psi \right. \tag{9}$$

$$\left. + \frac{1}{2}[\bar{b} \cdot \nabla(u - u^I)]\psi - \frac{1}{2}[\bar{b} \cdot \nabla\psi](u - u^I) \right\}\, dx\, dy.$$

Three of the terms are essentially straightforward to handle however the third term (the one containing the gradient of $(u - u^I)$) needs some further investigations. We estimate them seperatly and get

**Lemma 5** *Assuming (L) and (I) of Theorem 1. Then*

$$\left| \sum_{i,j=0}^{N-1} \int_{\Omega_{ij}} \frac{1}{2}[\bar{b} \cdot \nabla(u - u^I)]\psi\, dx\, dy \right| \le C\|\psi\|_\varepsilon\, h^{1/2}.$$

The proof starts with the rewriting

$$\sum_{i,j=0}^{N-1} \int_{\Omega_{ij}} \frac{1}{2}[\bar{b} \cdot \nabla(u - u^I)]\psi\, dx\, dy = -\sum_{i,j=0}^{N-1} \int_{\Omega_{ij}} \frac{1}{2}[\bar{b} \cdot \nabla\psi](u - u^I)\, dx\, dy \tag{10}$$

$$+ \sum_{i,j=0}^{N-1} \int_{y_j}^{y_{j+1}} \frac{1}{2}\left[ b_1^{i-1,j}(u - u^I)(x_i^-, y)\psi(x_i^-, y) - b_1^{ij}(u - u^I)(x_i^+, y)\psi(x_i^+, y) \right] dy$$

$$+ \sum_{i,j=0}^{N-1} \int_{x_i}^{x_{i+1}} \frac{1}{2}\left[ b_2^{i,j-1}(u - u^I)(x, y_j^-)\psi(x, y_j^-) - b_2^{ij}(u - u^I)(x, y_j^+)\psi(x, y_j^+) \right] dx.$$

The main idea then consists in adding some terms to the boundary integral sums getting terms of order $O(h^{1/2})$ and furthermore one term in each sum containing $(u_h^I - u^I)$ instead of $(u - u^I)$ without any other small factors (here $u_h^I$ denotes the special continuous interpolant corresponding to $u^I$). Essentially using a corollary of a well-known Lemma of Gartland (see [4],Theorem 1.1.) we were able to estimate the absolut value of this difference on the grid edges in a way which finally guarantees the desired order for the whole expression. For the exact proof see again [3], Lemma 21.

Using standard techniques for the other terms of (9) we end with

**Theorem 4 (approximation error)** *Assuming (L) and (I) of Theorem 1. Then we have*

$$|a_h(u^I - u, \psi)| \le C\|\psi\|_\varepsilon\, h^{1/2}.$$

After establishing uniformly convergent error estimates of order $O(h^{1/2})$ for the interpolation error, for the consistency error and for the approximation error in the three previous sections we did everything what we announced to do at the end of section 2. So the proof of our main result – Theorem 1 – is complete.

# 6 Numerical Results

The aim of this section is to show, that the method described yields reasonable numerical results, i.e., that our nonconforming method is numerically stable and the order of convergence obtained is optimal.

We tested, for instance, the problem

$$-\frac{1}{100}\Delta u + (2 + x - y)u_x + (3 - x + y^2)u_y + u = 2y + 3x - x^2 - y^2 + xy^2 + 2xy$$

with zero boundary conditions on a unit square, which satisfies the conditions (L) and (I). Note that $u_0(x, y) = xy$ is the solution of the reduced problem ($\varepsilon = 0$), so you get an impression how the discretized solution should look like.

As usual (see, e.g. [8]) we use the double mesh principle to determine the numerical order of convergence :

$$p = \frac{1}{\ln 2} \ln \frac{\|u_h - u_{h/2}\|_\varepsilon}{\|u_{h/2} - u_{h/4}\|_\varepsilon}\ .$$

The following table contains numerical convergence rates as well in the $\|.\|_\varepsilon$ – norm as in the $L_2$ – norm and the maximum norm.

Table 1 : Numerical convergence rates

| $h$ | numerical order of convergence | | | |
|---|---|---|---|---|
|  | $\|\cdot\|_\varepsilon$ – norm | $L_2$ – norm | discr. $L_2$ – norm | $L_\infty$ – norm |
| 1/16 | 0.61044 | 1.18559 | 0.67377 | 0.71073 |
| 1/32 | 0.56242 | 1.16409 | 0.83266 | 0.81058 |
| 1/64 | 0.54695 | 1.20125 | 0.94807 | 0.87308 |

The following figures show the results of our nonconfirming method and a standard streamline diffusion method.

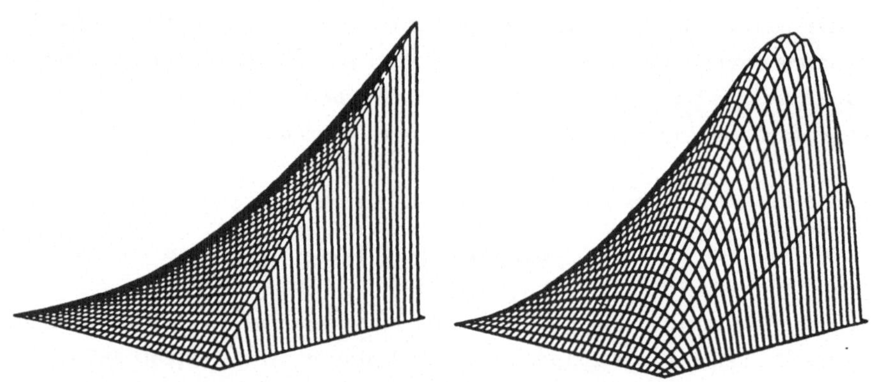

Figure 1 : Comparison between our nonconforming method (left)
and a standard streamline diffusion method (right)

*Acknowledgment*: We would like to thank Dr. L. Angermann (Erlangen) for providing us with the numerical results for solving several test problems based on the streamline diffusion method and an upwind finite volume technique.

# References


[1] D. Adam, A. Felgenhauer, H.-G. Roos, M. Stynes, *A nonconforming finite element method for a singularly perturbed boundary value problem*, technical report, University College Cork, Ireland, 1992.

[2] D. Adam, H.-G. Roos, *A nonconforming exponentially fitted finite element method I: The interpolation error*, report MATH-NM-06-1993, Technical University Dresden, 1993.

[3]  D. Adam, A.Felgenhauer, H.-G. Roos, *A nonconforming exponentially fitted finite element method II: The discretization error*, report MATH-NM-14-1993, Technical University Dresden, 1993.

[4]  E. Gartland Jr., *An analysis of a uniformly convergent finite difference / finite element scheme for a model singular-perturbation problem* Math. Comp. 51, 1988, 99–106.

[5]  K.Glasshoff, B.Werner, *Inverse Monotonicity of Monotone L-Operators with Applications to Quasilinear and Free Boundary Value Problems*, J. Math. Anal. Appl. 72, 1979, 89–105.

[6]  D. Gilbarg, N.S. Trudinger, *Elliptic partial differential operators of second order*, Springer, Berlin, Heidelberg, New York, 1983.

[7]  Ch. Grossmann, H.-G. Roos, *Numerik partieller Differentialgleichungen*, Teubner Verlag, Stuttgart, 1992.

[8]  T. J. R. Hughes, A. N. Brooks , *A multidimensional upwind scheme with no crosswind diffusion*, in: Finite element methods for convection dominated flows , ed. T. J. R. Hughes, AMP v.34 ASME, New York, 1980, pp. 19–35.

[9]  E. O'Riordan, M. Stynes, *A uniformly convergent difference scheme for an elliptic singular perturbation problem*, in: Discretization Methods of Singular Perturbation and Flow Problems, ed. L. Tobiska, Technical University "Otto von Guericke" Magdeburg, Magdeburg, 1989, pp. 48–55.

[10]  E. O'Riordan, M. Stynes, *A globally uniformly convergent finite element method for a singularly perturbed elliptic problem in two dimensions*, Math. Comp. 57, 1991, 47–62.

# AN EFFICIENT PARALLEL SOLUTION TECHNIQUE FOR THE INCOMPRESSIBLE NAVIER-STOKES EQUATIONS


*M. Schäfer, E. Schreck, K. Wechsler*

Lehrstuhl für Strömungsmechanik
Universität Erlangen-Nürnberg
Cauerstr. 4, D-91058 Erlangen



## SUMMARY

In this paper a parallel finite volume solution method for the Navier-Stokes equations is presented. The solution procedure is based on collocated blockstructured grids, iterative pressure-velocity coupling of SIMPLE type, multigrid methods, and automatic grid partitioning. A high numerical efficiency is ensured by the nonlinear multigrid method. The parallelization follows the message passing strategy allowing the code to be run on a wide range of actual parallel hardware platforms without much porting effort. By considering several 2-d and 3-d flow problems aspects concerning the numerical and parallel performance of the solver are discussed for different parallel architectures.


## 1. INTRODUCTION

The numerical simulation of practically relevant flows often involves the handling of complex geometries and complex physical and chemical phenomena requiring the use of very fine grids and small time steps in order to achieve the necessary accuracy. In this paper a numerical solution method for the incompressible Navier-Stokes equations is presented, which combines efficient numerical techniques and parallel computing giving the possibility to treat such problems within a reasonable amount of computing time.

The solution method is based on blockstructured grids, a fully conservative finite volume discretization, an iterative pressure-correction method of SIMPLE type, a nonlinear multigrid method, and a grid partitioning technique. Using the blockstructuring approach is advantageous for several reasons:

- complex geometries can be modelled easily,
- numerically efficient "structured" algorithms can be used within each block,
- a natural basis for the parallelization of the solution methods is provided,
- regions with different material properties (e.g. solid/liquid problems) can be handled in a straightforward way.

The parallelization of the method is achieved by a grid partitioning technique based on the blockstructured grids. According to the number of available processors an automatic load balancing procedure performs the mapping of the geometrical blockstructure to a parallel blockstructure such that the resulting subdomains can be assigned to individual processors. Two strategies for this automatic load balancing are presented.

The parallel multigrid method is implemented globally, i.e. without being affected by the grid partitioning. This ensures a close coupling of the subdomains and, therefore, only a slight deterioration of the numerical efficiency compared to the corresponding sequential algorithm can be observed.

Several numerical examples for 2-d and 3-d problems illustrate the capabilities of the considered approach. These results include investigations of the interdependence of the multigrid algorithm, the grid partitioning, and the performance data of the parallel computer with respect to numerical and parallel efficiency, as well as comparisons of several parallel computer platforms.

The results indicate that parallel computers, combined with advanced numerical methods, can yield the computational performance required for an efficient, accurate, and reliable solution of practical flow problems in engineering and science.

## 2. BASIC NUMERICAL METHOD

The solution method used in this study is described in detail by Demirdžić and Perić [2], so only a summary of the main features will be given here. The method is of finite volume type and uses non-orthogonal boundary-fitted grids with a collocated arrangement of variables. The continuity equation is used to obtain a pressure-correction equation according to the SIMPLE algorithm (Patankar and Spalding [5]). Second order discretization is used for all terms (central differences, linear interpolation). The part of diffusion fluxes which arises from grid non-orthogonality is treated explicitly. For the convection fluxes the so called "deferred correction" approach is used: only the part which corresponds to the first order upwind discretization is treated implicitly, while the difference between the central differencing and upwind fluxes is treated explicitly. The effect of non-orthogonality is also taken into account explicitly in the pressure-correction equation. In the first step, the pressure-correction equation is solved with these terms excluded (which suffices if the grid is not severely non-orthogonal). In a second step the non-orthogonal contribution is evaluated, and a second pressure-correction equation is solved.

Equations for the Cartesian velocity components $U$, $V$ and $W$, pressure correction $P'$ and temperature $T$ are discretized and solved one after another. The linear algebraic equation systems are solved iteratively using the ILU approach of Stone [7]. These inner iterations are stopped either after reducing the absolute sum of residuals over all CVs by a factor of five to ten, or after a prescribed number of iterations has been performed. Outer iterations are performed to take into account the non-linearity, coupling of variables and effects of grid non-orthogonality. Under-relaxation is used to improve the convergence of the outer iterations. The computation is stopped when at the begining of an outer iteration the sum of absolute residuals over all CVs in all equations becomes 4 to 5 orders of magnitude smaller than the initial values. This corresponds roughly to a 4–5 digit accuracy. Typical values of under-relaxation factors range between 0.5 and 0.8 for the velocity components and temperature. The fraction of pressure correction added to pressure after solving the pressure-correction equation is typically equal to $1 - \alpha_u$, where $\alpha_u$ is the under-relaxation factor for velocities. This choice leads to a nearly-optimum convergence rate.

The number of outer iterations increases linearly with the number of CVs, leading to a quadratic increase in computing time. For this reason a multigrid method is implemented, which keeps the number of outer iterations approximately independent of the number of CVs. The method is based on the so called "full approximation scheme" (FAS) with V-cycles and, for steady problems, it is implemented in the so called "full multigrid" (FMG) fashion. The equations solved on the coarse grids within a multigrid cycle contain an additional source term which describes the current solution and the residuals of the next finer grid. The employed multigrid technique is described in detail in Hortmann *et al.* [4].

For handling the coupling of the blockstructured grids auxiliary control volumes containing the corresponding boundary values of the neighbouring block are introduced along the block interfaces. The coupling is then ensured by the interchange of these boundary values during the iterative solution algorithm (e.g. Durst *et al.* [3]). This especially is important for the parallel implementation in order to reduce the communication effort.

## 3. GRID PARTITIONING

The parallelization of the solution algorithm described in the previous section is based on a grid partitioning technique directly related to the employed blockstructuring approach. The blockstructure used for modelling the problem geometry (geometrical blockstructure) is modified into a blockstructure suitable for parallel processing (parallel blockstructure). For this parallel blockstructure, which is created by an automatic mapping process, several requirements with respect to obtain an efficient parallel implementation can be formulated:

- similar block sizes to ensure a good load balancing,

- small number of neighbouring blocks to have few communication processes,

- short block interfaces to have a small number of data to transfer,

- block topology close to processor topology to avoid communication between physically non-neighboring processors (not for all parallel platforms of importance),

- avoiding of steep flow gradients across interfaces in order to retain a good coupling of the subdomains for a high numerical efficiency,

- mapping process not too much time consuming compared to the flow computation.

It is obvious that, for general flow problems, not all of these requirements can optimally be fulfilled simultaneously. Therefore, some compromise has to be found. In our approach the automatic mapping is based on the simple criteria that the number of control volumes in the different blocks differ in size as little as possible. Topological and flow specific aspects are not taken into account. This approach, which is very fast and ensures good load balancing, turns out to be quite efficient for a wide range of applications.

In general, for the mapping process two situations have to be distinguished. If the number of geometrical blocks is larger than the number of processors, the geometrical blocks are suitably grouped together, such that the sum of the block sizes in each group differ as little as possible. The resulting groups are then assigned to the individual processors.

In the other case, i.e. if the number of geometrical blocks is smaller than the number of processors, the geometrical blocks have to be partitioned in order to obtain finally a block structure with as many blocks as processors. For this partitioning two different strategies are considered:

- Direct decomposition:
  The processors are assigned to the geometrical blocks according to the number of control volumes in the blocks and the blocks are subdivided one-dimensionally in the coordinate direction with the largest number of control volumes.

- Recursive decomposition:

  The block with the largest number of control volumes is halved in the direction with the largest number of control volumes resulting in a new blockstructure. The process is repeated until the number of blocks equals the number of processors.

The principles of these two strategies are illustrated in Figs. 1 and 2. Which strategy is preferable depends on the problem geometry, the blockstructure used to model it, and the number of available processors. An advantage of the recursive decomposition is that, at least theoretically, as much processors as control volumes are on the coarsest grid can be used for the parallel computation. If direct decomposition is used, the number of processors assigned to a block may not be larger than the largest number of control volumes in one direction in this block. (However, as will become clear from the efficiency considerations given in the next section, from the numerical point of view it is not reasonable to use too many processors for a grid of given size.)

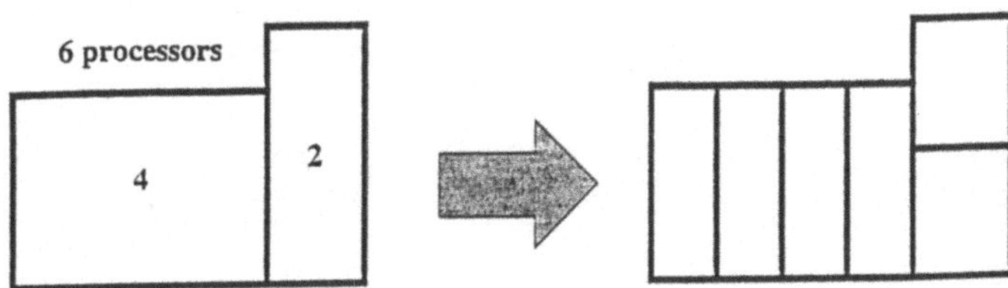

**Figure 1:** *Automatic grid partitioning by direct mapping of geometrical blockstructure to parallel blockstructure.*

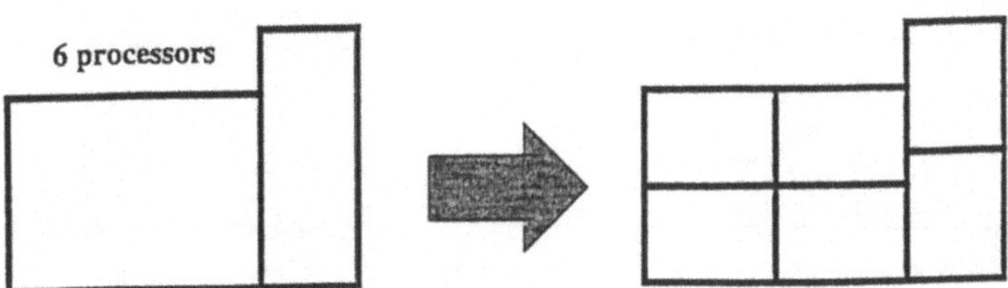

**Figure 2:** *Automatic grid partitioning by recursive mapping of geometrical blockstructure to parallel blockstructure.*

To illustrate the automatic mapping process and the dependence of the solution algorithm on the partitioning we consider two examples for complex geometries. The first is a non-isothermal flow in a complex channel system and the second concerns a buoyancy driven flow in a cavity with a complex obstacle. In Fig. 3 the geometries for the two problems are shown together with the computed velocity fields.

For the channel problem there are several inlets, where the fluid comes in with different flow rates and temperatures, and one outlet at the right bottom. For the cavity problem the walls are at low temperature, the obstacle walls are at high temperature, resulting in a Rayleigh number $Ra = 500$ based on the minimum distance between the cavity wall and the obstacle. In Figs. 4,5 the partitionings resulting from direct and recursive mapping are given (48 processors for the channel system, 32 processors for the cavity problem).

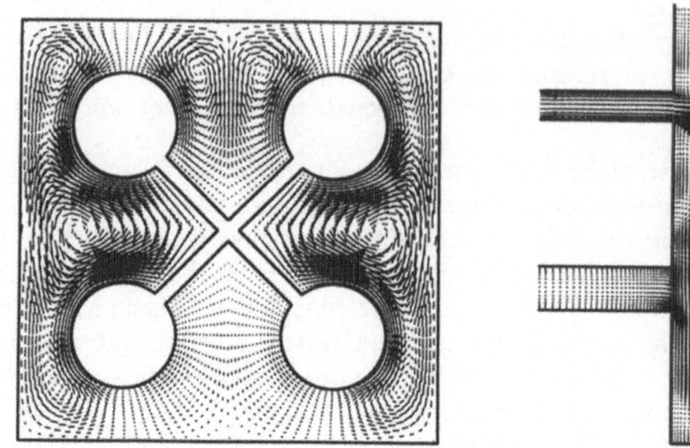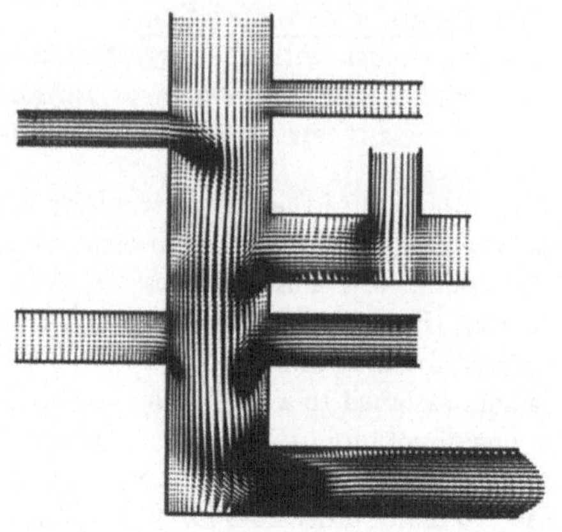

**Figure 3:** *Geometries and computed velocity fields for the buoyancy driven flow in a cavity with a complex obstacle and the flow in the complex channel system.*

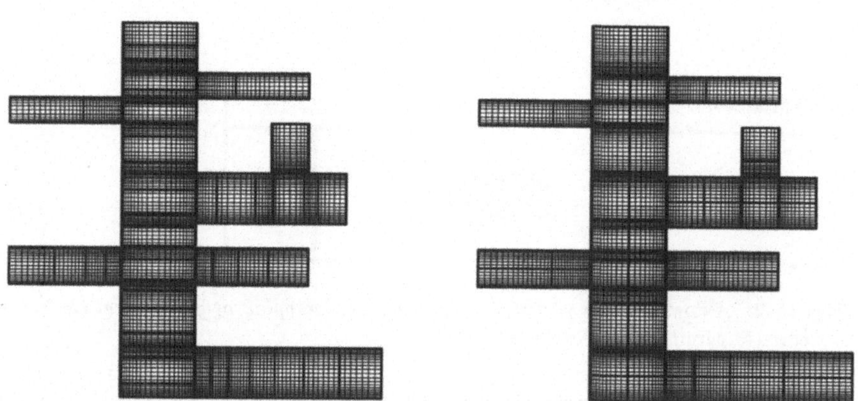

**Figure 4:** *Parallel blockstructures resulting from direct and recursive decomposition with 48 processors for the channel problem (coarse grid with 6.600 CV).*

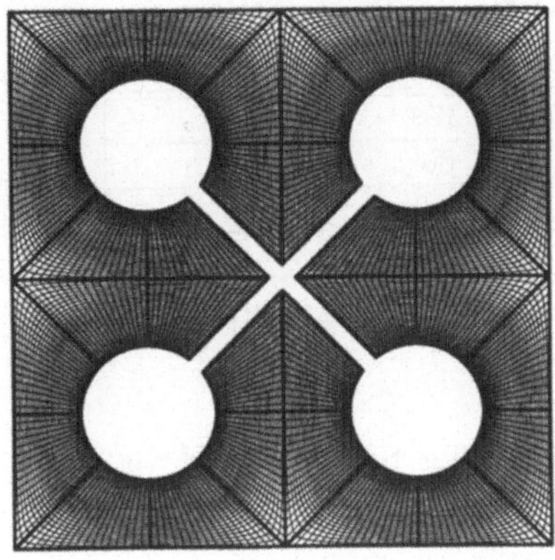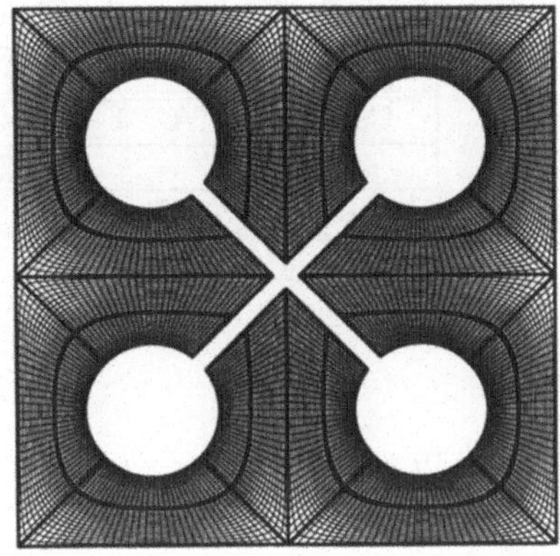

**Figure 5:** *Parallel blockstructures resulting from direct and recursive decomposition with 32 processors for the cavity problem.*

In Table 1 the number of iterations and the corresponding computing times on a Parsytec MultiCluster with T805 transputers under the operating system Parix are given for the two cases. One can see that for the channel flow the recursive mapping results in a better performance, while for the cavity flow the situation is opposite. The differences in the number of iterations between the direct and recursive approach are due to decoupling effects in the parallel ILU linear system solver. For the channel flow this difference is also the major reason for the different computing times (ratio of iterations is 0.85 and ratio of computing times is 0.82). For the cavity flow there is an additional overhead with the recursive mapping due to the larger number of communication processes for the recursive mapping (ratio of iterations is 0.84 and ratio of computing times is 0.69).

**Table 1:** *Iteration numbers and computing times for the channel and cavity flow with direct and recursive mapping.*

| Decomposition | Channel (26.400 CV) | | Cavity (16.386 CV) | |
|---|---|---|---|---|
| | Iterations | CPU-time (s) | Iterations | CPU-time (s) |
| Direct | 163 | 813 | 36 | 220 |
| Recursive | 139 | 666 | 43 | 319 |

## 4. RESULTS FOR STEADY FLOWS

As a test case for comparing different parallel architectures the buoyancy driven flow in an inclined cavity, which is one of the benchmark flows presented by Demirdžić *et al.* [1], is considered. All cavity walls are of the same area, the horizontal walls are adiabatic, and the inclined walls (angle 45°) are isothermal but at different temperatures. The temperature difference, cavity size, and fluid properties are chosen such that a Prandtl number $Pr = 0.1$ and a Rayleigh number $Ra = 10^6$ are obtained. The coarsest grid

Table 2: *Characteristic data of computers used for the calculations.*

| Computer | $\tau$ (MFlops) | $t^{st}(\mu s)$ | $R_{tr}$ (MBs) |
|---|---|---|---|
| Convex META | 20 | 370 | 8 |
| Intel IPSC | 2.5 | 80 | 2.8 |
| KSR-1 | 5 | 110 | 7 |
| Sun SPARC | 7 | 1400 | 1 |
| Parsytec SC | 0.55 | 84 | 1.5 |

consists of $10 \times 10$ CV, and six grid levels, corresponding to a finest grid with $320 \times 320$ CVs, are used.

The following five computer platforms are included in the comparison:

- a Parsytec Supercluster with T805 transputers (30 MHz) and Parix communication software,

- an Intel IPSC/860 with i860 processors (40 MHz) and NX/2 communication

- a virtual shared memory KSR-1 with proprietary processors (40 MFlops peak performance) and TCGMSG communication software,

- a Convex META with HP PA 1.1 processors (i.e. HP 9000/735 workstations with FDDI connection) and TCGMSG communication software,

- a cluster of SUN Sparc 10/20 workstations with Ethernet connection and TCGMSG communication software.

The main software and hardware parameters effecting the performance of a parallel algorithm (e.g. start-up or latency time $t^{st}$, data transfer rate $R_{tr}$, and computing speed $\tau$) are given in Table 2 for the different systems. The given MFlops rate is the sustained performance achieved with our application program.

In Tab. 3 the total computing times for solving the problem and the corresponding efficiencies are given for the various machines with different numbers of processors. As reference value the computing time on a single SUN Sparc 10/20 is also indicated. Of course, the computing times are mainly determined by the MFlop rates of the node processors. The crucial factor for the efficiencies are the start-up times in relation to the MFlop rate. The transfer rate is here of minor importance. While for the Parsytec SC good efficiencies also for large processor numbers are obtained, due to the very low processor performance, the computing times are not competitive. The advantage of an FDDI connection compared to an Ethernet can be seen from the efficiencies for the Sun cluster and the Convex META.

An important aspect for the performance of a parallel CFD method based on grid partitioning is the numerical efficiency, which is a measure for the decrease of the convergence rate due to the decoupling of the flow domain. With respect to this issue the employed multigrid algorithm behaves very well. The number of V-cycles for the finest grid only slightly depends on the number of processors and the grid partitioning (see [6]). Therefore, the efficiencies given in Tab. 3 are mainly determined by the communication effort. It

**Table 3:** *Computing times and efficiencies for the finest grid for the inclined cavity problem.*

| Computer | Processors | | CPU time(s) | Efficiency (%) |
|---|---|---|---|---|
| SUN Sparc 10/20 | 1 | Sparc V8 | 589 | - |
| SUN Sparc 10/20 | 5 | Sparc V8 | 243 | 49 |
| Convex META | 5 | HP PA 1.1 | 80 | 64 |
| KSR-1 | 5 | KSR-1 | 157 | 95 |
| KSR-1 | 10 | KSR-1 | 88 | 85 |
| Parsytec SC | 25 | T805 | 278 | 90 |
| Intel IPSC/860 | 25 | i860 | 73 | 76 |
| Parsytec SC | 100 | T805 | 120 | 52 |

should also be noted that using the multigrid method the computing time for the considered test case is more than 100 times shorter compared to the single grid computation, despite in terms of efficiency, due to the lower communication effort, for the single grid calculation higher efficiencies are possible (e.g. 96% with 100 processors on the Parsytec SC).

As a three dimensional test case the external turbulent flow around a ship hull with a Reynolds number of $10^6$ is considered. For the turbulence modelling a standard $k$-$\varepsilon$-model with wall functions is employed. For the computation a finest grid with $56 \times 24 \times 44$ CVs is used. The grid structure is shown in Fig. 6 (coarse grid with $28 \times 12 \times 22$ CVs) together with the predicted isobars. In Tab. 4 the numbers of iterations and computing times for a single grid calculation on the finest grid are indicated together with the corresponding numerical and total efficiencies for the Parsytec SC (48 processors) and the KSR-1 (18 processors). For reference the values for one HP 9000/735 workstation are also indicated. Here, as against for the multigrid method, one can observe the decrease in the numerical efficiency with increasing processor number. Concerning the total efficiencies and the computing times the situation is similar to the laminar 2d case discussed above. Again the latency time and the MFlop rate are the crucial factors.

**Table 4:** *Iteration numbers, computing times, and efficiencies for the 3d case for different hardware platforms.*

| Computer | # Procs. | Iter. | CPU-time (s) | num. Eff. | tot. Eff. |
|---|---|---|---|---|---|
| HP 9000/735 | 1 | 239 | 1995 | 100% | 100% |
| KSR 1 | 18 | 247 | 646 | 97% | 70% |
| Parsytec SC | 48 | 309 | 4641 | 77% | 49% |

## 5. RESULTS FOR UNSTEADY FLOW

The two-dimensional computation of the external flow around a cylinder has been chosen as an example for the determination of computation times on different parallel machines

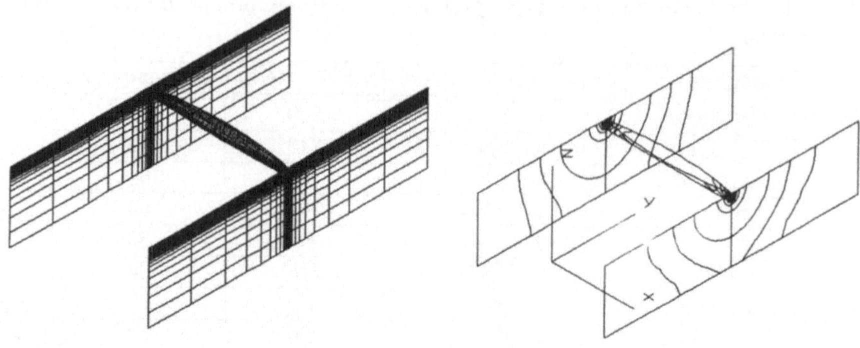

**Figure 6:** *Numerical grid and predicted isobars for the flow around a ship hull.*

for unsteady calculations. The size of the computational domain is about 22 diameters of the cylinder in streamwise direction and approximately 7 diameters in crossflow direction. Fluid properties and the diameter of the cylinder result in a Reynolds number Re=200 with respect to the diameter of the cylinder. The coarsest grid consists of 384 CVs, and unsteady calculations were performed with four and five grid levels using the multigrid technique (24576 CVs and 98304 CVs, respectively). For time discretization, we use first order, fully implicit, Euler method.

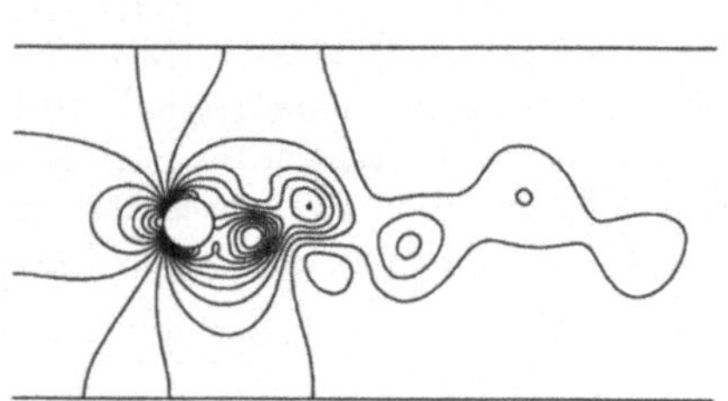

**Figure 7:** *Pressure distribution at $t = 50s$.*

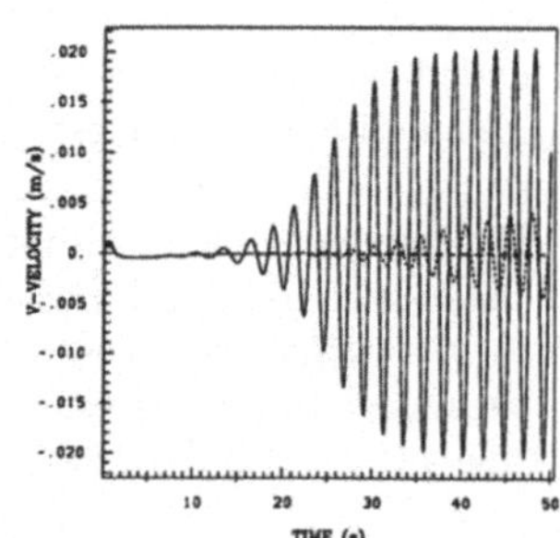

**Figure 8:** *Velocity at monitoring location.*

Figure 7 shows the instantaneous pressure distribution around the cylinder 50 seconds after the sudden start of the cylinder, calculated with a time step of $\Delta t = 0.05s$. The direction of the flow is from left to right, and symmetry boundary conditions are set on the top and at the bottom of the computational domain. The development of vortex shedding and therefore the unsteadiness of the problem (having stationary boundary conditions) can clearly be seen.

Figure 8 shows the time variation of a velocity component at a monitoring location in the wake of the flow behind the cylinder compared for two different time step sizes in the calculation (solid line for $\Delta t = 0.05s$ and dashed line for $\Delta t = 0.125s$). Obviously, chosing a higher time-step results in suppression of the development of the instationarity of the flow, a fact which is due to the strong damping properties of the implicit Euler

Table 5: *Computing times for 4 time steps in unsteady calculation.*

| Computer | Proc. | CPU time(s) 24576 CVs | CPU time(s) 98304 CVs |
|---|---|---|---|
| Sun Sparc 10/20 | 3 | 461 | 1343 |
| KSR-1 | 3 | 316 | 1145 |
|  | 6 | 186 | 610 |
|  | 12 | - | 383 |
|  | 24 | - | 379 |
| Convex META | 3 | 136 | 426 |
|  | 6 | 90 | 241 |
|  | 12 | - | 151 |

scheme for large time steps. While harmonic steady state has been reached at 40 seconds for $\Delta t = 0.05s$, velocity is still increasing and significantly lower for $\Delta t = 0.125s$.

Computation times, which are required for the first four time steps in the unsteady calculation are given in Table 5 obtained with the parallel machines Sun Sparc 10/20 workstation cluster, KSR-1, and Convex META, which are described in the previous chapter. Due to the high latency time of the Ethernet, computation on three processors of the KSR-1 is faster than on 3 processors of the Sun Sparc workstation cluster, although computation speed of one Sparc V8 processor is higher then that of one KSR-1 processor. Moreover, it is interesting to compare computation times for four grids on three processors and five grids on twelve processors. In this case, the number of CVs and the number of processors have been increased by a factor of four. This results in a relatively small increase of the computation time of 21% for the KSR-1 and 11% for the Convex Meta. But, this scalability was not achieved by using more processors, as can be seen for the computation times for six processors/four grid levels and 24 processors/five grid levels. Obviously, there is a relaxation of computation time while increasing the number of processors due to the enlarged communication requirements. We expect that this behaviour can be improved by the application and investigation of different mapping strategies.

## 6. CONCLUSIONS

We have presented a parallel implicit finite volume multigrid algorithm for the numerical prediction of flows in complex geometries. It has been shown that the employed parallelization approach based on blockstructured grid partitioning results in an efficient parallel implementation mostly retaining the high numerical efficiency of the powerful sequential multigrid solution procedure and providing very good portability properties.

The underlying partitioning approach for general geometries is a very complicated task, but for a wide range of applications relatively simple strategies give satisfactory results. Scalability mainly depends on the ratio of arithmetic performance and latency time of the parallel machine. By several numerical examples adequate efficiencies on actual parallel computers and high acceleration compared to serial flow computations are achieved.

ACKNOWLEDGMENTS

The authors would like to thank Prof. F. Durst for his support and many helpful discussions, as well as Ž. Lilek for the common work concerning the three dimensional calculation. The work was financed by the "Bayerische Forschungsstiftung" in the "Bavarian Consortium of High-Performance Scientific Computing (FORTWIHR)", the "Deutsche Forschungsgemeinschaft" in the special programme "Flow Simulation with High-Performance Computers", and the SFB 182 " Multiprocessor and Network Configurations". This support is gratefully acknowledged.

## REFERENCES

[1] I. Demirdžić, Ž. Lilek, and M. Perić. Fluid flow and heat transfer test problems for non-orthogonal grids: Bench-mark solutions. *Int. J. Num. Meth. in Fluids*, 15:329–354, 1992.

[2] I. Demirdžić and M. Perić. Finite volume method for prediction of fluid flow in arbitrary shaped domains with moving boundaries. *Int. J. Num. Meth. in Fluids*, 10:771–790, 1990.

[3] F. Durst, M. Perić, M. Schäfer, and E. Schreck. Parallelization of Efficient Numerical Methods for Flows in Complex Geometries. In *Flow Simulation with High-Performance Computers I*, Notes on Numerical Fluid Mechanics, pages 79–92. Vieweg Verlag, 1993.

[4] M. Hortmann, M. Perić, and G. Scheuerer. Finite volume multigrid prediction of laminar natural convection: Benchmark solutions. *Int. J. Num. Meth. in Fluids*, 11:189–207, 1990.

[5] S. V. Patankar and D. B. Spalding. A calculation procedure for heat, mass and momentum transfer in three dimesional parabolic flows. *Int. J. Heat Mass Transfer*, 15:1787–1806, 1972.

[6] M. Perić, M. Schäfer, and E. Schreck. Numerical simulation of complex fluid flows on MIMD computers. In *Parallel Computational Fluid Dynamics 92*, pages 311–324, Amsterdam, 1993. Elsevier.

[7] H. Stone. Iterative solution of implicit approximations of multi-dimensional partial differential equations. *SIAM Journal on Numerical Analysis*, 5:530–558, 1968.

# COUPLING OF TWO DIMENSIONAL VISCOUS AND INVISCID INCOMPRESSIBLE STOKES EQUATIONS


K.Schenk

Institut für Mathematik, Technische Universität Cottbus
Postfach 101344, 03013 Cottbus, Germany

F.K.Hebeker

IBM Scientific Center Heidelberg
Vangerowstraße 18, 69115 Heidelberg, Germany


## SUMMARY


We develop a method for the coupling of a generalized Stokes problem and a problem
with vanishing viscosity for an incompressible flow in a bounded region in two dimensions. Correct transmission conditions are derived, and an iterative procedure involving
the successive resolution of two subproblems is suggested.


## INTRODUCTION

A domain decomposition approach usually consists in the decomposition of the possibly complex domain into subdomains of simpler shape. Then the original problem is
reduced to a sequence of subproblems for the same partial differential equation which
can be solved independently to some extent [7, 10, 11, 14, 16].

Another very interesting area for the application of domain decomposition procedures is the coupling of partial differential equations of different type each used in a
suitable subregion of the whole domain. A typical situation is the investigation of an
incompressible flow around an obstacle. All the interesting features of the flow occur
near the boundary of the obstacle due to the important role of viscosity effects in this
area. In the region far away from the obstacle one can neglect viscosity effects. For
numerical and theoretical reasons it would be interesting to combine the solution to the
Navier-Stokes equations in an inner subregion near the boundary of the obstacle and
the solution to the Euler equations in an outer region far away from the obstacle and
to find correct transmission conditions.

The coupling of different fluid dynamical equations also occurs in connection with
the problem of missing boundary conditions at the outflow boundary of the computational domain [19].

We consider a stationary and linearized incompressible flow around an obstacle
described by generalized Stokes equations in a bounded domain in two dimensions. Two
problems of this type have to be solved at every time step of an operator splitting time
discretization scheme of the full incompressible Navier-Stokes equations [3, 8, 20]. To
reduce the amount of computations we want to solve this generalized Stokes problem
only in an inner subregion of the whole domain. In the outer subregion we use an
appropriate model with vanishing viscosity. We derive correct interface conditions at
the artificial interface boundary and suggest an iterative procedure for the solution to
the coupled problem.

In the literature one can find several approaches for similar problems. In [8] the coupling between the Navier-Stokes equations for unsteady incompressible viscous flows with the Laplace equation modeling inviscid incompressible potential flows is done through a domain decomposition procedure with overlapping.

For a simplified model problem a nonlinear perturbation of the original Navier-Stokes equations is introduced in [9]. This method ensures smooth coupling, and the choice of the interface boundary on the basis of experience and physical considerations is not necessary.

We use some theoretical means developed by A. Quarteroni et al for the investigation of the coupling of elliptic and hyperbolic equations [12, 13, 17, 21] and for compressible flows [15].

The full-length proofs of the presented results are contained in [22].


## THE REGULARIZED PROBLEM

Starting point of our considerations is a generalized Stokes problem for an incompressible viscous flow in a two dimensional domain $\Omega$ around an obstacle:

$$
\begin{aligned}
k\,\mathbf{z} - \mu\,\Delta\mathbf{z} + \mathbf{grad}\,r &= \mathbf{f} && \text{in} \quad \Omega \\
\operatorname{div}\mathbf{z} &= 0 && \text{in} \quad \Omega \\
\mathbf{z} &= \mathbf{0} && \text{on} \quad \Gamma_b \\
\mathbf{z} &= \mathbf{g}_\infty^- && \text{on} \quad \Gamma_\infty^-
\end{aligned}
\tag{1}
$$

$$+ \text{ natural boundary conditions on } \Gamma_\infty^+ .$$

By $\mathbf{z}$ we denote the velocity of the fluid, by $r$ the pressure and by $\mathbf{f}$ a known body force density, $k$ and $\mu$ are positive constants. At the boundary $\Gamma_b$ we suppose non-slip boundary conditions. The outer boundary $\Gamma_\infty$ consists of two disjoint parts. On $\Gamma_\infty^-$ an onset flow $\mathbf{g}_\infty^-$ is given. On the remaining part $\Gamma_\infty^+$ we have no physical data. To overcome this difficulty we use the natural boundary condition on $\Gamma_\infty^+$ corresponding to a suitable variational formulation of our problem [18].

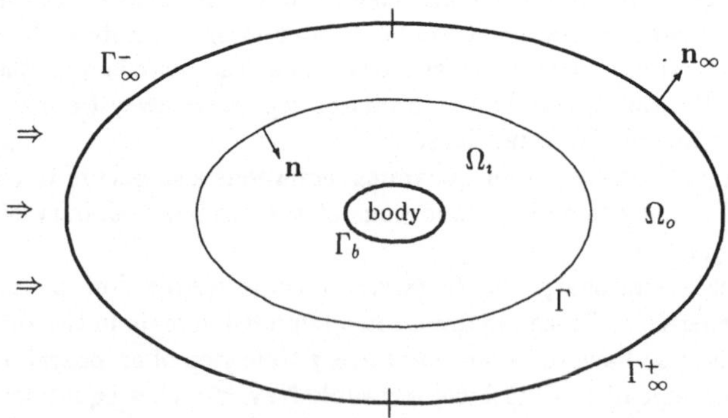


Figure 1: Notations

We introduce an artificial boundary $\Gamma$ that divides the domain $\Omega$ into an internal subdomain $\Omega_i$ near the body and an outer subdomain $\Omega_o$ far away from the obstacle. By $\mathbf{n}$ and $\mathbf{n}_\infty$ we denote the normal unit vectors to $\Gamma$ and $\Gamma_\infty$. We want to solve a generalized Stokes problem for an incompressible flow in the inner subdomain and a corresponding problem with vanishing viscosity in the outer subdomain. Of course we need suitable conditions at the different parts of the boundary.

To get these conditions as well as the coupled problem in a natural way by a vanishing viscosity argument we consider a regularized problem and the behaviour of the solution to a corresponding weak formulation for vanishing viscosity in $\Omega_o$:

$$
\begin{aligned}
k\,\mathbf{z}_\varepsilon - \mu_\varepsilon\,\Delta\mathbf{z}_\varepsilon + \mathbf{grad}\,r_\varepsilon &= \mathbf{f} && \text{in} \quad \Omega \\
\operatorname{div}\mathbf{z}_\varepsilon &= 0 && \text{in} \quad \Omega \\
\mathbf{z}_\varepsilon &= \mathbf{0} && \text{on} \quad \Gamma_b \\
\mathbf{z}_\varepsilon &= \mathbf{g}_\infty^- && \text{on} \quad \Gamma_\infty^-
\end{aligned}
\tag{2}
$$

$$+ \text{ natural boundary conditions on } \Gamma_\infty^+$$

with

$$
\mu_\varepsilon = \left\{ \begin{array}{ll} \mu & \text{in} \ \Omega_i \\ \varepsilon & \text{in} \ \Omega_o \end{array} \right. \ ,\ 
\mathbf{z}_\varepsilon = \left\{ \begin{array}{ll} \mathbf{w}_\varepsilon & \text{in} \ \Omega_i \\ \mathbf{u}_\varepsilon & \text{in} \ \Omega_o \end{array} \right. \ ,\ 
r_\varepsilon = \left\{ \begin{array}{ll} p_\varepsilon & \text{in} \ \Omega_i \\ q_\varepsilon & \text{in} \ \Omega_o \end{array} \right. .
\tag{3}
$$

We suppose that the boundaries are at least Lipschitz continuous and furthermore $\mathbf{f} \in \mathbf{L}^2(\Omega)$. For the given onset flow $\mathbf{g}_\infty^-$ at $\Gamma_\infty^-$ we assume:

$$
\exists\ \mathbf{g}_\infty \in \mathbf{H}^{1/2}(\Gamma_\infty) : \mathbf{g}_{\infty|\Gamma_\infty^-} = \mathbf{g}_\infty^- \quad \text{and} \quad \int_{\Gamma_\infty} \mathbf{g}_\infty \cdot \mathbf{n}_\infty\,ds = 0 .
\tag{4}
$$

This assumption is necessary for the solvability of our problem.

Furthermore, we use the abbreviations

$$
a_\varepsilon(\mathbf{z}, \mathbf{v}) = \mu_\varepsilon \sum_{j=1}^{2} (\mathbf{grad}\,z_j\,,\,\mathbf{grad}\,v_j)_{0,\Omega}, \qquad \mathbf{z}, \mathbf{v} \in \mathbf{H}^1(\Omega),
\tag{5}
$$

$$
\mathbf{H}(\operatorname{div};\Omega_o) = \left\{ \mathbf{v} \in \mathbf{L}^2(\Omega_o) : \operatorname{div}\mathbf{v} \in L^2(\Omega_o) \right\} ,
\tag{6}
$$

$$
\mathbf{V}_0(\Omega) = \left\{ \mathbf{v} \in \mathbf{H}^1(\Omega) : \operatorname{div}\mathbf{v} = 0,\ \mathbf{v}_{|\Gamma_\infty^-} = \mathbf{0},\ \mathbf{v}_{|\Gamma_b} = \mathbf{0} \right\} ,
\tag{7}
$$

$$
\mathbf{V}_1(\Omega) = \left\{ \mathbf{v} \in \mathbf{H}^1(\Omega) : \operatorname{div}\mathbf{v} = 0,\ \mathbf{v}_{|\Gamma_\infty^-} = \mathbf{g}_\infty^-,\ \mathbf{v}_{|\Gamma_b} = \mathbf{0} \right\} ,
\tag{8}
$$

$$
\mathbf{H}_{00}^{1/2}(\Gamma_\infty^+) = \left\{ \mathbf{g} \in \mathbf{L}^2(\Gamma_\infty^+) : \exists\,\mathbf{w} \in \mathbf{H}^{1/2}(\Gamma_\infty) \text{ with } \mathbf{w}_{|\Gamma_\infty^+} = \mathbf{g},\ \mathbf{w}_{|\Gamma_\infty^-} = 0 \right\},
\tag{9}
$$

$$
\mathbf{H}_{00,*}^{1/2}(\Gamma_\infty^+) = \left\{ \mathbf{v} \in \mathbf{H}_{00}^{1/2}(\Gamma_\infty^+) : \int_{\Gamma_\infty^+} \mathbf{v} \cdot \mathbf{n}_\infty\,ds = 0 \right\} \qquad (\text{see } [6]).
\tag{10}
$$

In general, the symbol $X'$ denotes the dual space of the space $X$. In some special cases we use different notations for dual spaces, for example $H_0^1(\Omega)' = H^{-1}(\Omega)$. Spaces of vector-valued functions and vector-valued functions itself are denoted by boldface type. By $\mathbf{grad}\,\mathbf{u}_\varepsilon$ we denote the matrix whose rows are the gradients of the components of $\mathbf{u}_\varepsilon$ and by $\mathbf{E}$ the identity matrix in $\mathbf{R}^2$, $(.\,,.)_{0,\Omega}$ is the inner product in $L^2(\Omega)$ and $\mathbf{L}^2(\Omega)$.

Now, let us consider the following variational problem:

$$\text{Find a function } \mathbf{z}_\varepsilon \in \mathbf{V}_1(\Omega)$$
$$\text{with} \tag{11}$$
$$k\,(\mathbf{z}_\varepsilon, \mathbf{v})_{0,\Omega} + a_\varepsilon(\mathbf{z}_\varepsilon, \mathbf{v}) = (\mathbf{f}, \mathbf{v})_{0,\Omega}, \qquad \forall\,\mathbf{v} \in \mathbf{V}_0(\Omega).$$

The first lemma states the relationship between the classical formulation (2) and the variational form (11) of the regularized problem in a distributional sense.

**Lemma 1:** For all $\varepsilon > 0$ problem (11) has a unique solution $\mathbf{z}_\varepsilon \in \mathbf{V}_1(\Omega)$. There exists a pressure $r_\varepsilon \in L^2(\Omega)$ unique up to an additive constant. For the corresponding restrictions to $\Omega_i$ and $\Omega_o$ we have :

$$k\,\mathbf{w}_\varepsilon - \mu\,\Delta\mathbf{w}_\varepsilon + \operatorname{grad} p_\varepsilon = \mathbf{f} \quad \text{in} \quad \mathbf{H}^{-1}(\Omega_i) \tag{12}$$

$$\operatorname{div}\mathbf{w}_\varepsilon = 0 \quad \text{in} \quad L^2(\Omega_i) \tag{13}$$

$$\mathbf{w}_{\varepsilon|\Gamma_b} = \mathbf{0} \quad \text{in} \quad \mathbf{H}^{1/2}(\Gamma_b) \tag{14}$$

$$k\,\mathbf{u}_\varepsilon - \varepsilon\,\Delta\mathbf{u}_\varepsilon + \operatorname{grad} q_\varepsilon = \mathbf{f} \quad \text{in} \quad \mathbf{H}^{-1}(\Omega_o) \tag{15}$$

$$\operatorname{div}\mathbf{u}_\varepsilon = 0 \quad \text{in} \quad L^2(\Omega_o) \tag{16}$$

$$\mathbf{u}_{\varepsilon|\Gamma_\infty^-} = \mathbf{g}_\infty^- \quad \text{in} \quad \mathbf{H}^{1/2}(\Gamma_\infty^-) \tag{17}$$

$$(-\varepsilon\operatorname{grad}\mathbf{u}_\varepsilon + q_\varepsilon\mathbf{E})\cdot\mathbf{n}_\infty = 0 \quad \text{in} \quad \mathbf{H}_{00,*}^{1/2}(\Gamma_\infty^+)' \tag{18}$$

$$(-\mu\operatorname{grad}\mathbf{w}_\varepsilon + p_\varepsilon\mathbf{E})\cdot\mathbf{n} = (-\varepsilon\operatorname{grad}\mathbf{u}_\varepsilon + q_\varepsilon\mathbf{E})\cdot\mathbf{n} \quad \text{in} \quad \mathbf{H}^{-1/2}(\Gamma) \tag{19}$$

$$\mathbf{w}_{\varepsilon|\Gamma} = \mathbf{u}_{\varepsilon|\Gamma} \quad \text{in} \quad \mathbf{H}^{1/2}(\Gamma)\,. \tag{20}$$

For the proof of this lemma [22] we used the Lax-Milgram Theorem, De Rham's Theorem, trace theorems and the generalized Stokes formula [2, 4, 5].

For sufficiently smooth velocity and pressure restrictions to $\Omega_0$ one can show that the natural boundary condition (18) may be interpreted in a classical way:

$$(-\varepsilon\operatorname{grad}\mathbf{u}_\varepsilon + q_\varepsilon\mathbf{E})\cdot\mathbf{n}_\infty = c_\varepsilon\cdot\mathbf{n}_\infty \quad \text{at } \Gamma_\infty^+, \qquad c_\varepsilon : \text{open constant}\,. \tag{21}$$

The interface condition (19) resembles a continuity condition for the stress vector. Both sides of this condition have the form of the left hand side of the natural boundary condition (18). However, the interface condition is a condition in the space $\mathbf{H}^{-1/2}(\Gamma)$ whereas the natural boundary condition holds in $\mathbf{H}_{00,*}^{1/2}(\Gamma_\infty^+)'$. The second interface condition (20) may be interpreted as a continuity condition for the velocity.

Now let us suppose that the functions $r_{c,\varepsilon}$ are just those unique pressures that have a prescribed common mean value $c$, for all $\varepsilon$.


## ASYMPTOTIC ANALYSIS

The next lemma describes the behaviour of the restrictions of the velocities $\mathbf{z}_\varepsilon$ and of the pressures $r_{c,\varepsilon}$ to $\Omega_i$ and $\Omega_o$ for vanishing $\varepsilon$.

**Lemma 2:** There exist functions $\mathbf{w} \in \mathbf{H}^1(\Omega_i), p \in L^2(\Omega_i), \mathbf{u} \in \mathbf{H}(\operatorname{div};\Omega_o), q \in L^2(\Omega_o)$ such that for $\varepsilon \to 0$, possibly taking subsequences,

$$\mathbf{w}_\varepsilon \rightharpoonup \mathbf{w} \quad \text{in} \quad \mathbf{H}^1(\Omega_i), \quad \operatorname{div}\mathbf{w} = 0, \tag{22}$$

$$\mathbf{u}_\varepsilon \rightharpoonup \mathbf{u} \quad \text{in} \quad \mathbf{H}(\operatorname{div};\Omega_o), \quad \operatorname{div}\mathbf{u} = 0, \tag{23}$$

$$r_{c,\varepsilon|\Omega_i} \rightharpoonup p \quad \text{in} \quad L^2(\Omega_i), \tag{24}$$

$$r_{c,\varepsilon|\Omega_o} \rightharpoonup q \quad \text{in} \quad L^2(\Omega_o), \quad q \in H^1(\Omega_o). \tag{25}$$

The proof of these propositions is based on a priori estimates for the restrictions of the solution to (11) uniform with respect to $\varepsilon$, on the possibility to extract a weakly convergent subsequence from every bounded sequence in a reflexive Banach space and on general properties of weakly convergent sequences [1]. Obviously, Lemma 2 only states the existence of limit functions and not their uniqueness. The pressure limit functions depend on $c$.

The next lemma deals with the properties of the limit functions on the boundaries $\Gamma_b, \Gamma$ and $\Gamma_\infty$.

**Lemma 3:** The functions $\mathbf{w} \in \mathbf{H}^1(\Omega_i)$, $\mathbf{u} \in \mathbf{H}(\mathrm{div}; \Omega_o)$, $p \in L^2(\Omega_i)$ and $q \in H^1(\Omega_o)$ from Lemma 2 fulfil the following boundary and interface conditions

$$\mathbf{w}_{|\Gamma_b} = \mathbf{0}, \qquad \text{in} \quad \mathbf{H}^{1/2}(\Gamma_b), \tag{26}$$

$$(-\mu \, \mathrm{grad}\, \mathbf{w} + p \cdot \mathbf{E}) \cdot \mathbf{n} = q \cdot \mathbf{n}, \qquad \text{in} \quad \mathbf{H}^{-1/2}(\Gamma), \tag{27}$$

$$\mathbf{w} \cdot \mathbf{n} = \mathbf{u} \cdot \mathbf{n}, \qquad \text{in} \quad H^{-1/2}(\Gamma), \tag{28}$$

$$\mathbf{u} \cdot \mathbf{n}_\infty = \mathbf{g}_\infty^- \cdot \mathbf{n}_\infty, \qquad \text{in} \quad H_{00}^{1/2}(\Gamma_\infty^-)', \tag{29}$$

$$q \cdot \mathbf{n}_\infty = 0, \qquad \text{in} \quad \mathbf{H}_{00,*}^{1/2}(\Gamma_\infty^+)'. \tag{30}$$

This lemma is proved using the fact that for a linear bounded operator $T \in L(X, Y)$, $X$ and $Y$ Banach spaces, the property $x_n \rightharpoonup x$ in $X$ implies $T(x_n) \rightharpoonup T(x)$ in $Y$.

We note that the transitions of the velocity and of the pressure at the coupling boundary are not continuous, in general. The first interface condition (27) states particularly that the jump of the pressure is proportional to the value of the viscosity at the coupling boundary. This relation looks similar as the corresponding condition (19) in the regularized problem, but the outer velocity does not occur on the right hand side of (27). The second interface condition (28) expresses the continuity of the normal component of the velocity in distinction to the continuity condition (20) for the velocity in the regularized problem.

The Dirichlet condition (17) on $\Gamma_\infty^-$ turns into condition (29) for the normal component for the outer velocity.

From the natural boundary condition (18) in the regularized problem we get relation (30). For $\Gamma_\infty^+$ sufficiently smooth, we can show that (30) is equivalent to a Dirichlet condition:

$$q_{|\Gamma_\infty^+} = \text{constant} \qquad \text{on} \quad \Gamma_\infty^+. \tag{31}$$

One can choose the common mean value $c$ of the pressure functions $r_{c,\varepsilon}$ in such a way, that (31) is just a homogeneous Dirichlet condition.

All these statements result in a theorem, that describes two coupled limit problems fulfilled by the limit pressures and by the inner velocity limit function.

**Theorem 4:** For $\mathbf{f} \in \mathbf{H}(\mathrm{div}; \Omega_o)$ and sufficiently smooth $\Gamma_\infty^+$, the functions $\mathbf{w} \in \mathbf{H}^1(\Omega_i)$, $p \in L^2(\Omega_i)$ and $q \in H^1(\Omega_o)$ from Lemma 2 fulfil the following coupled boundary value problems:

Mixed boundary value problem for the Poisson equation in $\Omega_o$:

$$\Delta q = \mathrm{div}\, \mathbf{f} \qquad \text{in} \quad D(\Omega_o)' \tag{32}$$

$$\frac{\partial q}{\partial \mathbf{n}_\infty} = (\mathbf{f} - k\, \mathbf{g}_\infty^-) \cdot \mathbf{n}_\infty \qquad \text{in} \quad H_{00}^{1/2}(\Gamma_\infty^-)' \tag{33}$$

$$q_{|\Gamma_\infty^+} = 0 \qquad \text{in} \quad H^{1/2}(\Gamma_\infty^+). \tag{34}$$

Coupling conditions at the interface boundary:

$$\frac{\partial q}{\partial \mathbf{n}} \;=\; (\mathbf{f} - k\,\mathbf{w})\cdot \mathbf{n} \qquad\qquad \text{in } H^{-1/2}(\Gamma) \tag{35}$$

$$(-\mu\,\mathrm{grad}\,\mathbf{w} + p\cdot \mathbf{E})\cdot \mathbf{n} \;=\; q\cdot \mathbf{n} \qquad\qquad \text{in } \mathbf{H}^{-1/2}(\Gamma)\,. \tag{36}$$

Generalized Stokes problem in $\Omega_i$:

$$k\,\mathbf{w} - \mu\,\Delta\mathbf{w} + \mathrm{grad}\,p \;=\; \mathbf{f} \qquad\qquad \text{in } \mathbf{D}(\Omega_i)' \tag{37}$$

$$\mathrm{div}\,\mathbf{w} \;=\; 0 \qquad\qquad \text{in } L^2(\Omega_i) \tag{38}$$

$$\mathbf{w}_{|\Gamma_b} \;=\; \mathbf{0} \qquad\qquad \text{in } \mathbf{H}^{1/2}(\Gamma_b)\,. \tag{39}$$

The outer velocity does not occur in this theorem, because it is possible to eliminate it by the difference of the force and the outer pressure. The additional assumption $\mathbf{f} \in \mathbf{H}(\mathrm{div},\Omega_o)$ is required for the definition of the right hand sides of relations (32), (33) and (35).

## ITERATIVE PROCEDURE

Until here we have only shown the existence of solutions to the whole limit problem. To prove uniqueness of the limit functions as well as of the solution to the limit problem and to provide a theoretical approach for the computation of these limit functions we establish the following iterative algorithm. We introduce the space

$$\mathbf{H}_*^{1/2}(\Gamma) = \left\{ \mathbf{v} \in \mathbf{H}^{1/2}(\Gamma) : \int_\Gamma \mathbf{v}\cdot \mathbf{n}\,ds = 0 \right\}. \tag{40}$$

**Step 1:**
Choose an initial function $\mathbf{s}^1 \in \mathbf{H}_*^{1/2}(\Gamma)$. Set j=1.
**Step 2:**
Solve a mixed boundary value problem for the Poisson equation in $\Omega_o$ using $\mathbf{s}^j \in \mathbf{H}_*^{1/2}(\Gamma)$ from the preceding step:

$$\Delta q^j \;=\; \mathrm{div}\,\mathbf{f} \qquad\qquad \text{in } D(\Omega_o)' \tag{41}$$

$$\frac{\partial q^j}{\partial \mathbf{n}_\infty} \;=\; (\mathbf{f} - k\,\mathbf{g}_\infty^-)\cdot \mathbf{n}_\infty \quad \text{in } H_{00}^{1/2}(\Gamma_\infty^-)' \tag{42}$$

$$q^j_{|\Gamma_\infty^+} \;=\; 0 \qquad\qquad \text{in } H^{1/2}(\Gamma_\infty^+) \tag{43}$$

$$\frac{\partial q^j}{\partial \mathbf{n}} \;=\; (\mathbf{f} - k\,\mathbf{s}^j)\cdot \mathbf{n} \qquad \text{in } H^{-1/2}(\Gamma)\,. \tag{44}$$

**Step 3:**
Solve a mixed boundary value problem for the generalized Stokes equations in $\Omega_i$ using $q^j \in H^1(\Omega_o)$ from the preceding step:

$$k\,\mathbf{w}^j - \mu\,\Delta\mathbf{w}^j + \mathrm{grad}\,p^j \;=\; \mathbf{f} \qquad\qquad \text{in } \mathbf{D}(\Omega_i)' \tag{45}$$

$$\mathrm{div}\,\mathbf{w}^j \;=\; 0 \qquad\qquad \text{in } L^2(\Omega_i) \tag{46}$$

$$\mathbf{w}^j_{|\Gamma_b} \;=\; \mathbf{0} \qquad\qquad \text{in } \mathbf{H}^{1/2}(\Gamma_b) \tag{47}$$

$$(-\mu\,\mathrm{grad}\,\mathbf{w}^j + p^j\cdot \mathbf{E})\cdot \mathbf{n} \;=\; q^j\cdot \mathbf{n} \qquad \text{in } \mathbf{H}^{-1/2}(\Gamma)\,. \tag{48}$$

**Step 4:**
Compute $s^{j+1} \in \mathbf{H}_*^{1/2}(\Gamma)$ using $s^j \in \mathbf{H}_*^{1/2}(\Gamma)$ and $\mathbf{w}^j \in \mathbf{H}^1(\Omega_i)$ by the following formula with any $0 < \theta < 1$:

$$s^{j+1} = R_\theta(s^j) = \theta\, \gamma_{\mathfrak{t},\Gamma}(\mathbf{w}^j) + (1-\theta)\, s^j. \tag{49}$$

**Step 5:**
Stop the iteration procedure if the $\mathbf{H}_*^{1/2}$-norm of the difference $(s^{j+1} - s^j)$ is small enough. Otherwise set $j = j + 1$ and go to Step 2.

Concerning this iteration, we have to prove the following statements:

- Existence and uniqueness of the solution to the mixed boundary value problem for the Poisson equation (Step 2).

- Existence and uniqueness of the solution to the mixed boundary value problem for the generalized Stokes equations (Step 3).

- Convergence of the sequence $\{s^j\}_{j=1}^{\infty}$ in $\mathbf{H}_*^{1/2}(\Gamma)$ for $j \to \infty$.

In order to be able to write the boundary conditions containing the body force density in an integral form instead of a functional form, we need $\mathbf{f} \in \mathbf{H}^1(\Omega_o)$.

With this assumption, we consider a first variational problem for a given function $s \in \mathbf{H}_*^{1/2}(\Gamma)$:

Find a function $q^s \in H_0(\Omega_o) = \left\{ \varphi \in H^1(\Omega_o) : \varphi_{|\Gamma_\infty^\pm} = 0 \right\}$ such that

$$(\mathbf{grad}\, q^s, \mathbf{grad}\, \varphi)_{0,\Omega_o} = -(\operatorname{div} \mathbf{f}, \varphi)_{0,\Omega_o} + \int_\Gamma (\mathbf{f} - k\,s) \cdot \mathbf{n} \cdot \varphi_{|\Gamma}\, ds + \tag{50}$$

$$+ \int_{\Gamma_\infty^-} (\mathbf{f} - k\,\mathbf{g}_\infty^-) \cdot \mathbf{n}_\infty \cdot \varphi_{|\Gamma_\infty^-}\, ds \quad \forall \varphi \in H_0(\Omega_o).$$

**Lemma 5:** Problem (50) has a unique solution $q^s \in H_0(\Omega_o)$. This solution is the unique solution to the mixed boundary value problem for the Poisson equation from Step 2 of the iteration process for $s^j = s$.

For a given function $q \in H^1(\Omega_o)$ and with the usual suppositions we establish the variational formulation of the boundary value problem from Step 3:

Find a function $\quad \mathbf{w}^q \in \mathbf{W}_0(\Omega_i) = \left\{ \mathbf{v} \in \mathbf{H}^1(\Omega_i) : \operatorname{div} \mathbf{v} = 0,\; \mathbf{v}_{|\Gamma_b} = \mathbf{0} \right\}$

$$\text{such that} \quad k\,(\mathbf{w}^q, \mathbf{v})_{0,\Omega_i} + \mu \sum_{j=1}^{2} (\mathbf{grad}\, w_j^q,\, \mathbf{grad}\, v_j)_{0,\Omega_i} = \tag{51}$$

$$= (\mathbf{f}, \mathbf{v})_{0,\Omega_i} + \int_\Gamma q_{|\Gamma} \cdot \mathbf{v}_{|\Gamma} \cdot \mathbf{n}\, ds \quad \forall \mathbf{v} \in \mathbf{W}_0(\Omega_i).$$

**Lemma 6:** Problem (51) has a unique solution $\mathbf{w}^q \in \mathbf{W}_0(\Omega_i)$. Furthermore, there exists a unique pressure $p^q \in L^2(\Omega_i)$ such that the pair $\mathbf{w}^q \in \mathbf{W}_0(\Omega_i)$, $p^q \in L^2(\Omega_i)$ is the unique solution to the mixed boundary value problem for the Stokes equations from Step 3 of our iteration for $q^j = q$.

Consequently our iteration process generates a well defined sequence $s^j$ in $\mathbf{H}_*^{1/2}(\Gamma)$ for every initial function $s^1$.

Furthermore, we show that the operators $R_\theta : \mathbf{H}_*^{1/2}(\Gamma) \to \mathbf{H}_*^{1/2}(\Gamma)$ from Step 4 have the same fixed points for all $\theta$ and that each solution of the limit problem (32)-(39) corresponds to a fixed point of $R_\theta$ for all $\theta$.

Now the crucial point is, that for sufficiently small $\theta$ the operators $R_\theta$ are contractions in a special norm in $\mathbf{H}_*^{1/2}(\Gamma)$. This norm is defined using an extension operator from $\mathbf{H}_*^{1/2}(\Gamma)$ to $\mathbf{W}_0(\Omega_i)$. To define this operator, we consider the following variational problem:

$$\text{For a given } \mathbf{s} \in \mathbf{H}_*^{1/2}(\Gamma) \text{ find a function } \mathbf{y} \in \mathbf{W}_0(\Omega_i) \text{ with } \mathbf{y}_{|\Gamma} = \mathbf{s}$$

$$\text{and} \quad k\,(\mathbf{y},\mathbf{v})_{0,\Omega_i} + \mu \sum_{j=1}^{2} (\mathbf{grad}\, y_j\,,\, \mathbf{grad}\, v_j)_{0,\Omega_i} = 0\,, \tag{52}$$

$$\forall\, \mathbf{v} \in \mathbf{H}_0^1(\Omega_i) \quad \text{with} \quad \operatorname{div} \mathbf{v} = 0\,.$$

It is easy to prove that this problem has a unique solution which represents a special extension of $\mathbf{s} \in \mathbf{H}_*^{1/2}(\Gamma)$. We denote it by $\mathbf{y} = E(\mathbf{s})$. Problem (52) is the weak formulation of a generalized Stokes problem in $\Omega_i$ with non-homogeneous Dirichlet data at $\Gamma$, homogeneous Dirichlet data at $\Gamma_b$ and vanishing body forces.

For $\mathbf{s}, \mathbf{t} \in \mathbf{H}_*^{1/2}(\Gamma)$ we set

$$(\mathbf{s},\mathbf{t})_* = k\,(E(\mathbf{s}), E(\mathbf{t}))_{0,\Omega_i} + \mu \sum_{j=1}^{2} (\mathbf{grad}\, E(\mathbf{s})_j\,,\, \mathbf{grad}\, E(\mathbf{t})_j)_{0,\Omega_i}\,. \tag{53}$$

**Lemma 7:**

1. Relation (53) defines an inner product $(\cdot,\cdot)_*$ and a corresponding norm $\|\cdot\|_*$ in $\mathbf{H}_*^{1/2}(\Gamma)$, whereby $\|\cdot\|_*$ and the usual $\mathbf{H}^{1/2}(\Gamma)$-norm are equivalent in the subspace $\mathbf{H}_*^{1/2}(\Gamma) \subset \mathbf{H}^{1/2}(\Gamma)$.

2. For an appropriate $\theta$-interval contained in $(0,1)$, the operators $R_\theta$ are contractions with regard to the $\|\cdot\|_*$-norm in $\mathbf{H}_*^{1/2}(\Gamma)$. The contractivity constant depends on $k$, $\mu$ and $\theta$. It tends to one from below for decreasing $\mu$.

Using Banach's fixed point theorem contractivity provides uniqueness of the limit functions as well as of the solution to the limit problem, and the iteration converges for suitable $\theta$:

**Theorem 8:** For each $\mathbf{s}^1 \in \mathbf{H}_*^{1/2}(\Gamma)$ and for sufficiently small $\theta$ our iteration process creates a sequence $\{\mathbf{s}^j\}_{j=1}^{\infty}$ with

$$\mathbf{s}^j \;\to\; \mathbf{w}_{|\Gamma} \quad \text{in} \quad \mathbf{H}_*^{1/2}(\Gamma)\,, \tag{54}$$

$$\mathbf{w}^j \;\to\; \mathbf{w}\,, \quad \text{in} \quad \mathbf{H}^1(\Omega_i)\,, \tag{55}$$

$$q^j \;\to\; q\,, \quad \text{in} \quad H^1(\Omega_o)\,. \tag{56}$$

The limit functions $\mathbf{w} \in \mathbf{H}^1(\Omega_i)$, $\mathbf{u} \in \mathbf{H}(\operatorname{div}; \Omega_o)$, $p \in L^2(\Omega_i)$ and $q \in H^1(\Omega_o)$ from Lemma 2 are the only solution to the coupled boundary value problem (32)-(39).

Thus we have proved all our statements. Of course these results are preliminary and this model is a first attempt to analyse coupling problems. It would be an interesting task to investigate the following open problems:

1. Estimates for the difference between the solution to the full Stokes problem and the solutions to the coupled problem as well as the influence of the location of the artificial boundary on these estimates

2. Numerical realization of the iteration algorithm, consisting of the successive solution to a mixed boundary value problem for the Poisson equation in the outer subregion using the variational formulation (50) and the solution to the mixed boundary value problem for the Stokes equations in the inner subregion by means of the variational formulation (51)

3. Generalization to the Stokes equations with an additional convective term of the form

$$\sum_{j=1}^{2} a_j(x) \frac{\partial \mathbf{z}}{\partial x_j} \quad , \qquad \mathbf{a} = (a_1, a_2) : \Omega \to \mathbf{R}^2 \,,$$

for a given divergence-free function $\mathbf{a}$.

For the second task we need an effective discretization method where the speed of convergence of the iteration does not depend on the discretization parameters. Furthermore, direct application of the weak formulation (51) in the frame of a finite element procedure would require divergence-free elements for the velocity in the inner subdomain.

## REFERENCES


[1] Zeidler, E.: "Vorlesungen über nichtlineare Funktionalanalysis II: Monotone Operatoren", B. G. Teubner Leipzig 1977.

[2] Temam, R.: "Navier-Stokes equations: theory and numerical analysis", North Holland, Amsterdam, Oxford 1984.

[3] Glowinski, R.: "Numerical methods for nonlinear variational problems", Springer-Verlag, Berlin, Heidelberg, New York, Tokyo 1984.

[4] Girault, V. and Raviart, P.-A.: "Finite element methods for the Navier-Stokes equations", Springer-Verlag, Berlin, Heidelberg, New York, Tokyo 1986.

[5] Hackbusch, W.: "Theorie und Numerik elliptischer Differentialgleichungen", B. G. Teubner Stuttgart 1986.

[6] Dautray, R., and Lions, J.-L.: "Mathematical analysis and numerical methods for science and technology, volume 2", Springer-Verlag, Berlin, Heidelberg, New York, London, Paris, Tokyo 1988.

[7] Cahouet, J.: "On some difficulties occurring in the simulation of incompressible fluid flows by domain decomposition", In: Chan, T. et al (eds.): Domain decomposition methods for partial differential equations, vol. I. SIAM, Philadelphia (1988), pp. 313-332.

[8] Dinh, Q.V., Glowinski, R., Periaux, J., and Terrasson, G.: "On the coupling of viscous and inviscid models for incompressible fluid flows via domain decomposition", In: Chan, T. et al (eds.): Domain decomposition methods for partial differential equations, vol. I. SIAM, Philadelphia (1988), pp. 350-369.

[9] Brezzi, F., Canuto, C., and Russo, A.: "A self-adaptive formulation for the Euler/Navier-Stokes coupling", Comp. Meth. Appl. Mech. Eng., 73 (1989) pp. 317-330.

[10] Pasciak, J.E.: "Two domain decomposition techniques for Stokes problems", In: Chan, T. et al (eds.): Domain decomposition methods for partial differential equations, vol. II. SIAM, Philadelphia (1989), pp. 419-430.

[11] Quarteroni, A.: "Domain decomposition algorithms for the Stokes equations", In: Chan, T. et al (eds.): Domain decomposition methods for partial differential equations, vol. II. SIAM, Philadelphia (1989), pp. 431-442.

[12] Gastaldi, F., Quarteroni, A.: "On the coupling of hyperbolic and parabolic systems: analytical and numerical approach", Appl. Numer. Math., 6 (1989) pp. 3-31.

[13] Gastaldi, F., Quarteroni, A. and Sacchi Landriani, G.: "On the coupling of two dimensional hyperbolic and elliptic equations: analytical and numerical approach", In: Chan, T. et al (eds.): Domain decomposition methods for partial differential equations, vol. III. SIAM, Philadelphia (1990), pp. 22-63.

[14] Quarteroni, A.: "Domain decomposition and parallel processing for the numerical solution of partial differential equations", Mathematics for Industry, 1 (1991) pp. 75-118.

[15] Quarteroni, A., Sacchi Landriani, G., and Valli, A.: "Coupling of viscous and inviscid Stokes equations via a domain decomposition method for finite elements", Numer. Math., 59 (1991) pp. 831-859.

[16] Quarteroni, A., Valli, A.: "Domain decomposition for a generalized Stokes problem", In: Manley, J. et al (eds.): Proceedings of the third European conference on mathematics in industry, Kluwer, Dortrecht, B.G.Teubner, Stuttgart (1991), pp. 59-74.

[17] Quarteroni, A., Pasquarelli, F., and Valli, A.: "Heterogeneous domain decomposition : principles, algorithms, applications", Preprint Nr. 40/P, Milano 1991.

[18] Heywood, J. G., Rannacher, R., and Turek, S.: "Artificial boundaries and flux and pressure conditions for the incompressible Navier-Stokes equations", Preprint Nr. 681, SFB 123, Universität Heidelberg 1992.

[19] Hebeker, F.-K., and Wilde, P.: "On missing boundary conditions with unsteady incompressible Navier-Stokes flow", Math. Meth. in the Appl. Sci., 15 (1992) pp. 421-432.

[20] Borchers, W.: "A Fourier spectral method for incompressible viscous flows past obstacles", In: Hirschel, E.H. et al (eds.): Flow simulation with high-performance computers I, Vieweg, Braunschweig (1993).

[21] Frati, A., Pasquarelli, F., and Quarteroni, A.: "Spectral approximation to advection-diffusion problems by the fictitious interface method", J. Comp. Phys., 107 (1993) pp. 201-212.

[22] Schenk, K., and Hebeker, F.-K.: "Coupling of two dimensional viscous and inviscid incompressible Stokes equations", Preprint, Universität Heidelberg Dezember 1993.

# ON THE ORDER OF TWO NONCONFORMING FINITE ELEMENT APPROXIMATIONS OF UPWIND TYPE FOR THE NAVIER–STOKES EQUATIONS


Friedhelm Schieweck

Institute for Analysis and Numerical Mathematics, University Magdeburg

PSF 4120, D-39016 Magdeburg, Germany



## Abstract

We study the discretization error for a finite element approximation of the stationary incompressible Navier–Stokes equations in the primitive variables $u$ and $p$ for velocity and pressure, respectively. Two types of nonconforming finite element pairs are considered: the triangular $P_1/P_0$ element of Crouzeix/Raviart and the quadrilateral $Q_1^{\text{rot}}/Q_0$ element of Rannacher/Turek. In order to stabilize the numerical scheme in the case of high Reynolds numbers, we use an upwind discretization of the convective term. For the corresponding $P_1/P_0$ discretization, an optimal error estimate in the two- and three-dimensional case is given. We discuss a numerical test problem which gives an impression about the quantity that the errors may have and about cases that are not covered by the theory.


# 1 Introduction

We are interested in the numerical solution of the stationary incompressible Navier–Stokes equations in primitive variables

$$- \nu \triangle u + (u \cdot \nabla)u + \nabla p = f, \quad \nabla \cdot u = 0 \quad \text{in } \Omega\,, \qquad u = 0 \quad \text{on } \partial\Omega \tag{1}$$

where $u$ and $p$ denote the unknown velocity and pressure field, respectively, in a bounded convex polyhedral domain $\Omega \subset \mathbf{R}^d$ with $d \leq 3$, $f$ a given body force and $\nu = 1/Re$ with the *Reynolds number Re*. For the discretization we consider two types of nonconforming finite element pairs - the $P_1/P_0$ element of Crouzeix/Raviart [1] and the rotated bilinear $Q_1^{\text{rot}}/Q_0$ element of Rannacher/Turek [3]. In order to ensure numerical stability also for the case of high Reynolds numbers we use an upwind discretization of the convective term. Both element pairs have shown to be attractive for practical computations [2],[7]. On the one hand they guarantee the Babuška-Brezzi stability uniformly with respect to the mesh size (where a sufficiently regular mesh has to be assumed for the $Q_1^{\text{rot}}/Q_0$ element) and on the other hand the total number of unknowns is relatively small compared to other stable element pairs of higher order. Furthermore, efficient and robust multigrid solvers can be created for both discretizations [2], [7].

The aim of this paper is to discuss the accuracy of these methods. We present an optimal theoretical result for the $P_1/P_0$ discretization and some experimental convergence results for the $Q_1^{\mathrm{rot}}/Q_0$ element which give an impression about the quantity of the discretization error and about its dependence on the Reynolds number.

## 2   Finite Element Approximation of Upwind Type

Using the function spaces $V:=(H_0^1(\Omega))^d$ and $Q:=L_0^2(\Omega)$ the weak formulation of (1) reads

$$\begin{aligned}
&\text{Find } (u,p) \in V \times Q \text{ such that}\\
&\nu\, a(u,v)+n(u,u,v)+b(p,v)+b(q,u) = (f,v) \quad \forall\, (v,q) \in V\times Q
\end{aligned} \tag{2}$$

where the bilinear and trilinear forms are defined by

$$a(u,v) := (\nabla u, \nabla v), \quad b(p,v) := -(p, \nabla\cdot v), \quad n(z,u,v) := ((z\cdot\nabla)u, v). \tag{3}$$

Let $\mathbf{T}_h$ be a regular decomposition of the domain $\Omega \subset \mathbf{R}^d$ into elements $K\in\mathbf{T}_h$ where the mesh parameter $h$ represents the maximum diameter of all elements $K\in\mathbf{T}_h$. We denote by $\Gamma_i$ the (d–1)–dimensional faces of the elements $K\in\mathbf{T}_h$ assuming inner faces for $i=1,\ldots,N$ and boundary faces $\Gamma_i\subset\partial\Omega$ for $i=N+1,\ldots,N+M$. $B_i$ is supposed to be the barycentre of $\Gamma_i$. Now we can define the discrete spaces $V_h\approx V$ and $Q_h\approx Q$ by

$$V_h := \left\{ \begin{array}{l} v_h \in (L^2(\Omega))^d \mid v_{h|K} \in (P(K))^d \ \ \forall K \in \mathbf{T}_h, \ \ v_h \text{ is continuous at}\\ B_i \ \ \forall i=1,\ldots,N \ \ \text{and} \ \ v_h(B_i)=0 \ \ \forall i=N+1,\ldots,N+M \end{array} \right\}, \tag{4}$$

$$Q_h := \{q_h \in L_0^2(\Omega) \mid q_{h|K} = \mathrm{const.}, \forall K \in \mathbf{T}_h\}. \tag{5}$$

Herein $P(K)$ is for the Crouzeix/Raviart element the set of all linear functions on $K$ and for the $Q_1^{\mathrm{rot}}/Q_0$ element the set

$$P(K) = Q_1^{\mathrm{rot}}(K) := \left\{\hat{q} \circ \psi_K^{-1} \mid \hat{q} \in \mathrm{span}\left(1, \hat{x}_i, \hat{x}_i^2 - \hat{x}_{i+1}^2, \ i = 1, ..., d\right)\right\}. \tag{6}$$

where $\psi_K : \hat{K} \to K$ denotes the multi-linear transformation between the reference element $\hat{K} = [-1,1]^d$ and the original element $K$. The degrees of freedom are the velocity values at the barycentres $B_i$ of the element faces $\Gamma_i$ and the pressure values within each element $K\in\mathbf{T}_h$. Note that a function $v\in V_h$ in general is discontinuous on the element faces $\Gamma_i$ which implies $V_h \not\subset V$. Therefore, we need for the discretization of the weak formulation (2) the elementwise defined bilinear forms

$$a_h(u_h,v_h) := \sum_{K\in\mathbf{T}_h} \int_K \nabla u_h \cdot \nabla v_h\, dx, \qquad b_h(q_h,v_h) := - \sum_{K\in\mathbf{T}_h} \int_K q_h \nabla\cdot v_h\, dx. \tag{7}$$

For the discretization of the trilinear form $n(\cdot,\cdot,\cdot)$ we do not use the elementwise version but apply an upwind technique proposed in [4]. Let $C_K$ be the barycentre of element $K$ and $S_{K,l}$ the d-dimensional pyramid contained in $K$ which has $\Gamma_l$ as its basis face and the point $C_K$ at the top. We denote by $\Lambda_l$ the set of all indices $k \neq l$ for which the nodes $B_k$ and $B_l$ belong to a common element $K$ and define in this case $\Gamma_{lk} := \partial S_{K,l} \cap \partial S_{K,k}$ as the common (d-1)-dimensional face of $S_{K,l}$ and $S_{K,k}$ (see Figure 1). For $k \in \Lambda_l$ we denote by $n_{lk}$ the outer unit normal on $\Gamma_{lk}$ with respect to $S_{K,l}$ where $K$ is the element

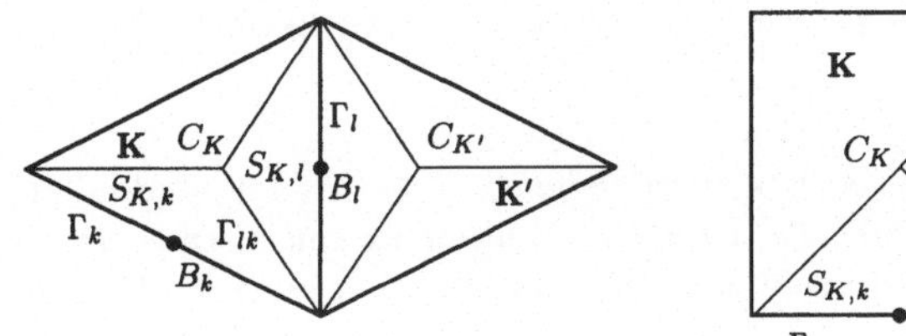 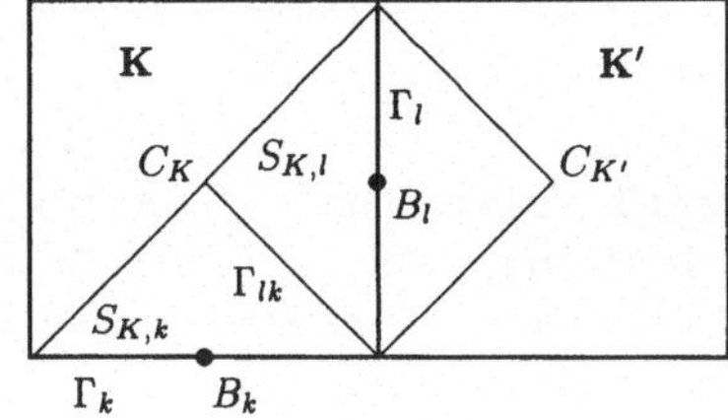

Figure 1: *illustration for the triangular and quadrilateral case*

that contains both nodes $B_l$ and $B_k$. Now, our upwind discretization of $n(\cdot,\cdot,\cdot)$ can be written as

$$\tilde{n}_h(z,u,v) := \sum_{l=1}^{N+M} \sum_{k\in\Lambda_l} \int_{\Gamma_{lk}} (z\cdot n_{lk})\, d\gamma\, (1 - \lambda_{lk}(z))\, \{\, u(B_k) - u(B_l)\,\}\, v(B_l) \qquad (8)$$

where the function $\lambda_{lk}(\cdot)$ is defined by

$$\lambda_{lk}(z) = \Phi(t) \quad \text{with} \quad t := \frac{1}{\nu}\int_{\Gamma_{lk}} z\cdot n_{lk}\, d\gamma\,. \qquad (9)$$

Possible choices for $\Phi(\cdot)$ which have been already used in practical computations (see [2],[7]) are

$$\Phi_1(t) \;=\; \begin{cases} 1\,, & \text{if } t \geq 0 \\ 0\,, & \text{if } t < 0 \end{cases} \qquad \text{and} \qquad \Phi_2(t) \;=\; \frac{1}{2} + \frac{t}{2(1 + |t|)} \qquad (10)$$

where $\Phi_1(\cdot)$ is called *simple* or *sharp upwind* and $\Phi_2(\cdot)$ *Samarskij upwind*.

Finally, our discrete Navier–Stokes problem reads:

$$\begin{aligned} &\text{Find } (u_h, p_h) \in V_h \times Q_h \text{ such that for all } (v_h, q_h) \in V_h \times Q_h \\ &\nu\, a_h(u_h, v_h) + \tilde{n}_h(u_h, u_h, v_h) + b_h(p_h, v_h) + b_h(q_h, u_h) = (f, v_h)\,. \end{aligned} \qquad (11)$$

## 3 An Error Estimate of Optimal Order

In this section, we will give a recent result on the discretization error of our upwind method applied to the Crouzeix/Raviart element. The corresponding error estimate is of optimal order and improves the results of some former work [4],[6].

We assume that the elements $K \in T_h$ are shape regular in the usual sense and quasi-uniform, i.e. $h/h_K \leq C$ for all $K \in T_h$ where $h_K$ denotes the diameter of the element $K$ and $h$ the maximum of all $h_K$ with $K \in T_h$. Let $\|\cdot\|_m$ be the usual norm in the Sobolev space $H^m(\Omega)$ or $(H^m(\Omega))^d$, respectively, and $\|\cdot\|_h$ the following discrete $H^1$-norm on $V + V_h$

$$\|v\|_h \;:=\; a_h(v,v)^{1/2}\,. \qquad (12)$$

For the weighting function $\Phi(\cdot)$ we need the assumptions:

$(A1) \quad \Phi(t) = 1 - \Phi(-t) \quad \text{and} \quad 0 \leq \Phi(t) \leq 1 \quad \forall\, t \neq 0$

$(A2) \quad \Phi(t) \geq \frac{1}{2} \quad \forall\, t \geq 0$

$(A3) \quad g(t) := t\Phi(t) \quad \text{is Lipschitz continuous.}$

These assumptions are satisfied for the *simple* and *Samarskij* upwind. Now, we can formulate the following result.

**Theorem 1** *Assume that (A1),(A2),(A3) are fulfilled, $f \in (L^2(\Omega))^d$ and $\nu \geq \nu_0$ with a sufficiently large $\nu_0 = \nu_0(\Omega, f) > 0$. Then both the continuous and discrete problem (2) and (11) have uniquely determined solutions $(u, p)$ and $(u_h, p_h)$, respectively. Under the additional regularity assumption $(u, p) \in (H^2(\Omega))^d \times H^1(\Omega)$ the error estimate*

$$\|u - u_h\|_h + \|p - p_h\|_0 \leq C h \tag{13}$$

*holds for space dimensions $d \leq 3$.*

For a detailed proof we refer to a forthcoming paper [5]. The estimate (13) shows that the discretization error is of optimal order for the velocity in the discrete $H^1$-norm and for the pressure in the $L^2$-norm. Moreover, we can derive from (13) an estimate for the velocity error in the $L^2$-norm of order $O(h)$. This is of course a nonoptimal result. However, for our upwind method, which looks (from the way of its construction) similar to other first order upwind schemes, one would not expect in general more than first order for the $L^2$-norm error of the velocity.

The $Q_1^{\mathrm{rot}}/Q_0$ element was analyzed for the Stokes problem by Rannacher and Turek [3]. Under a regularity assumption on the mesh (which is fulfilled for uniformly refined multilevel grids) they obtained optimal error estimates. An analysis for our upwind discretization (11) of the Navier–Stokes problem is not yet available. However, such an analysis (if it would exist) probably would have the same lack as for the Crouzeix/Raviart element which will be discussed in the next section.

# 4 Critical Evaluation of the Theory

Although the theoretical result in Theorem 1 looks quite nice it has an essential drawback from the practical point of view. This theory assumes that $\nu \geq \nu_0$ with a sufficiently large $\nu_0$ such that the more interesting cases of high Reynolds numbers are not covered. There are several open questions left from theory. How does the constant C in the estimate (13) depend on the Reynolds number $Re$ and how do the error norms $\|u - u_h\|_h$ and $\|p - p_h\|_0$ behave if we keep the mesh size $h$ fixed and increase $Re$ ? What is the order of these error norms with respect to $h$ if we choose a high Reynolds number which violates the theoretical assumption $\nu \geq \nu_0$ ?

In such a situation, where the existing theoretical convergence analysis cannot answer some important questions, we can try to make some *experimental convergence analysis* on the computer in order to get first answers at all. That means we take simple test examples where we can compute all interesting quantities. Then we can answer our questions for these examples. Concerning the discretization error we get a first impression about the real quantity of the error norms and about the numerical order of convergence.

# 5 Experimental Convergence Analysis

We consider as a test example the 2D Navier–Stokes problem (1) on the unit square $\Omega = (0,1)^2$ where we define the right hand side by $f := -\nu\Delta u + (u \cdot \nabla)u + \nabla p$ with the following prescribed exact solution

$$u = \begin{pmatrix} \psi_y \\ -\psi_x \end{pmatrix} \quad \text{with} \quad \psi(x,y) = x^2(x-1)^2 y^2 (y-1)^2, \qquad p(x,y) = x^3 + y^3 - 0.5.$$

For this example, all assumptions of the theory concerning the domain and the smoothness of the data are satisfied. Moreover, the exact solution is very smooth and does not depend on the Reynolds number. In contrast to more practice-oriented problems, where the exact solution in general is unknown, we can calculate here the accurate error norms. However, for high Reynolds numbers we have to suppose as a hypothesis that there is only one exact solution which cannot be guaranteed by the theory.

In all numerical calculations we have used the simple upwind $\Phi=\Phi_1$ from (10). Table 1 shows for the $P_1/P_0$ and the $Q_1^{\text{rot}}/Q_0$ finite element approximation the error norms for different Reynolds numbers $Re$ and different mesh sizes $h$. The elements $K\in\mathbf{T}_h$ in

Table 1: *comparison between the $P_1/P_0$ and $Q_1^{\text{rot}}/Q_0$ element*

| $Re$ | $h$ | $\|u - u_h\|_h$ | | $\|u - u_h\|_0$ | | $\|p - p_h\|_0$ | |
|---|---|---|---|---|---|---|---|
| | | $Q_1^{\text{rot}}/Q_0$ | $P_1/P_0$ | $Q_1^{\text{rot}}/Q_0$ | $P_1/P_0$ | $Q_1^{\text{rot}}/Q_0$ | $P_1/P_0$ |
| 10 | 1/8 | 0.67 +0 | 0.41 +0 | 0.30 -1 | 0.15 -1 | 0.72 -1 | 0.14 -1 |
| | 1/16 | 0.35 +0 | 0.21 +0 | 0.80 -2 | 0.71 -2 | 0.36 -1 | 0.67 -2 |
| | 1/32 | 0.18 +0 | 0.11 +0 | 0.21 -2 | 0.36 -2 | 0.17 -1 | 0.35 -2 |
| | 1/64 | 0.91 -1 | 0.69 -1 | 0.53 -3 | 0.17 -2 | 0.86 -2 | 0.19 -2 |
| | 1/128 | 0.45 -1 | 0.34 -1 | 0.13 -3 | 0.83 -3 | 0.43 -2 | 0.91 -3 |
| 100 | 1/8 | 0.39 +1 | 0.37 +1 | 0.17 +0 | 0.12 +0 | 0.73 -1 | 0.14 -1 |
| | 1/16 | 0.31 +1 | 0.21 +1 | 0.69 -1 | 0.35 -1 | 0.36 -1 | 0.60 -2 |
| | 1/32 | 0.18 +1 | 0.11 +1 | 0.20 -1 | 0.11 -1 | 0.17 -1 | 0.26 -2 |
| | 1/64 | 0.90 +0 | 0.53 +0 | 0.52 -2 | 0.65 -2 | 0.86 -2 | 0.13 -2 |
| | 1/128 | 0.45 +0 | 0.27 +0 | 0.13 -2 | 0.31 -2 | 0.43 -2 | 0.67 -3 |
| 1000 | 1/8 | 0.60 +1 | 0.98 +1 | 0.27 +0 | 0.31 +0 | 0.75 -1 | 0.18 -1 |
| | 1/16 | 0.78 +1 | 0.10 +2 | 0.17 +0 | 0.18 +0 | 0.36 -1 | 0.11 -1 |
| | 1/32 | 0.87 +1 | 0.88 +1 | 0.97 -1 | 0.74 -1 | 0.18 -1 | 0.23 -2 |
| | 1/64 | 0.73 +1 | 0.51 +1 | 0.41 -1 | 0.22 -1 | 0.87 -2 | 0.12 -2 |
| | 1/128 | 0.44 +1 | 0.26 +1 | 0.13 -1 | 0.89 -2 | 0.43 -2 | 0.60 -3 |
| 2000 | 1/8 | 0.62 +1 | 0.10 +2 | 0.28 +0 | 0.34 +0 | 0.75 -1 | 0.23 -1 |
| | 1/16 | 0.83 +1 | 0.12 +2 | 0.18 +0 | 0.21 +0 | 0.36 -1 | 0.10 -1 |
| | 1/32 | 0.10 +2 | 0.12 +2 | 0.11 +0 | 0.12 +0 | 0.18 -1 | 0.84 -2 |
| | 1/64 | 0.11 +2 | 0.78 +1 | 0.60 -1 | 0.49 -1 | 0.88 -2 | 0.39 -2 |
| | 1/128 | 0.80 +1 | 0.47 +1 | 0.23 -1 | 0.15 -1 | 0.43 -2 | 0.87 -3 |

the $Q_1^{\text{rot}}/Q_0$ case are defined by the quadrilateral cells $K=[(i-1)h, ih]\times[(j-1)h, jh]$, $i,j = 1, 2, ...$, and in the $P_1/P_0$ case by dividing each of these quadrilaterals (by means of the diagonals) into four triangles. That means the triangular mesh is somewhat finer than the quadrilateral one. Let us note that for fixed $h$ the total number of unknowns for the $P_1/P_0$ elements is more than three times larger than for the $Q_1^{\text{rot}}/Q_0$ elements.

In the following, we will consider only the $Q_1^{\text{rot}}/Q_0$ finite element approximation. Figure 2 shows for different Reynolds numbers $Re$ the reduction of the error norms if we increase the grid level $l$, i.e. if we decrease the mesh size $h = 2^{-l}$. The line marked by "int_err" represents the norm of the corresponding interpolation error (e.g. $\|u - i_h u\|$ for the velocity where $i_h u \in V_h$ is the usual finite element interpolation of the exact solution $u$). In contrast to the velocity, the $L^2$-norm error of the pressure shows the optimal order $O(h)$ and does not depend on $Re$. For moderate Reynolds numbers, the velocity error decreases with the order $O(h)$ in the discrete $H^1$-norm and with $O(h^2)$ in the $L^2$-norm. However, for high Reynolds numbers these orders seem to be reached only when the mesh size $h$ is very small whereas for coarser meshes the behavior of the error looks like $O(1)$ in the discrete $H^1$-norm and $O(h)$ in the $L^2$-norm.

Another observation is that the error norms increase if we keep the mesh size fixed and increase the Reynolds number. This effect can be seen more clearly in Figure 3 for the discrete $H^1$-norm and in Figure 4 for the $L^2$-norm where the error behaves like $O(Re)$ in some range of $Re$. Figure 5 shows, for the fixed mesh size $h=1/64$, how strong the velocity errors for increasing Reynolds numbers affect the streamlines of the computed discrete solutions. It is amazing that this can happen even if the exact solution $u$ does not depend on $Re$.

# 6 Conclusions

Although our upwind method has proved by some practical computations to be reasonable for the numerical solution of the stationary and nonstationary Navier–Stokes problem [2],[7], the simple example in the previous section shows that for high Reynolds numbers the discretization error can really deteriorate strongly and that there is not only a lack in the theoretical analysis. Therefore, an important conclusion for practical computations in the high Reynolds number case is that we have to be careful with the accuracy of our computed solution and that we have to verify it in a suitable way.

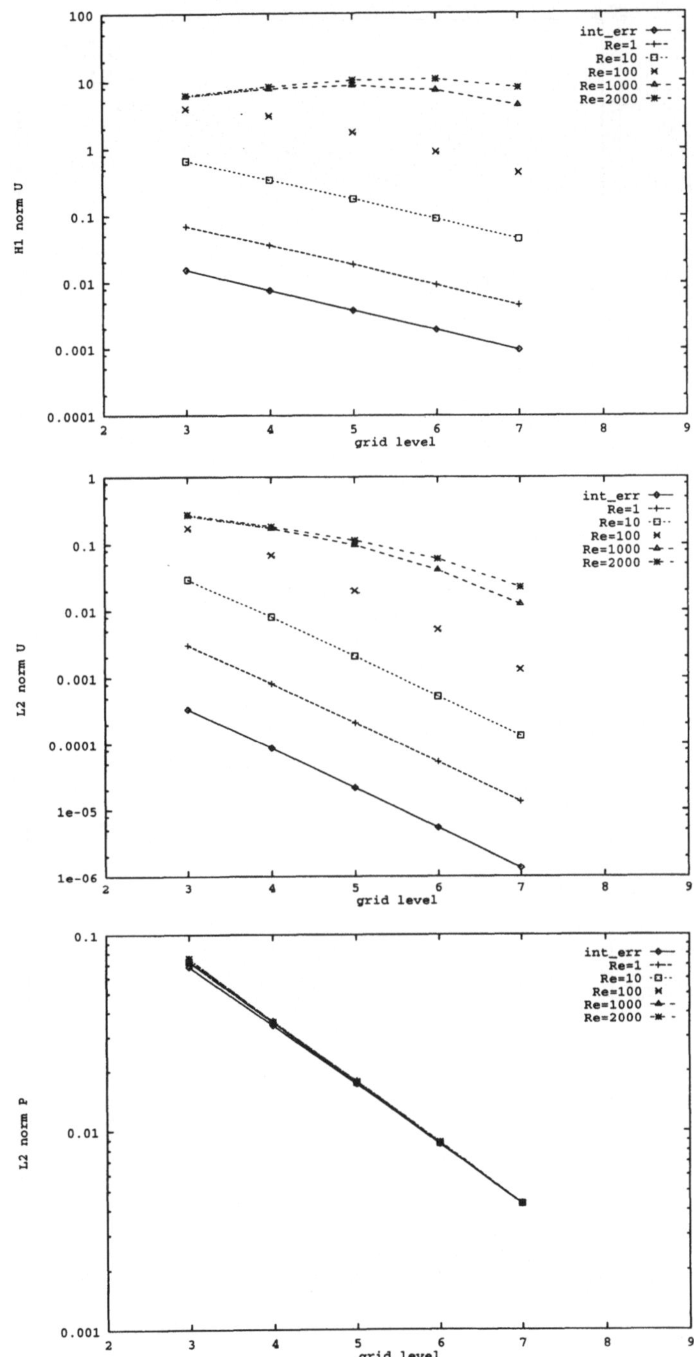

Figure 2: *error reduction for* $h \to 0$

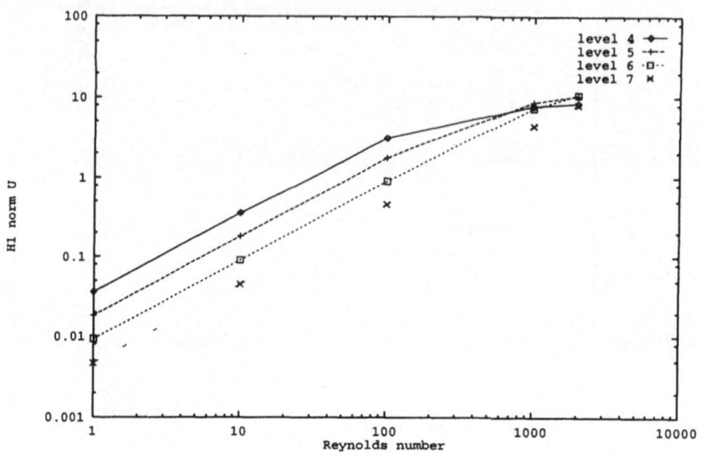

Figure 3: *discrete $H^1$-norm error of the velocity for increasing Re*

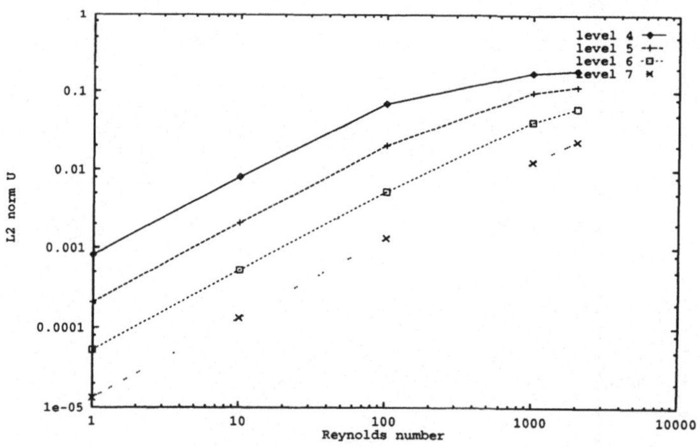

Figure 4: *$L^2$-norm error of the velocity for increasing Re*

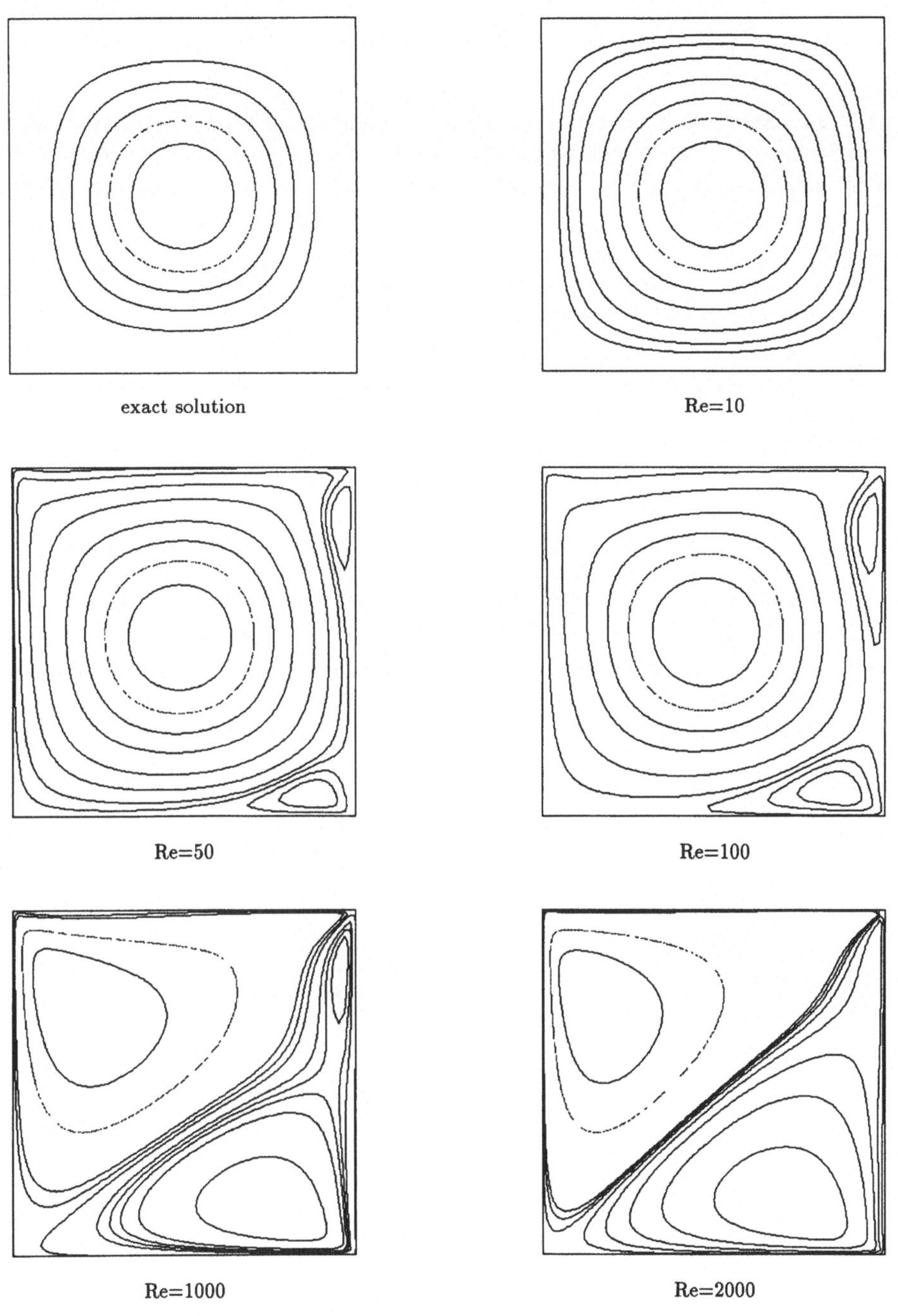

Figure 5: *streamlines for h=1/64*

# References


[1] M. Crouzeix, P.A. Raviart, Conforming and Nonconforming Finite Element Methods for Solving the Stationary Stokes Equations. *RAIRO Numer. Anal.* **3**, 33-76 (1973).

[2] W. Grambow, U. Risch, F. Schieweck, *Experiences with the Multigrid Method Applied to High Reynolds Number, Steady, Incompressible Flow.* Preprint Math 6/90, TU Magdeburg (1990).

[3] R. Rannacher, S. Turek, Simple Nonconforming Quadrilateral Stokes Element. *Numer. Methods Partial Differential Equations* **8**, 97-111 (1992).

[4] F. Schieweck, L. Tobiska, A nonconforming finite element method of upstream type applied to the stationary Navier–Stokes equation. *RAIRO Modél. Math. Anal. Numér.* **23**, 627–647 (1989).

[5] F. Schieweck, L. Tobiska, *An error estimate of optimal order for a nonconforming upwind finite element discretization of the Navier–Stokes equations.* Preprint TU Magdeburg, in preparation.

[6] L. Tobiska, A three-dimensional nonconforming finite element method of upstream type and its application to the Navier Stokes equations. In *"Proceedings of the workshop International Seminar on Applied Mathematics 1991"*, (H.-G. Roos, A. Felgenhauer, L. Angermann, eds.), TU–Dresden 1991.

[7] S. Turek, *Ein robustes und effizientes Mehrgitterverfahren zur Lösung der instationären, inkompressiblen, 2D Navier–Stokes-Gleichungen mit diskret divergenzfreien finiten Elementen.* Thesis, Preprint Nr. 642, SFB 123, Universität Heidelberg (1991).

# AN ACCURATE AND EFFICIENT IMPLICIT UPWIND SOLVER FOR THE NAVIER-STOKES EQUATIONS


E. Schöll, H.-H. Frühauf
Institut für Raumfahrtsysteme, Universität Stuttgart
Pfaffenwaldring 31
70550 Stuttgart, Germany


## SUMMARY


A new computer code for solving the Navier-Stokes equations for perfect gas and nonequilibrium flows is introduced. An unfactored implicit scheme has been developed which employs an efficient solution algorithm for the resulting system of linear equations. The nearly unrestricted stability of the scheme allows for large time steps and is therefore adequate for the resolution of the largely different time scales in complex viscous flow simulations. Small numerical dissipation is introduced in the calculation of the inviscid fluxes through the careful implementation of a Godunov-type upwind scheme which maintains uniform second order accuracy on locally skewed and stretched meshes. Results showing the accuracy and the convergence properties of this newly developed implicit Navier-Stokes solver are presented for inviscid and viscous channel and cascade flows.


## INTRODUCTION

The application of a Navier-Stokes code as a routine engineering tool depends on its robustness, its stability and the total time necessary to obtain a solution, as well as on the accuracy of the computed results for a complex flow problem. Another desirable feature is the applicability of a Navier-Stokes solver in a wide Mach and Reynolds number range.

In the calculation of the inviscid terms a minimum of inherent numerical dissipation is indispensable for a sufficiently accurate prediction of losses or drag in viscous flow computations, because this numerical dissipation can interfere, or even dominate, the physical diffusive effects. Godunov-type methods have shown to produce excellent results for viscous simulations while retaining their favourable shock capturing capability due to their low 'build-in' dissipation [8]. Furthermore, they have been applied to a wide range of flow regimes.

Implicit schemes guarantee nearly unrestricted stability especially in the presence of mesh and source term stiffness, and allow for much larger time steps and consequently faster convergence compared to explicit schemes. On the other hand, however, implicit schemes require the inversion of a large banded matrix leading to a high computational work per time step. The use of upwind methods permits the utilization of unfactored implicit schemes relying on line relaxation techniques for which very efficient solution algorithms can be developed.

The purpose of this paper is to introduce a newly developed Navier-Stokes solver employing the above mentioned numerical methods and to show the accuracy and the convergence behaviour for practical flow problems.

## GOVERNING EQUATIONS

The governing equations are the two-dimensional, compressible Navier-Stokes equations which may be written in dimensionless, integral form as

$$\frac{\partial}{\partial t} \int_V \vec{Q}\, dV + \oint_A \left( \vec{\varphi} - \frac{\vec{\varphi}_v}{Re_{ref}} \right) dA = 0 \quad ; \quad d\vec{A} = \vec{n} \cdot dA = \begin{bmatrix} n_x \\ n_y \end{bmatrix} \cdot dA \qquad (1)$$

where $\vec{\varphi}$ denotes the flux vector pointing in the normal direction of a surface element $d\vec{A}$. The cartesian flux components are related to $\vec{\varphi}$ by the relation $\vec{\varphi}_{(v)} = \vec{E}_{(v)} \cdot n_x + \vec{F}_{(v)} \cdot n_y$ with $n_x$ and $n_y$ being the components of the unit normal vector $\vec{n}$. The vector of the conserved variables $\vec{Q}$ and the cartesian fluxes are given by:

$$
\vec{Q} = \begin{bmatrix} \rho \\ \rho u \\ \rho v \\ \rho e_t \end{bmatrix} \qquad
\vec{E} = \begin{bmatrix} \rho u \\ \rho u^2 + p \\ \rho u v \\ \rho u h_t \end{bmatrix} \qquad
\vec{F} = \begin{bmatrix} \rho v \\ \rho v u \\ \rho v^2 + p \\ \rho v h_t \end{bmatrix}
$$

$$
\vec{E}_v = \begin{bmatrix} 0 \\ \tau_{xx} \\ \tau_{xy} \\ u\tau_{xx} + v\tau_{xy} - \left(\frac{\gamma}{\gamma-1}\frac{1}{Pr}\right)_{ref} \dot{q}_x \end{bmatrix} \qquad
\vec{F}_v = \begin{bmatrix} 0 \\ \tau_{xy} \\ \tau_{yy} \\ u\tau_{xy} + v\tau_{yy} - \left(\frac{\gamma}{\gamma-1}\frac{1}{Pr}\right)_{ref} \dot{q}_y \end{bmatrix} \tag{2}
$$

The pressure is defined by the equation of state for a perfect gas

$$
p = (\gamma - 1) \cdot \rho \cdot \left[ e_t - \frac{u^2 + v^2}{2} \right] \tag{3}
$$

and the shear stresses by

$$
\tau_{xx} = 2\mu\frac{\partial u}{\partial x} + \lambda\left(\frac{\partial u}{\partial x} + \frac{\partial v}{\partial y}\right) \;\; ; \;\;
\tau_{xy} = \mu\left(\frac{\partial u}{\partial y} + \frac{\partial v}{\partial x}\right) \;\; ; \;\;
\tau_{yy} = 2\mu\frac{\partial v}{\partial y} + \lambda\left(\frac{\partial u}{\partial x} + \frac{\partial v}{\partial y}\right) \tag{4}
$$

using Stokes' hypothesis $\lambda = -2/3 \cdot \mu$. The heat fluxes are determined by Fourier's law:

$$
\dot{q}_x = -\kappa \cdot \frac{\partial T}{\partial x} \qquad ; \qquad \dot{q}_y = -\kappa \cdot \frac{\partial T}{\partial y} \; . \tag{5}
$$

Turbulence is modelled by the eddy viscosity approach employing the algebraic turbulence model of Baldwin and Lomax [1]. The preceeding equations have been nondimensionalized by reference values of density $\rho_{ref}$, velocity $v_{ref}$, viscosity $\mu_{ref}$, heat conductivity $\kappa_{ref}$ and a reference length $l_{ref}$. The quantities $Re_{ref}$ and $Pr_{ref}$ are the Reynolds number and the Prandtl number calculated with the reference values, respectively.

## NUMERICAL METHOD

### Spatial Discretization

The spatial discretization of (1) is performed by the cell-centered Finite Volume method for a structured mesh. It can be shown, that the Finite Volume discretization of the Navier-Stokes equations transformed into the curvilinear computational space (see [12]) is equivalent to the standard Finite Volume discretization using the control volume described in the physical space. The prerequisite is the computation of the metric relations between the cartesian $(x, y)$ and the curvilinear $(\xi, \eta)$ coordinates with standard central difference formulas. This fact allows for the following combined approach in calculating the fluxes at the cell faces:

$$
V_{i,j}\frac{\partial \vec{Q}_{i,j}}{\partial t} + \left(\vec{\mathcal{F}} - \vec{\mathcal{F}}_v\right)_{i,j} = 0
$$

$$
\vec{\mathcal{F}}_{i,j} = (\vec{\varphi}\,\Delta A)_{i+1/2,j} - (\vec{\varphi}\,\Delta A)_{i-1/2,j} + (\vec{\varphi}\,\Delta A)_{i,j+1/2} - (\vec{\varphi}\,\Delta A)_{i,j-1/2} \tag{6}
$$

$$
(\vec{\mathcal{F}}_v)_{i,j} = \frac{1}{Re_{ref}}\left[(\hat{\vec{E}}_v)_{i+1/2,j} - (\hat{\vec{E}}_v)_{i-1/2,j} + (\hat{\vec{F}}_v)_{i,j+1/2} - (\hat{\vec{F}}_v)_{i,j-1/2}\right] .
$$

The inviscid fluxes $\vec{\varphi}$ are formulated in the physical space thus avoiding the transformation of the eigenvalues and eigenvectors. The determination of the viscous fluxes in the computational space, indicated by the hat sign, guarantees a high accuracy on stretched and skewed grids as well as a straightforward implementation of wall boundary conditions in a formal manner.

### Time Integration

The temporal discretization of (6) is done by the implicit Euler backward scheme

$$
V_{i,j}\frac{\vec{Q}_{i,j}^{n+1} - \vec{Q}_{i,j}^{n}}{\Delta t} + \left[\vec{\mathcal{F}} - \vec{\mathcal{F}}_v\right]_{i,j}^{n+1} = \vec{\mathcal{R}}_{i,j} = 0 \qquad\qquad t = n \cdot \Delta t \tag{7}
$$

where $\vec{\mathcal{R}}_{i,j}$ denotes the Residual at each grid point. The solution of this nonlinear equation system is obtained by the application of the iterative Newton method with underrelaxation:

$$\left[\frac{\partial \vec{\mathcal{R}}}{\partial \vec{Q}}\right]^{\alpha} \cdot \Delta \vec{Q}^{\alpha} = -\vec{\mathcal{R}}(\vec{Q}^{\alpha}) \quad ; \quad \vec{Q}^{\alpha+1} = \vec{Q}^{\alpha} + \omega \cdot \Delta \vec{Q}^{\alpha} \quad ; \quad 0 < \omega \leq 1 \cdot \tag{8}$$

Currently only the steady state solution is of interest, thus, the time index $n$ is used as Newton iteration index $\alpha$ [4]. This yields the following system of linear equations:

$$\frac{V_{i,j}}{\Delta t} \cdot \Delta \vec{Q}^{n}_{i,j} + \sum_{k,l} \frac{\partial}{\partial \vec{Q}_{k,l}} \left[\vec{\mathcal{F}} - \vec{\mathcal{F}}_{v}\right]^{n}_{i,j} \cdot \Delta \vec{Q}^{n}_{k,l} = -\left[\vec{\mathcal{F}} - \vec{\mathcal{F}}_{v}\right]^{n}_{i,j} \cdot \tag{9}$$

The summation sign indicates that a Jacobian matrix has to be calculated for each grid point the discretized flux vector depends on. This is in contrast to the approach where the Jacobians are determined analytically and are subsequently discretized. To accelerate convergence to the steady state, a local time step for each volume cell is determined from a prescribed CFL number:

$$\frac{V_{i,j}}{\Delta t} = \frac{(\Delta s_{max})_{i,j} \cdot (c_{i,j} + |\vec{v}|_{i,j})}{CFL} \cdot \tag{10}$$

## Calculation of the Inviscid Terms

A flux difference splitting scheme employing Roe's approximate Riemann solver [13] is used to calculate the inviscid fluxes normal to the cell faces:

$$\vec{\bar{\varphi}}(\vec{Q}^{R}, \vec{Q}^{L}) = \frac{1}{2} \cdot \left[\vec{\varphi}(\vec{Q}^{R}) + \vec{\varphi}(\vec{Q}^{L}) - \bar{\bar{R}} \, | \, \bar{\bar{\Lambda}} \, | \, \bar{\bar{R}}^{-1} \cdot \left(\vec{Q}^{R} - \vec{Q}^{L}\right)\right] . \tag{11}$$

$\bar{\bar{\Lambda}}$ is the diagonal matrix of the eigenvalues and $\bar{\bar{R}}$ the matrix of the corresponding right eigenvectors, whereby the matrix elements are computed using Roe's averages. A spatially first order scheme is obtained if the neighbouring cell center values are considered as left and right states, e.g. $\vec{\varphi}_{i+1/2,j} = \vec{\bar{\varphi}}\left(\vec{Q}_{i+1,j}, \vec{Q}_{i,j}\right)$. The extension to second order spatial accuracy is achieved by a variable extrapolation in each mesh direction:

$$\vec{Q}_{i,j}(\vec{s}|^{i}) \approx \vec{Q}(\vec{s}_{i,j}) + \left.\frac{\partial \vec{Q}}{\partial \vec{s}}\right|^{i}_{i,j} \cdot (\vec{s}|^{i} - \vec{s}_{i,j}) + O(\Delta s^{2}) . \tag{12}$$

The slopes in conservative variables are determined via limited calculation of variations in characteristic variables, which characterize the direction of information propagation:

$$\left.\frac{\partial \vec{Q}}{\partial \vec{s}}\right|^{i}_{i,j} = \left.\delta \vec{Q}\right|^{i}_{i,j} \approx R^{-1}_{i,j} \cdot \left.\delta \vec{W}\right|^{i}_{i,j} \quad ; \quad \delta \vec{W}^{i} = f_{lim}\left(\delta^{+}\vec{W}^{i}, \delta^{0}\vec{W}^{i}, \delta^{-}\vec{W}^{i}\right) . \tag{13}$$

The expressions $\delta^{+}\vec{W}^{i}, \delta^{-}\vec{W}^{i}, \delta^{0}\vec{W}^{i}$ stand for the forward, backward and central differences formulated for nonequidistant mesh cells, respectively [3]. The limiter function $f_{lim}$ selects the smoothest difference or becomes zero at discontinuities to prevent oscillations. In the code the limiter formulation of Van Leer [15] and the differentiable function of Eberle [7] based on the sensor of Van Albada et al. [14] are implemented.

The determination of the Jacobian matrices for the Roe fluxes follows the considerations of Barth [2] and is done approximately by neglecting the derivatives of the Roe matrices in (11)

$$\frac{\partial \vec{\varphi}_{i+1/2,j}}{\partial \vec{Q}_{i+1,j}} = \frac{1}{2} \cdot \left[\Phi(\vec{Q}^{R}_{i+1/2,j}) - \left(\bar{\bar{R}} \, | \, \bar{\bar{\Lambda}} \, | \, \bar{\bar{R}}^{-1}\right)_{i+1/2,j}\right]$$

$$\frac{\partial \vec{\varphi}_{i+1/2,j}}{\partial \vec{Q}_{i,j}} = \frac{1}{2} \cdot \left[\Phi(\vec{Q}^{L}_{i+1/2,j}) + \left(\bar{\bar{R}} \, | \, \bar{\bar{\Lambda}} \, | \, \bar{\bar{R}}^{-1}\right)_{i+1/2,j}\right] \tag{14}$$

with the analytical Jacobian matrix $\Phi(\vec{Q}) = \partial \vec{\varphi}/\partial \vec{Q}$.

<u>**Calculation of the Viscous Terms**</u>

The viscous fluxes in curvilinear coordinates are obtained from the cartesian flux components by the well-known metric relationships (see [12])

$$\hat{\vec{E}}_v = \frac{1}{J}\cdot\left[\frac{\partial\xi}{\partial x}\vec{E}_v + \frac{\partial\xi}{\partial y}\vec{F}_v\right] \quad ; \quad \hat{\vec{F}}_v = \frac{1}{J}\cdot\left[\frac{\partial\eta}{\partial x}\vec{E}_v + \frac{\partial\eta}{\partial y}\vec{F}_v\right] \tag{15}$$

where $J$ denotes the metric Jacobian of the transformation. Also the equations (4) and (5) defining the shear stresses and the heat fluxes have to be transformed into the computational space.

The metric derivatives at the cell faces are calculated from the cartesian coordinates of the cell vertices by means of standard central difference formulas. In a similar manner the derivatives with respect to $\xi$ and $\eta$ appearing in the viscous fluxes are discretized with difference formulas of second order spatial accuracy, e.g. at the cell face $(i+1/2, j)$:

$$\left(\frac{\partial\alpha}{\partial\xi}\right)_{i+\frac{1}{2},j} \approx \alpha_{i+1,j} - \alpha_{i,j} \quad ; \quad \left(\frac{\partial\alpha}{\partial\eta}\right)_{i+\frac{1}{2},j} \approx \frac{1}{4}\cdot(\alpha_{i+1,j+1} + \alpha_{i,j+1} - \alpha_{i+1,j-1} + \alpha_{i,j-1}) \; . \tag{16}$$

Variables at the cell faces are determined by the arithmetic mean of the centroid values.

For the derivation of the viscous Jacobian matrices the Thin-Layer approximation is applied. So only the fluxes in $\eta$-direction with neglected $\xi$-derivatives are discretized using central differences leading to the dependency $\hat{\vec{F}}^{*}_{v;\ i,j+1/2} = f(\vec{Q}_{i,j+1}, \vec{Q}_{i,j})$. The viscous Jacobians are derived from these Thin-Layer fluxes under the assumption of locally constant transport coefficients $\mu$ and $\kappa$ and the resulting balance of the viscous Jacobians can be written as:

$$\frac{\partial(\hat{\vec{F}}_v)_{i,j}}{\partial\vec{Q}} = \sum_{m=j-1,j,j+1}\left[\frac{\partial(\hat{\vec{F}}^{*}_v)_{i,j+1/2}}{\partial\vec{Q}_{i,m}} - \frac{\partial(\hat{\vec{F}}^{*}_v)_{i,j-1/2}}{\partial\vec{Q}_{i,m}}\right]\cdot\Delta\vec{Q}_{i,m} \; . \tag{17}$$

<u>**Boundary Conditions**</u>

An implicit characteristic approach [9] is used to prescribe the values at the inflow and the outflow boundaries. At solid walls the pressure is extrapolated from the interior for inviscid flow simulations and for viscous solutions the no-slip condition, adiabatic wall or constant wall temperatur and $\partial p/\partial\vec{n}$ are formulated as functions of the conservation vector $\vec{Q}$ and are implemented implicitly:

$$\vec{B}(\vec{Q}) = 0 \quad \rightarrow \quad \frac{\partial\vec{B}(\vec{Q})}{\partial\vec{Q}}\cdot\Delta\vec{Q} = -\vec{B}(\vec{Q}) \; . \tag{18}$$

<u>**Solution Procedure**</u>

With the just described schemes for the determination of the fluxes and the Jacobians, equation (9) becomes a blockpentadiagonal system of linear equations:

$$\left[M^{0,0}\right]_{i,j}\cdot\Delta\vec{Q}_{i,j} + \left[M^{1,0}\right]_{i,j}\cdot\Delta\vec{Q}_{i+1,j} + \left[M^{-1,0}\right]_{i,j}\cdot\Delta\vec{Q}_{i-1,j}$$
$$+ \left[M^{0,1}\right]_{i,j}\cdot\Delta\vec{Q}_{i,j+1} + \left[M^{0,-1}\right]_{i,j}\cdot\Delta\vec{Q}_{i,j-1} = -\vec{\mathcal{R}}_{i,j} \; . \tag{19}$$

This equation system is solved approximately by the Jacobi line relaxation method employing a subiteration scheme. First the equation system is preconditioned by multiplying each equation with the inverse of the main diagonal matrix $M^{0,0}_{i,j}$ thus obtaining unit matrices on the main diagonal. The solution procedure for (19) then consists of four steps:

1) Starting value is the solution of the last global iteration: $\quad \Delta\vec{Q}^{n,\nu=0} = \Delta\vec{Q}^{n-1}$ .

2) Set $\nu = \nu+1$ and solve the following blocktridiagonal equation system for every $j = const.$:

$$M^{0,0}_{i,j}\Delta\vec{Q}^{n,\nu}_{i,j} + M^{1,0}_{i,j}\Delta\vec{Q}^{n,\nu}_{i+1,j} + M^{-1,0}_{i,j}\Delta\vec{Q}^{n,\nu}_{i-1,j} = -\vec{\mathcal{R}}_{i,j} - M^{0,1}_{i,j}\Delta\vec{Q}^{n,\nu-1}_{i,j+1} - M^{0,-1}_{i,j}\Delta\vec{Q}^{n,\nu-1}_{i,j-1} \; . \tag{20}$$

3) Set $\nu = \nu+1$ and solve the following blocktridiagonal equation system for every $i = const.$:

$$M_{i,j}^{0,0}\Delta\vec{Q}_{i,j}^{n,\nu} + M_{i,j}^{0,1}\Delta\vec{Q}_{i,j+1}^{n,\nu} + M_{i,j}^{0,-1}\Delta\vec{Q}_{i,j-1}^{n,\nu} = -\vec{\mathcal{R}}_{i,j} - M_{i,j}^{1,0}\Delta\vec{Q}_{i+1,j}^{n,\nu-1} - M_{i,j}^{-1,0}\Delta\vec{Q}_{i-1,j}^{n,\nu-1}. \quad (21)$$

4) Repeat step 2 and 3 until subconvergence $\Delta\vec{Q}^{n,\nu} \approx \Delta\vec{Q}^{n,\nu-1}$ or a maximum number of subiterations is reached.

The LU decomposition of the blocktridiagonal matrices in (20) and (21) is done only at the beginning of the subiterations and is simplified by the fact, that the preconditioning results in unit matrices on the main diagonal. Thus, equation (19) can be solved efficiently with arbitrary accuracy.

## RESULTS

In Fig.1 the convergence behaviour of the code for <u>inviscid channel flows</u> is depicted to demonstrate the influence of the approximate computation of the inviscid Jacobians. The three cases shown are all calculated with $CFL = 100$ starting from a constant initial distribution. In the subsonic case the last term in (11) and thus the influence of the approximate determination of the inviscid Jacobians becomes rather small and the nearly quadratic convergence of Newton's method is obtained. The appearance of a discontinuity in the transonic flow increases the inconsistency between the approximate inviscid Jacobians and the inviscid flux terms on the right-hand-side. Furthermore, the process of shock capturing and positioning degrades the convergence rate and thus more iterations are required to drop the residual to the same order of magnitude as in the subsonic case. For completely supersonic inviscid flows the information has to be propagated only in one direction resulting in a faster convergence compared with the transonic case despite the complex shock system.

The <u>laminar flat plate flow</u> with $Ma_1 = 0.3$ and $Re_1 = 10^5$ has been calculated on a $121 \times 61$ grid with highly resolved leading and trailing edges. The computed velocity profile at the middle station of the plate coincides very well with the theory of Blasius (Fig.2a). Also the skin friction coefficient is in good agreement with the theory of Dijkstra et al. [6] which accounts for leading and trailing edge effects (Fig.2b). The influence of the Thin-Layer approximation for the viscous Jacobians is investigated in Fig.2c for laminar flat plate flows on a $36 \times 93$ grid without considering the region in front of the leading and behind the trailing edge. Increasing the consistency by using the Thin-Layer approximation also for the viscous fluxes on the right-hand-side leads in the subsonic case with $Ma_1 = 0.3$ and $Re_1 = 10^5$ to a residual drop to machine accuracy. Opposed to that the convergence becomes asymptotic after 200 iterations if the full viscous terms are considered on the right-hand-side. For the hypersonic laminar flow with $Ma_1 = 10$ and $Re_1 = 3 \cdot 10^5$ a shock is generated at the leading edge resulting in an additional inconsistency between the inviscid terms on the left- and right-hand-side. Therefore, the "asymptotic residual" is located at a higher order of magnitude. All computations were performed with a CFL number of 100 and constant initial conditions.

As a practical test case the convergence behaviour and the accuracy for <u>inviscid cascade flows</u> is studied for the <u>VKI-1 turbine cascade flow</u> on a $161 \times 25$ C-grid in Fig.3 and 4. In Fig.3 the accuracy is demonstrated for inviscid cascade flows. First the cascade is computed with a wedge-type trailing edge which is appropriately resolved. Solutions for the transonic flow with $Ma_{2is} = 1.21$ were calculated with the new solver and with the Finit Difference ADI solver of Lecheler [10] on the identical grid with the same boundary conditions. The Mach contours exhibit an improved shock resolution and a better representation of the contact discontinuity emanating from the trailing edge in the solution of the new upwind code. Furthermore, the upwind code generates less numerical entropy in the outer flow field and convects the shock-produced entropy in a less diffusive way. This is an essential requirement for the correct prediction of drag or losses in Navier-Stokes simulations. Fig.4a shows the Mach number contours for a subsonic ($Ma_{2is} = 0.68$) and a transonic ($Ma_{2is} = 1.21$) inviscid flow through the VKI-1 cascade with the original round trailing edge which was resolved by only 12 grid points. The corresponding convergence rates for both subsonic and transonic flows are nearly identical (Fig.4b) and are obtained with a maximum CFL number of 100. For the transonic flow case the profile isentropic

Mach number distribution is depicted in Fig.4c and compares well with the experiment. Deviations are due to the inviscid flow model used.

In Fig.5 the simulation of the <u>turbulent transonic flow</u> through the <u>MTU T5.1 turbine cascade</u> with $Ma_{2is} = 0.9$ and $Re_2 = 3 \cdot 10^5$ on a $201 \times 46$ C-grid is shown. The trailing edge was resolved by only 8 points. The Mach contours are compared to the solution obtained by Zimmermann [16] on a $148 \times 26$ C-grid. The new Navier-Stokes code gives a higher peak Mach number as well as a sharper resolution of the shock on the suction side of the blade (Fig.5a). Fig.5b shows that the total pressure losses outside the wake in the inviscid flow region are correctly determined to zero due to the low numerical dissipation of the new upwind scheme. To predict the total pressure loss in the wake more accurately the trailing edge resolution has to be refined further. The convergence of the turbulent calculation (max. CFL=50) in contrast to an inviscid calculation (max. CFL=100) for the same cascade and flow conditions is represented in Fig.5c. The turbulence modelling introduces additional inconsistencies due to the strong nonlinear dependence of the eddy viscosity on the conserved variables, which is neglected in the computation of the Jacobians, and thus the residual drop is smaller than in the inviscid computation.

A version of the Navier-Stokes solver has been extended and validated for the simulation of nonequilibrium high-temperature flows including a 11-species, 5-temperatures air model [5, 9]. The basic equations are solved fully coupled. Stiffness may arise from the source terms of the species continuity, vibrational and electron energy equations. Complex nonequilibrium flows can be computed with a convergence rate comparable to perfect gas flow simulations if the source term Jacobians are determined exactly. This is illustrated in Fig.6 for a $Ma_\infty = 35$, $Re_\infty = 3000$ flow around a 0.1 m diameter sphere.

## REFERENCES


[1] BALDWIN B.S., LOMAX H., Thin Layer Approximation and Algebraic Model for Separated Turbulent Flow, *AIAA 78-0257*.

[2] BARTH T.J., Analysis of Implicit Local Linearization Techniques for Upwind and TVD Algorithms, *AIAA 87-0595*.

[3] BORSBOOM M., Numerical Modelling of Navier-Stokes Equations, *VKI Lecture Series 1986-02*.

[4] CHAKRAVARTHY S.R., Relaxation Methods for Unfactored Implicit Upwind Schemes, *AIAA 84-0165*.

[5] DAISS A., SCHÖLL E., FRÜHAUF H.-H., KNAB O., Validation of the URANUS Navier-Stokes Code for High-Temperature Nonequilibrium Flows, *AIAA 93-5070*.

[6] DIJKSTRA D., KÜRTEN J.G.M., An Easy Test-Case for a Navier-Stokes Solver, *Proceedings of the First European CFD Conference, Brussels, 1992, Vol.2, Hirsch Ch. et al. (Eds.)*.

[7] EBERLE A., SCHMATZ M.A., BISSINGER N.C., Generalized Flux Vectors for Hypersonic Shock-Capturing, *AIAA 90-0390*.

[8] HIRSCH CH., Numerical Computation of Internal and External Flows, Volume 2, *John Wiley & Sons, Chichester, 1990*.

[9] JONAS S., Implizites Godunov-Typ-Verfahren zur voll gekoppelten Berechnung reibungsfreier Hyperschallströmungen im thermo-chemischen Nichtgleichgewicht, *Ph.D. Thesis, University of Stuttgart, 1993*.

[10] LECHELER S., Ein vollimplizites 3D Euler-Verfahren zur genauen und schnellkonvergenten Strömungsberechnung in Schaufelreihen von Turbomaschinen, *Ph.D. Thesis, University of Stuttgart, 1992*.

[11] SIEVERDING C., Experimental Data on Two Transonic Turbine Blade Sections and Comparioson with Various Theoretical Methods, *VKI Lecture Series 59, Transonic Flows in Turbomachines, 1973*.

[12] PULLIAM T.H., Efficient Solution Methods for the Navier-Stokes Equations, *VKI Lecture Series 1986-02*.

[13] ROE P.L., Approximate Riemann Solvers, Parameter Vectors, and Difference Schemes, *Journal of Computational Physics, Vol.43, pp.357-372, 198*

[14] VAN ALBADA G.D., VAN LEER B., ROBERTS JR. W.W., A Comparative Study of Computational Methods in Cosmic Gas Dynamics, *Astronomy and Astrophysics, Vol.108, pp.76-84, 1982*.

[15] VAN LEER B., Towards the Ultimate Conservative Difference Scheme. IV. A New Approach to Numerical Convection, *Journal of Computational Physics, Vol.23, pp.276-299, 1977*.

[16] ZIMMERMANN H., Berechnung von 2- und 3-dimensionalen Strömungen in einem transsonischen Turbinengitter unter besonderer Berücksichtigung der Verluste, *ZLR-Report 90-03, Technical University of Braunschweig, 1990*.

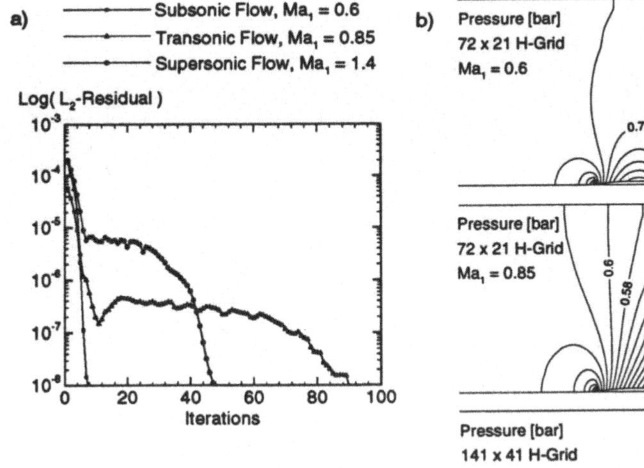


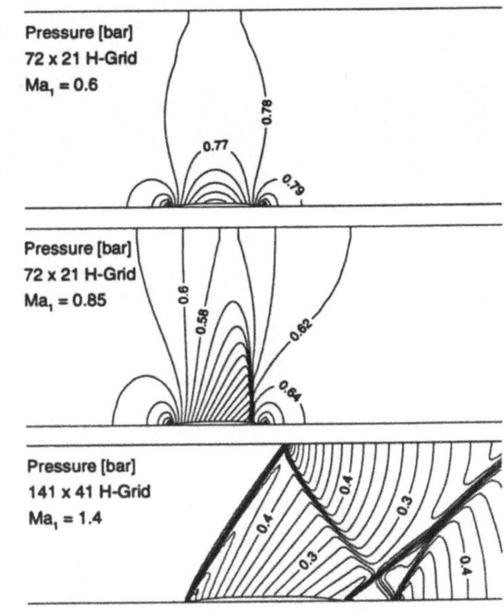


Fig.1: Inviscid Channel Flows
Convergence Behaviour

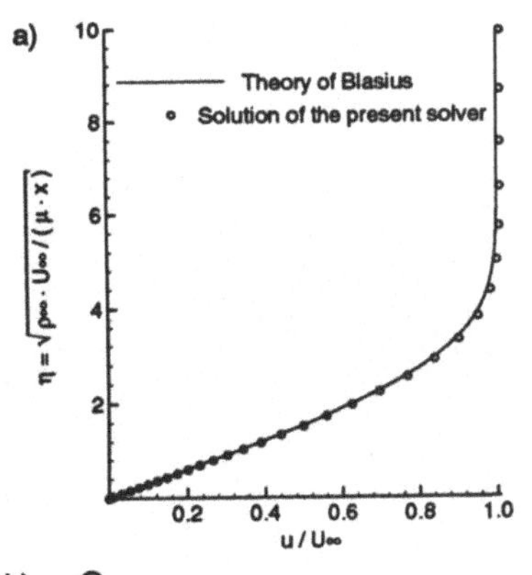


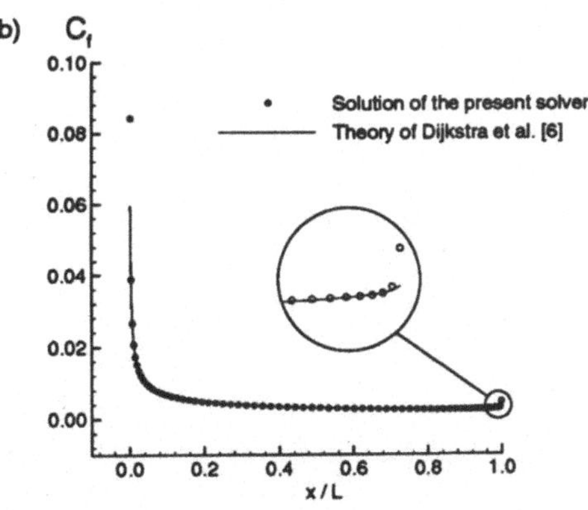


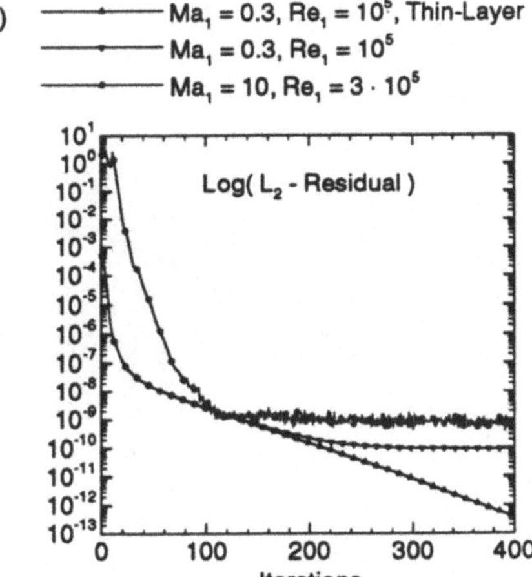


Fig.2: Laminar Flat Plate Flows
a) Velocity Distribution in the
   Boundary Layer
b) Skin Friction Distribution
   along the Plate
   for $Ma_1 = 0.3$, $Re_1 = 10^5$
c) Convergence Behaviour

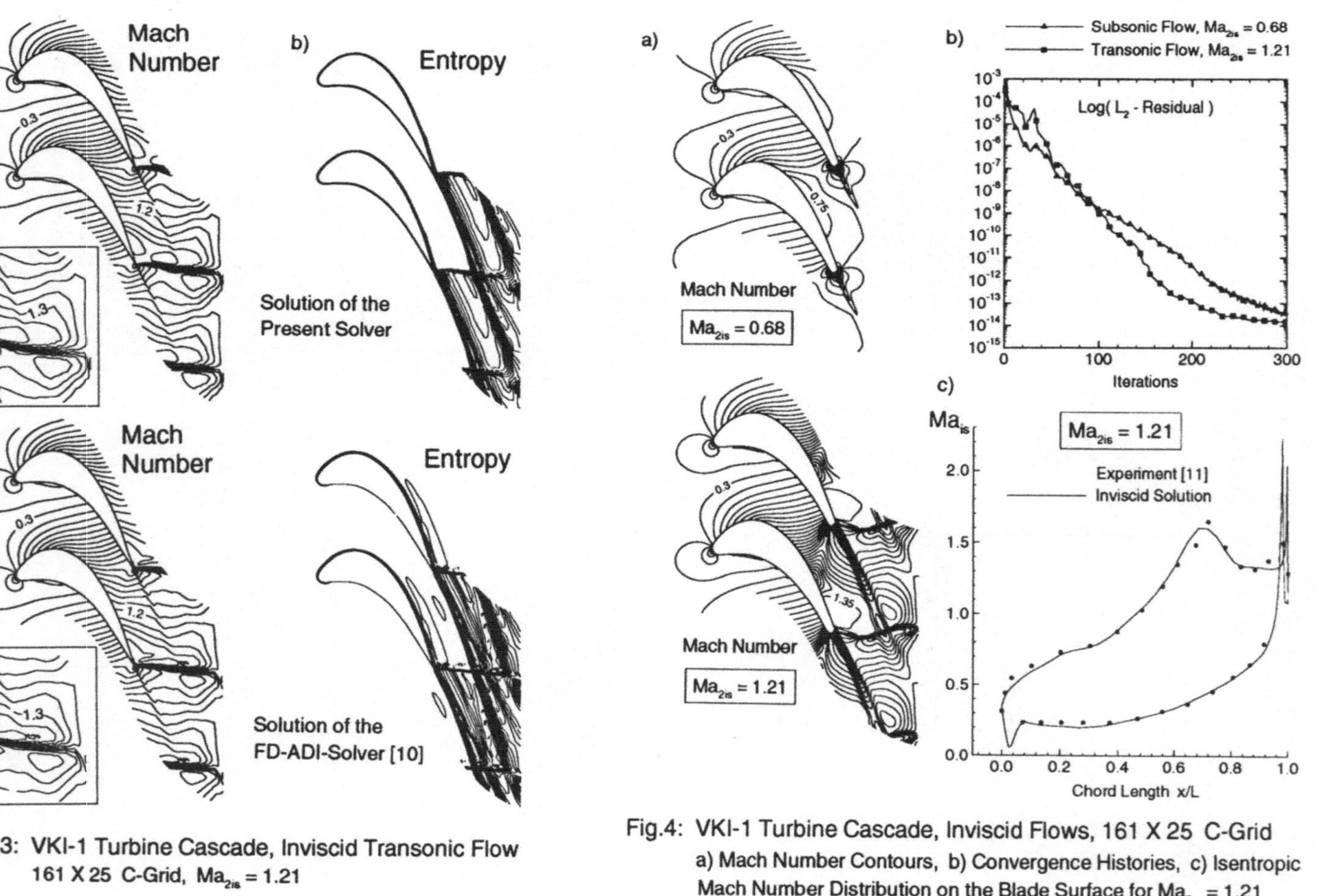


Fig.4: VKI-1 Turbine Cascade, Inviscid Flows, 161 X 25 C-Grid
a) Mach Number Contours, b) Convergence Histories, c) Isentropic
Mach Number Distribution on the Blade Surface for $Ma_{2is} = 1.21$

Fig.3: VKI-1 Turbine Cascade, Inviscid Transonic Flow
161 X 25 C-Grid, $Ma_{2is} = 1.21$

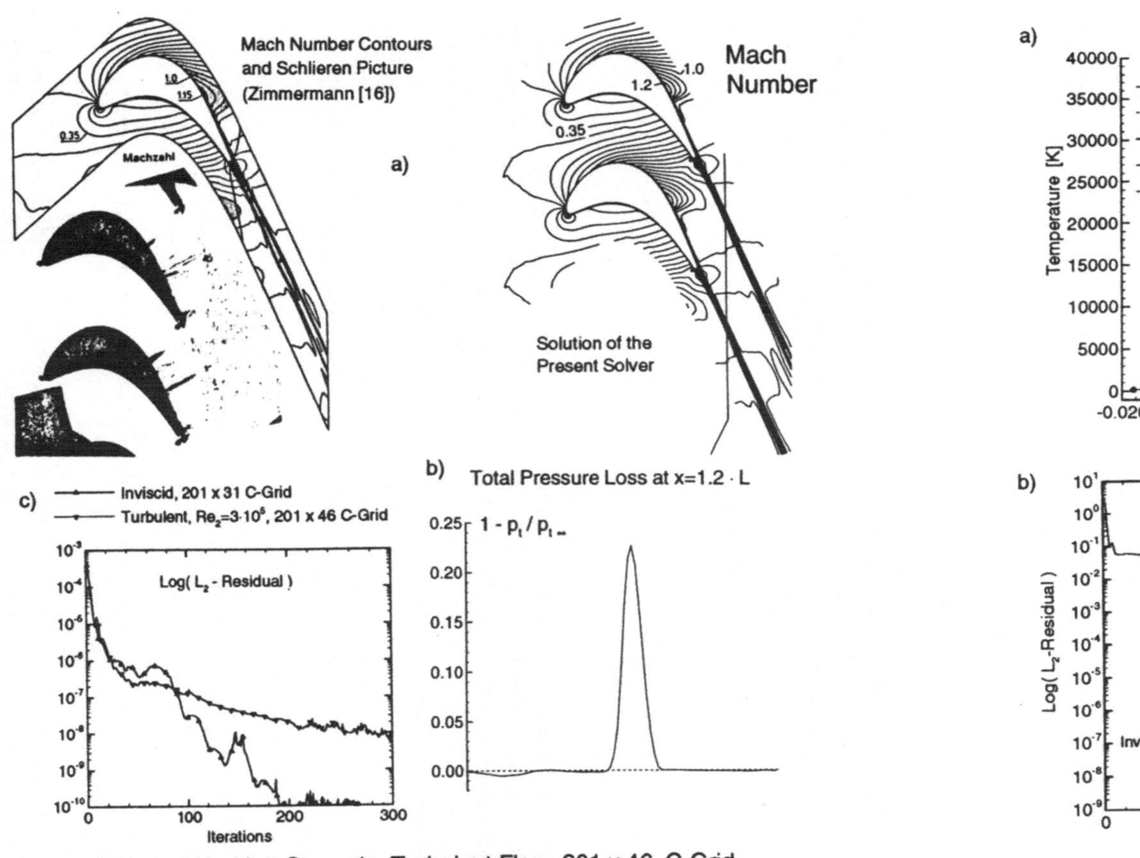

Fig.5: MTU T5.1 Turbine Cascade, Turbulent Flow, 201 x 46 C-Grid
a) Mach Number Contours, b) Total Pressure Loss in the Wake,
c) Convergence Histories

Fig.6: Laminar Nonequilibrium Flow around a
Sphere, Ma$_\infty$ = 35, Re$_\infty$ = 3000, 25 x 62 Grid

# An efficient finite element solver for the nonstationary incompressible Navier–Stokes equations in two and three dimensions


Peter Schreiber, Stefan Turek

Institut für Angewandte Mathematik, Universität Heidelberg
Im Neuenheimer Feld 294, 69120 Heidelberg, Germany


## Summary


We develop simulation tools for the nonstationary incompressible Navier–Stokes equations. The most important components of the finite element code are: The Fractional-step $\vartheta$–scheme, which is of second order accuracy and strongly A–stable, for the time discretization; a fixed point defect correction method with adaptive step–length control for the nonlinear problems (stationary Navier–Stokes equations); a modified upwind discretization of higher order accuracy for the convective terms. Finally, the resulting nonsymmetric linear subproblems are treated by a special multigrid algorithm which is adapted to the nonconforming finite elements. For the graphical postprocess we use a fully nonstationary and interactive particle tracing method. With extensive test calculations we show that our method may be a candidate for a "Black Box" solver.


## 1. Introduction

We consider the nonstationary incompressible Navier–Stokes equations,

$$\mathbf{u}_t - \nu\Delta\mathbf{u} + (\mathbf{u}\cdot\nabla)\mathbf{u} + \nabla p = \mathbf{f}\,,\ \nabla\!\cdot\!\mathbf{u} = 0\,,\ \text{in}\ \Omega \times (0,T)\,,$$

for a given force $\mathbf{f}$ and viscosity $\nu$, with prescribed boundary values on $\partial\Omega$ and an initial condition at $t = 0$. The variables $\mathbf{u}$ and $p$ describe the velocity and the pressure of a viscous incompressible flow in a bounded region $\Omega \subset \mathbf{R}^n, n = 2, 3$. These fundamental equations are of interest to both more theoretical scientists like mathematicians or physicists, and more applied ones like engineers or industrial users. Therefore, we try to derive solution methods which satisfy both parties by combining our mathematical methods and experiences (like accurate discretizations and efficient solution techniques) with techniques coming from the field of scientific computing. Methods for code optimization and visualization techniques on workstations come together with mathematical models and, furthermore, are improved by an adaption to physical and "real life"configurations. The result of this research is a family of codes: The 2D divergence–free version **NSDIV2D** ([5]), its primitive variable counterpart **NSP2D** (developed by Schieweck/Turek) and the 3D analogon **NSP3D**. This software package is just in development by Schreiber/Turek.

What are the theoretical aspects needed to develop an algorithm to solve efficiently these equations? Our mathematical research led us to examine the following problems:

Efficient treatment of the nonlinearity, error estimates for spatial and time discretizations, construction of fast linear solvers, modelling of boundary conditions, and many others.

It is very important not only to solve these problems in a theoretical way, but also to see the results in "real life", via computer simulations. On the other hand, engineers are also interested in an efficient solution method for these equations. The simulation of some complex fluid structures, the development of new models for "real" problems and, not to forget, the saving of money, by doing a computer simulation rather than an expensive laboratory test, these are all wishes, which appear every day. Now, our aim is to develop and really to implement a solution method, which satisfies both parties, this means we would like to construct a so called "Black Box solver". This term is often used, and in many cases its meaning is not clear, so we must define it. What we cannot offer is a "black universe method", we only present a method which attempts to give the user a "fast", "robust" and "accurate" solution. Then, the problem is to determine whether the computed "solution" is physically relevant. But our method, by itself, does not necessarily produce the "right solution", satisfying a prescribed accuracy corresponding to the real flow. In this sense, we try to find a good compromise between "fast", "robust" and "accurate", since it seems to be impossible to satisfy all points in practical.

Our implemented algorithm shall produce the results in a "short" amount of time, working on a workstation. We prefer this class of machine, because at today's prices (nearly) every mathematical institute can afford such computers. Additionally, these machines are very fast, complete as well as robust, and the future trend indicates that these will only improve. By fast we mean that we want to compute a fully developed nonstationary flow in hours (2D) or days (3D), and this with an accuracy, which is essentially determined by the available RAM memory. Another aspect of the term "fast" involves being able to change and to easily implement new ideas by adding subroutines to the existing code. Hence, we need a very modular programming structure and a very clever data management. The basis of all our programs is the finite element package **FEAT** ([1]), which was developed by our group in Heidelberg. This basis enables several people to work and to coorporate in the same field. Finally, another important aspect of "fast" is the fast graphical presentation of the computed data. Most workstations have very fine graphic features and enable, for example, one to compute and visualize at the same time. Hence, we also have to provide the corresponding graphical algorithms.

If we say our method is stable, we think to several aspects. First, the implemented algorithm shall be independent of given "data", like the shape of the domain (not convex, not polygonal, several boundary components), the mesh (not equidistant, refined in some parts), and the size of force, viscosity and boundary values. So by "robust" we mean that our finite element discretization works for (nearly) all meshes, that our linear solvers (almost) always produce the same convergence rates, and that the nonlinear convergence rates are not becoming too bad for higher Reynolds numbers. What we do not claim is that our algorithm works completely independently of all these "parameters", since, for instance, it is clear that the stationary Navier–Stokes solver cannot converge for higher Reynolds numbers, if there is no stationary solution, and it is also clear that the time step size is dependent on the physical flow structure. The aim is to develop a method, which requires only defining the domain, the right hand side, the boundary values and the viscosity, and then, the algorithm attempts to produce results in an "optimal" way. We think, the proposed method is a step in this direction. Another aspect is the robustness of the implementation. The program should be organized in a very modular structure in order to simplify and to control possible errors when making changes to the program.

We also emphasize that "robustness" includes the aspect, as to whether the algorithm is analysable, and in a certain sense "deterministic" for the user. Therefore, for instance, variational formulations and discretizations by finite elements are used, because this allows, at least for some simpler model problems, a strong analysis, and can be helpful to explain some "funny" results of the implemented code. So, only well known "standard methods" are used, like stable finite element pairs for the discretization, fast multigrid algorithms, fixed point defect correction methods for the nonlinearity, and well known time discretization schemes for the nonstationary processes. Finally, also our graphical support should work (in a robust way) independently of the above mentioned "data", something which is not fulfilled by all existing graphic packages.

The question of accuracy is the most critical and difficult point of our method. As mentioned above we cannot present a "black universe method", which tells the user the quality of the computed solutions compared with the exact ones. What we can hope to present is a "solution" with the "best accuracy" reachable for a given machine and time. Here, one has to consider that the more accurate methods are sometimes not the fastest ones. Therefore, only very simple finite elements (of second order), upwind schemes for the convective parts and time discretization schemes of second order are used, since then corresponding solution methods which are very fast can be constructed (divergence–free finite elements, multigrid methods). For the grid construction process the best method would be a fully adaptive method which we do not have implemented right now. On the other hand our "semi–adaptive" method leads to a very regular data structure, which gives not the most accurate results, but leads at least to a very efficient multigrid method. However, the question of adaptivity will be central for further development.

## 2. Components of the solver

We want to introduce the three main parts of our solution method: The preprocessing part, the postprocessing part and the finite element solvers.

### Input/Preprocessing part

The perfect method would be to define only the computational domain (perhaps in a graphical way), the input data (like force term, viscosity and boundary values), and then have the algorithm produce the most accurate solution, depending on the available memory and proposed time limit. Unfortunately, at this time, we are unable to accomplish this goal. At present our realized method needs, beside the description of the domain (by parametrization or given boundary points), also a "coarse" grid, which can be already slightly adapted to the expected solution (for example after small test calculations, or after comparing with existing theoretical and practical results). Then, this "coarse" grid is systematically refined (corresponding to the available memory), and the computation may begin. For nonstationary problems, one still has to define the time step size, because (presently) we have not implemented an adaptive control. But, since our implicit time discretization scheme is unconditionally stable, this parameter can be chosen depending solely on physical reasons, and not on (purely) mathematical stability conditions. These non–self adaptive methods do not guarantee the most accurate "solution", but do give a nearly optimal "solution" in a short amount of time.

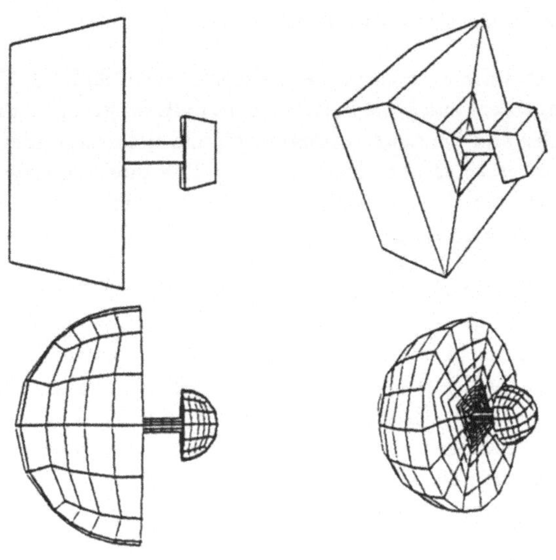

Figure 1: Coarse "semi-adaptive" grid and two times refined grid

## Visualization/Postprocessing part

We use standard graphic packages like MOVIE.BYU (for 2D) and AVS for 3D applications. Additionally, we developed some modules for nonstationary flows, for example a fully interactive 2D particle tracing with the corresponding data structure. These "movies" can also be transfered directly from screen to video tape. We think, this nonstationary postprocess modul is the most important one, since only with its aid one is really able to examine nonstationary flows, because these graphical results differ essentially from the streamline or vector plot snapshots (see Figure 2).

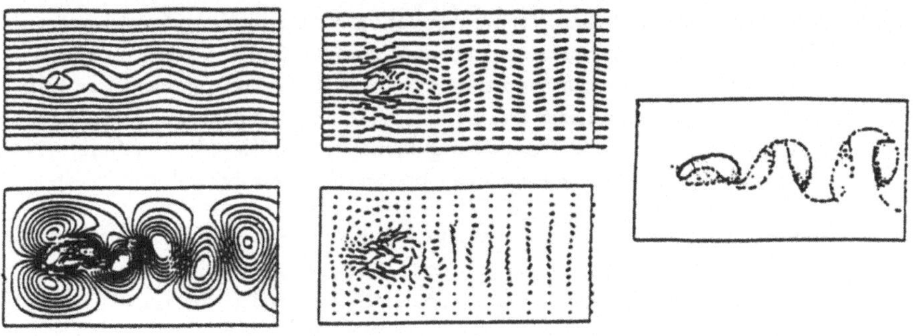

Figure 2: Different visualization techniques for 2D flow around an ellipse .

## The fully modular finite element solvers

They are based on the finite element package **FEAT** (in FORTRAN 77) with pseudody-namic memory management and basic finite element tools for grid generation, stiffness ma-trices, right hand sides and boundary components, and a corresponding **BLAS**–package with basic linear algebra tools and linear solvers. The most important components of our finite element code are the following. We delay a more detailed description to the subsequent papers in the reference list.

We use nonconforming "rotated bilinear", resp., "rotated trilinear" finite element spaces for the velocity (generated by $\langle x^2-y^2, x^2-z^2, x, y, z, 1\rangle$) at the midpoints of the edges/areas of the triangulation, and a piecewise constant pressure approximation. In this case an explicit construction of the discretely divergence–free subspaces satisfying $\int_T \nabla\cdot v_h^d \, dx = 0$ for all elements $T$ is possible. Then, the pressure is eliminated, and can be calculated in a postprocess by a marching process. This pair of finite element spaces fulfills the LBB–condition, and guarantees $h^2/h$ approximation properties. The amount of storage in 3D is about 750 Byte per cell, independent of the used mesh. The corresponding 2D

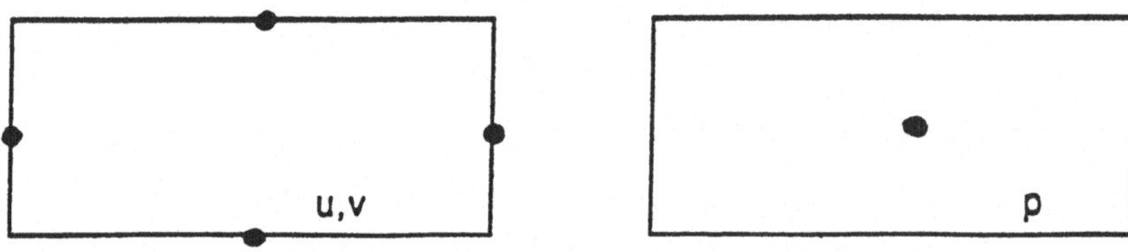


Figure 3: D.o.f.'s for the used finite elements in 2D

discretely divergence–free basis functions are living in the midpoints (approximating tan-gential velocities) and vertices (for streamfunction values). As a summary we can state the advantages: Elimination of the pressure, reduction of the number of unknowns, sym-metric positive definite stiffness matrices and approximation of the streamfunction (for graphics). The disadvantages are a condition number for stiffness matrices like $O(h^{-4})$, a condition number for mass matrices like $O(h^{-2})$ and the hard process of implementing the basis functions. For several boundary components no additional basis functions are needed, since so–called projection methods can be used for iterative solution methods.

Corresponding to these special finite elements we developed a suited multigrid algo-rithm with "makro-elementwise full interpolation". After a careful implementation this multigrid method with simple Jacobi/Gauss–Seidel/ILU–iteration for smoothing in the divergence–free case or a block Gauss–Seidel method in the primitive formulation , an adaptive step length control for the correction and the "full" interpolation leads to a very efficient and robust Stokes–solver with a numerical effort which is comparable to scalar Poisson–solvers for conforming quadratic finite elements.

For an efficient treatment of the Oseen–equations we use for the convective parts

$$b(\mathbf{u}, \mathbf{v}, \mathbf{w}) := \int_\Omega \mathbf{u}_i D_i \mathbf{v}_j \mathbf{w}_j \, dx$$

a central or an upstream discretization with lumping regions $\Lambda_l$ around each midpoint $B_l$ ($N$ number of midpoints)

$$\sum_{l=1}^{N} \sum_{k\in\Lambda_l} \int_{\Gamma_{lk}} \mathbf{un}^{lk} d\gamma \, (1 - \lambda_{lk}(\mathbf{u})) \, (\mathbf{v}_j(B_k) - \mathbf{v}_j(B_l)) \, \mathbf{w}_j(B_l).$$

Here, we have several choices for $\lambda_{lk}$ (with "local Re–number" $x := \dfrac{1}{\nu} \int_{\Gamma_{lk}} \mathbf{un}^{lk} d\gamma$):

$$\lambda_{lk}(\mathbf{u}) := \left\{ \begin{array}{ll} 1 & \text{if } x \geq 0 \\ 0 & \text{else} \end{array} \right\} \quad \text{or} \quad \lambda_{lk}(\mathbf{u}) := \left\{ \begin{array}{ll} \dfrac{\frac{1}{2}+\alpha x}{1+\alpha x} & \text{if } x \geq 0 \\[2mm] \dfrac{1}{2(1-\alpha x)} & \text{else} \end{array} \right\}.$$

Choice 1 (equivalent to "$\alpha = \infty$") leads to a very stable discretization scheme with M–matrices, but only of first order accuracy, while the second one (also called Samarski–scheme) has better approximation properties, but no complete theory in 3D is known.

We solve the steady nonlinear Navier–Stokes problems

$$A(\mathbf{u})\mathbf{u} := -\nu\Delta\mathbf{u} + (\mathbf{u} \cdot \nabla)\mathbf{u} + \nabla p = \mathbf{f}$$

by a fixed point–defect correction–method with adaptive step length control:

$$\mathbf{u}^{n+1} = \mathbf{u}^n + \omega^n [\bar{A}(\mathbf{u}^n)]^{-1}(\mathbf{f} - A(\mathbf{u}^n)\mathbf{u}^n).$$

This means, in each step an Oseen equation with unknown $\mathbf{u}$ and known coefficient function $\tilde{\mathbf{u}}$ must be solved:

$$-\nu\Delta\mathbf{u} + (\tilde{\mathbf{u}} \cdot \nabla)\mathbf{u} + \nabla p = \text{right hand side}, \ \nabla\cdot\mathbf{u} = 0.$$

This can be done if we use for $\bar{A}(\mathbf{u}^n)$ one of our proposed upwind–schemes, since the resulting matrices are similiar to M–matrices, and, therefore, can be easily inverted by our proposed multigrid–solver. The "optimal" value for $\omega^n$ can be calculated by a simple approximated defect minimization.

For solving the unsteady Navier–Stokes equations we use two different kinds of time discretization schemes:

<u>1) The one step $\theta$–scheme</u>

$$\frac{\mathbf{u}^{n+1} - \mathbf{u}^n}{\Delta t} = \theta A^{n+1}\mathbf{u}^{n+1} + (1 - \theta)A^n\mathbf{u}^n + \theta\mathbf{f}^{n+1} + (1 - \theta)\mathbf{f}^n.$$

These methods correspond to the usual explicit Euler–scheme ($\theta = 0$, first order), Crank-Nicolson ($\theta = 1/2$, implicit, A–stable), and the implicit Euler–scheme ($\theta = 1$, strongly A–stable).

2) The fractional step $\theta$-scheme

Choosing $\theta \in (0,1)$, $\theta' = 1 - 2\theta$ , and $\alpha \in [0.5,1]$, $\beta = 1 - \alpha$ , the time step $t_n \to t_{n+1}$ is split into three substeps as follows:

$$[I + \alpha\theta\Delta t A^{n+\theta}] \ \mathbf{u}^{n+\theta} \quad = \quad [I - \beta\theta\Delta t\, A^n]\mathbf{u}^n + \theta\Delta t\mathbf{f}^n$$

$$[I + \beta\theta'\Delta t A^{n+1-\theta}] \ \mathbf{u}^{n+1-\theta} \ = \ [I - \alpha\theta'\Delta t\, A^{n+\theta}]\mathbf{u}^{n+\theta} + \theta'\Delta t\, \mathbf{f}^{n+1-\theta}$$

$$[I + \alpha\theta\Delta t A^{n+1}] \ \mathbf{u}^{n+1} \quad = \quad [I - \beta\theta\Delta t\, A^{n+1-\theta}]\mathbf{u}^{n+1-\theta} + \theta\Delta t\, \mathbf{f}^{n+1} .$$

For the special choice $\theta = 1 - \sqrt{2}/2$ this scheme is of second order and strongly A-stable. For $\alpha = (1 - 2\theta)/(1 - \theta), \beta = \theta/(1 - \theta)$ the coefficient matrices are the same in all substeps. Therefore we think that this scheme is able to combine the advantages of the implicit Euler and Crank–Nicolson–scheme, without more additional numerical amount comparing with 3 one step $\theta$-scheme steps. These schemes applied to the Navier–Stokes equations lead to the following type of problems ($M$ mass matrix)

$$[M + \alpha\Delta t\, A(\mathbf{u}^{n+1})]\mathbf{u}^{n+1} = [M - (1 - \alpha)\Delta t\, A(\mathbf{u}^n)]\mathbf{u}^n + \alpha\Delta t\, \mathbf{f}^{n+1} + (1 - \alpha)\Delta t\, \mathbf{f}^n .$$

As boundary conditions we can prescribe unsteady velocity profiles, unsteady fluxes, unsteady pressure drops and unsteady tangential velocities (for cylinder rotations). Additionally we can calculate with unsteady forces or start from rest. At "free" boundaries (this means where we don't know what to prescribe, for instance at outflows), we use our Neumann-like "do nothing" conditions (see [3]).

## 3. Numerical examples and summary

By the numerical examples given we hope to show that our proposed components work well, even in really interesting tests which also have "real life" applications. The conditions on the robustness and fastness seem to be satisfied, only the accuracy should be improved, due to the fact that a rather simple upwinding and only linear elements are used and no adaptive strategies are implemented. So, we can state that we are developing a candidate for a "Black Box" solver, working efficiently in very general 2D and 3D domains, at least for midrange Reynolds numbers up to the size of $O(10^3)$ - $O(10^4)$ .

## References


[1] Blum, H., Harig, J., Müller, S., Turek, S.: **FEAT2D**. *Finite element analysis tools. User Manual. Release 1.3*, Technical report, University Heidelberg, 1992 .

[2] Girault, V., Raviart, P.A.: *Finite Element Methods for Navier–Stokes equations*, Springer, Berlin–Heidelberg 1986 .

[3] Heywood, J., Rannacher, R., Turek, S.: *Artificial boundaries and flux and pressure conditions for the incompressible Navier–Stokes equations*, Technical Report 681, SFB 123, 1992 .

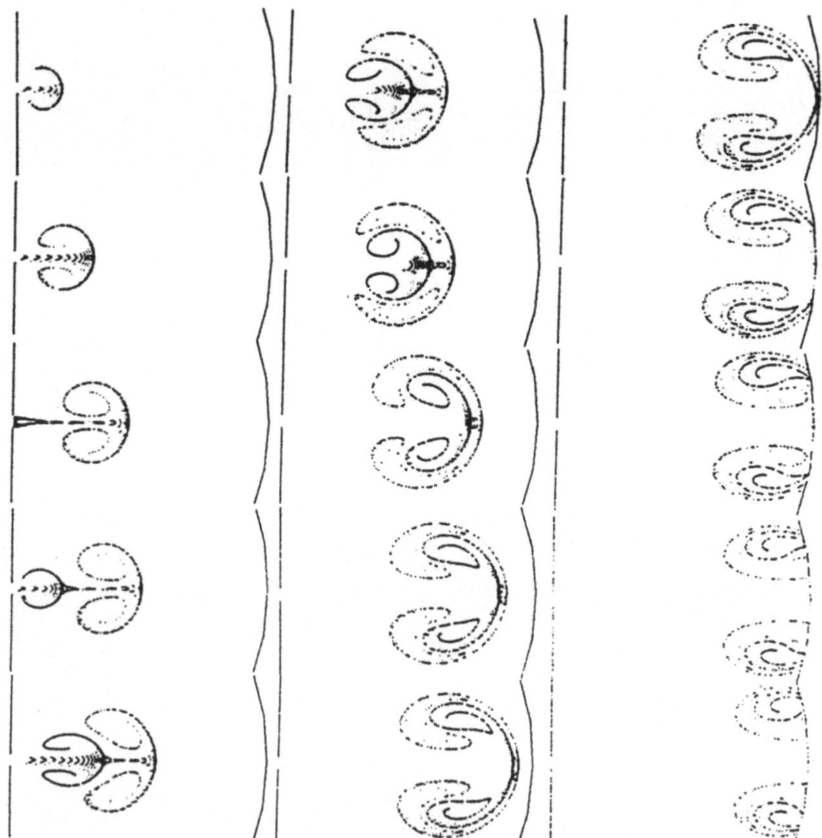

Figure 4: 2D particle tracing representation at different times of a flow through a hole ("Leapfrogging of two smoke rings"). The flow is driven by time dependent pressure drop boundary conditions. The Re–number is of order $O(10^2)$.


[4] Rannacher, R., Turek, S.: *A simple nonconforming quadrilateral Stokes element*, Numerical Methods for Partial Differential Equations, John Wiley and Sons, Inc., 8, 97–111 (1992).

[5] Turek, S.: *Tools for simulating non–stationary incompressible flow via discretely divergence–free finite element models*, Int. J. for Num. Meth. in Fluids, 18, 71–107 (1994).

$$P_- \equiv \lim_{\substack{|x|\to\infty \\ x_1<0}} p \qquad s \qquad P_+ \equiv \lim_{\substack{|x|\to\infty \\ x_1>0}} p$$

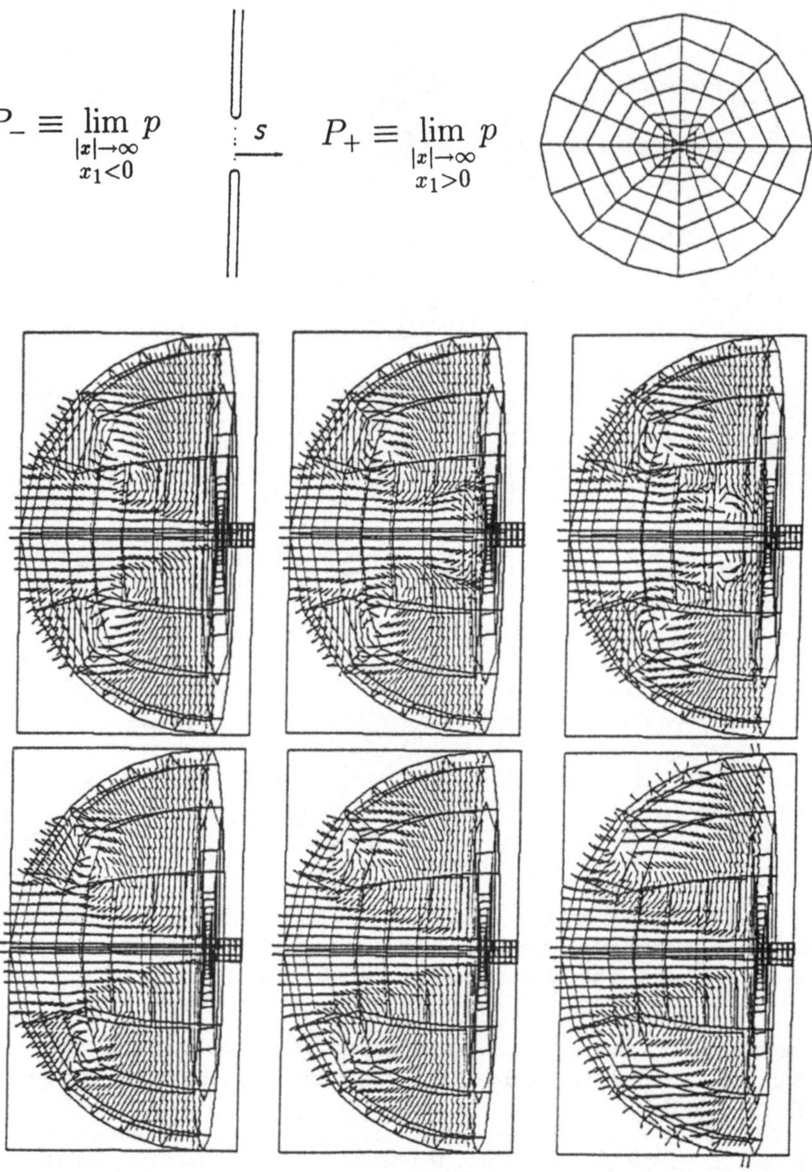

Figure 5: Vector plot representation of the same flow in 3D.

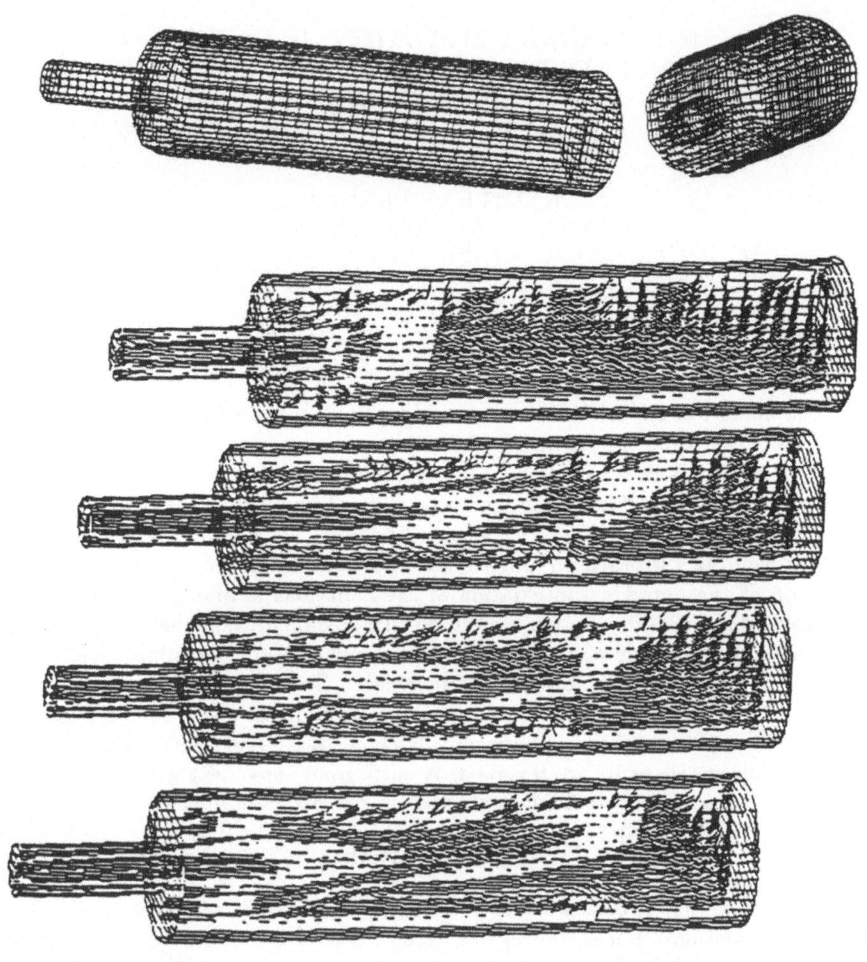

Figure 6: Vector plot representation at different times of a flow in a 3D pipe with five slits on the top. The flow is driven by time dependent pressure drop boundary conditions. The Re–number is of order $O(10^2)$.

# NUMERICAL SIMULATION OF UNSTEADY COMPRESSIBLE VISCOUS FLOW AROUND NACA 0012 AIRFOIL


C. Tenaud, Ta Phuoc Loc

L.I.M.S.I. - UPR CNRS 3251
BP.133, 91403 ORSAY Cedex, France


## Summary


The numerical simulation of the compressible flow passed a NACA 0012 airfoil has been undertaken using the Naviers-Stokes and total energy equations. The resolution of the governing equations is based on a Roe's approximate Riemann solver. The integration of these equations is performed by means of an implicit upwind Total Variation Diminishing (TVD) scheme, developed by Harten and Yee. The resolution of the set of the governing equations is based on an Alternating Direction Implicit (ADI) formulation. A description of the numerical integration of the equations is first given. The resolution of the governing equations is then described. Two configurations has been investigated. First, this numerical code was checked on the calculation of the compressible flow in the vicinity of a NACA 0012 airfoil without incidence. Secondly, the interaction between a vortex and a NACA 0012 airfoil has been investigated. In these two calculations, the Mach and Reynolds numbers are respectively equal to 0.6 and 22000., the $CFL$ number is 20. The numerical results are reviewed. A quite good agreement with numerical results obtain by other authors on similar configuration has been obtained.


## Introduction

The interaction of concentrated vortices with lifting surfaces occurs in many aerodynamic applications. For instance, this kind of interaction could be encountered in an helicopter rotor flow field, which is of particular interest. It is well known that in such a flow field, the interaction of trailing vortex wake with the oncomming rotor blades can induce unsteady blade loading and aerodynamic noise. One of the mechanisms which is responsible of noise generation is due to the instantaneous modification of the airfoil lift induced by the interaction with the vortex-wake. According to Srinivasan and McCroskey [3], there exist two limiting configurations to be encountered :

- the vortex axis is parallel to the leading edge of the airfoil, this is a two dimensional configuration which is mainly unsteady ;

- on the opposite, the vortex axis is normal to the leading edge, this is a quasi-steady three dimensional configuration.

It seems that, following several experimental results, the first configuration provides with the highest pressure variations which play an important role in the noise generation. Of course, this impulsive noise is mainly sensitive to the compressibility of the flow. This is the reason why, in this present study, we are mainly interested in the interaction of a concentrated vortex with a NACA 0012 airfoil. Therefor, in order to understand the mechanisms involved in the interaction of an incomming vortex with an airfoil, the numerical simulation of the compressible flow passed a NACA 0012 airfoil has been undertaken using the Naviers-Stokes and total energy equations. In order to obtain a well estimated solution, the numerical method should at least be conservative, have as weak as possible numerical dissipation and the integration scheme should be sufficiently accurate in both time and space. The resolution of the governing equations is based on a Roe's approximate Riemann solver [2]. The integration of these equations is performed by means of an implicit upwind Total Variation Diminishing (TVD) scheme, developed by Harten [1] and Yee [4]. The resolution of the set of the governing equations is based on an Alternating Direction Implicit (ADI) formulation. A description of the numerical integration of the equations is first given. The resolution of the governing equations is then described. Finally, the results, calculated for a peculiar configuration, are discussed and compared with results on rather similar configuration obtained by other authors.

### Numerical approach

**Basic equations.** The equations of the flow are the Navier-Stokes and total energy equations written in conservative form in a cartesian coordinate system as follows :

$$\frac{\partial Q}{\partial t} + \frac{\partial F}{\partial x} + \frac{\partial G}{\partial y} + \frac{\partial F_v}{\partial x} + \frac{\partial G_v}{\partial y} = 0. \tag{1}$$

$x$ and $y$ represent respectively, the streamwise and the vertical directions. $Q$ is the solution vector and $F$ et $G$ are the convective fluxes and $F_v$ and $G_v$ stand for the viscous fluxes :

$$Q = \begin{bmatrix} \rho \\ \rho u \\ \rho v \\ \rho E \end{bmatrix} \quad ; \quad F = \begin{bmatrix} \rho u \\ \rho u^2 + P \\ \rho u v \\ \rho u E + u P \end{bmatrix} \quad ; \quad G = \begin{bmatrix} \rho v \\ \rho u v \\ \rho v^2 + P \\ \rho v E + v P \end{bmatrix}$$

$$F_v = \begin{bmatrix} 0 \\ -\sigma_{xx} \\ -\sigma_{xy} \\ -u\,\sigma_{xx} - v\,\sigma_{xy} + \Phi_x \end{bmatrix} \quad ; \quad G_v = \begin{bmatrix} 0 \\ -\sigma_{xy} \\ -\sigma_{yy} \\ -u\,\sigma_{xy} - v\,\sigma_{yy} + \Phi_y \end{bmatrix}$$

where the stress tensor components are :

$$\sigma_{xx} = 2\mu\frac{\partial u}{\partial x} - \frac{2}{3}\Theta \quad ; \quad \sigma_{yy} = 2\mu\frac{\partial v}{\partial y} - \frac{2}{3}\Theta \quad ; \quad \sigma_{xy} = 2\mu\left(\frac{\partial u}{\partial x} + \frac{\partial v}{\partial y}\right).$$

$\Theta$ is the dilatation and $\Phi_x$, $\Phi_y$ the heat fluxes :

$$\Theta = \left(\frac{\partial u}{\partial x} + \frac{\partial v}{\partial y}\right) \quad ; \quad \Phi_x = -\lambda\frac{\partial T}{\partial x} \quad ; \quad \Phi_y = -\lambda\frac{\partial T}{\partial y}.$$

The static pressure $P$ and temperature $T$ are calculated through the use of the state equation written for a perfect gas. The dynamic viscosity ($\mu$) as well as the heat conductivity ($\lambda$) are supposed to be only function of the mean temperature. For the dynamic viscosity, two different laws are employed depending on the local mean temperature value :

$$\begin{cases} \mu = \mu_\infty \left(\frac{T}{T_\infty}\right)^{3/2} \frac{T_\infty + 110.}{T + 110.} & if \quad T \geq 120.K \\ \mu = \mu_{120}\frac{T}{120.} & if \quad T < 120.K. \end{cases}$$

In these relationships, $\mu_\infty$ and $T_\infty$ represent the values of viscosity and temperature at infinity ; $\mu_{120}$ is calculated from the Sutherland's law for $T = 120.K$ to ensure continuity.

The heat conductivity is expressed through the use of a constant Prandtl number assumption : $\lambda = \frac{\mu C_p}{\mathcal{P}}$ with $\mathcal{P} = 0.725$.

**Grid and coordinate transformation**  The grid employed is a C-grid which fits to the body. The mesh is tightened near the wall in the normal to the wall direction ($\eta$) as well as in the vicinity of the leaing and trailling edge of the airfoil in the direction parallel to the wall ($\xi$). Therefor, a typical mesh size is about 300 points in the $\xi$ direction and 100 points following $\eta$. To solve the governing equations, a generalized coordinate transformation ($\xi = \xi(x,y)$ and $\eta = \eta(x,y)$) is used which preserves the conservative form of equations (1) written as follow :

$$J\frac{\partial Q}{\partial t} + \frac{\partial \widehat{F}}{\partial \xi} + \frac{\partial \widehat{G}}{\partial \eta} + \frac{\partial \widehat{F_v}}{\partial \xi} + \frac{\partial \widehat{G_v}}{\partial \eta} = 0. \tag{2}$$

where $J = x_\xi y_\eta - x_\eta y_\xi$ is the jacobian of the coordinate transformation and the fluxes in $\xi$ and $\eta$ directions are written respectively :

$$\widehat{F} = (y_\eta F - x_\eta G) \quad ; \quad \widehat{G} = (-x_\eta F + x_\xi G) \, .$$

Let $\widehat{A} = \partial\widehat{F}/\partial Q$ and $\widehat{B} = \partial\widehat{G}/\partial Q$ be the jacobians of the Euler flux function ; they can be written as :

$$\widehat{A} = (y_\eta A - x_\eta B) \quad ; \quad \widehat{B} = (-x_\eta A + x_\xi B) \, .$$

$\widehat{A}$ and $\widehat{B}$ have a complete set of 4 eigenvalues ($a_\xi^l$ and $a_\eta^l$) and eigenvectors ; we denote $R_\xi$ and $R_\eta$ the matrices whose columns are those eigenvectors.

**Time and space discretization.**  The resolution of the Navier-Stokes and total energy equations (2) has been performed by means of a finite difference method. If we consider a grid point where $\xi = i.\Delta\xi$ and $\eta = j.\Delta\eta$, the implicit discretisation of equations (2) is written :

$$
\begin{aligned}
\Delta^n Q_{ij} \; &+ \; \frac{\Delta t}{\Delta\xi}\theta\left[\overline{F}^{n+1}_{i+1/2,j} - \overline{F}^{n+1}_{i-1/2,j} + \overline{F_v}^{n+1}_{i+1/2,j} - \overline{F_v}^{n+1}_{i-1/2,j}\right] \\
&+ \; \frac{\Delta t}{\Delta\eta}\theta\left[\overline{G}^{n+1}_{i,j+1/2} - \overline{G}^{n+1}_{i,j-1/2} + \overline{G_v}^{n+1}_{i,j+1/2} - \overline{G_v}^{n+1}_{i,j-1/2}\right] \\
&= \; -\frac{\Delta t}{\Delta\xi}(1-\theta)\left[\overline{F}^n_{i+1/2,j} - \overline{F}^n_{i-1/2,j} + \overline{F_v}^n_{i+1/2,j} - \overline{F_v}^n_{i-1/2,j}\right] \\
&- \; \frac{\Delta t}{\Delta\eta}(1-\theta)\left[\overline{G}^n_{i,j+1/2} - \overline{G}^n_{i,j-1/2} + \overline{G_v}^n_{i,j+1/2} - \overline{G_v}^n_{i,j-1/2}\right]
\end{aligned}
\tag{3}
$$

where $\Delta^n Q_{ij} = \left(Q^{n+1}_{ij} - Q^n_{ij}\right)$ If $\theta$ is not equal to zero, we have an implicit scheme. Here, $\theta = 1/2$ and this scheme is second order accurate in time.

As regards the convective terms, their integration is based on a Roe's approximate Riemann solver [2]. These equations are projected along the characteristic direction ; a set of 4 characteristic equations is then obtained. Therefor, in order to integrate these equations, a scalar scheme could be applied on. This integration is performed by means of a shock capturing finite difference method based on an implicit upwind Total Variation

Diminishing (TVD) scheme, developed by Harten [1] and Yee [4]. The numerical flux function $\overline{F}_{i+1/2,j}$ is expressed following :

$$\overline{F}_{i+1/2,j} = \frac{1}{2}\left(\widehat{F}_{i+1,j} + \widehat{F}_{i,j} - R_{\xi_{i+1/2,j}}\Phi_{\xi_{i+1/2,j}}\right).$$

$\Phi_{\xi_{i+1/2,j}}$ is a 4x4 matrix for which the elements are denoted $\phi^l_{\xi_{i+1/2,j}}$ ; in order to ensure that this scheme is a second order upwind TVD scheme, the $\phi$ elements must be written [1], [4] :

$$\phi^l_{\xi_{i+1/2,j}} = \sigma\left(a^l_{\xi_{i+1/2,j}}\right)\left(g^l_{\xi_{i+1,j}} + g^l_{\xi_{i,j}}\right) - \psi\left(a^l_{\xi_{i+1/2,j}} + \gamma^l_{\xi_{i+1/2,j}}\right)\alpha^l_{\xi_{i+1/2,j}}$$

with $\alpha_{\xi_{i+1/2,j}} = R^{-1}_{\xi_{i+1/2,j}}(Q_{i+1,j} - Q_{i,j})$ is the forward difference of the local charcateristic variables in the $\xi$ direction

$$\psi(z) = |z| \qquad ; \qquad \sigma(z) = \frac{1}{2}\left(\psi(z) - \frac{\Delta t}{\Delta\xi}z^2\right)$$

$$\gamma_{\xi_{i+1/2,j}} = \sigma\left(a^l_{\xi_{i+1/2,j}}\right)\begin{cases}\left(g_{\xi_{i+1,j}} - g_{\xi_{i,j}}\right)/\alpha_{\xi_{i+1/2,j}} & if \quad \alpha_{\xi_{i+1/2,j}} \neq 0 \\ 0 & if \quad \alpha_{\xi_{i+1/2,j}} = 0.\end{cases}$$

And the limitor function is expressed using the relationship proposed by Van-Leer [5] :

$$g^l_{\xi_{i,j}} = \left(\alpha_{\xi_{i+1/2,j}}\alpha_{\xi_{i-1/2,j}} + \left|\alpha_{\xi_{i+1/2,j}}\alpha_{\xi_{i-1/2,j}}\right|\right)/\left(\alpha_{\xi_{i+1/2,j}} + \alpha_{\xi_{i-1/2,j}}\right).$$

In order to evaluate the terms at the point (i+1/2,j), we employ a mass-weighted averaging technique introduce by Roe [2] between states $Q_{i+1,j}$ and $Q_{i,j}$. Note that this scheme is second order accurate in both time and space.
Similar formulations as in streamwise direction are used to express the numerical flux functions in the $\eta$ direction.

Concerning the diffusive flux function ($\widehat{F}_v$ and $\widehat{G}_v$), a central differencing scheme is applied giving a second order accurate in space.


**Resolution and numerical procedure.** We need to solve a set of nonlinear equations in order to know $Q^{n+1}$. Concerning the convective flux functions, we must linearize the implicit operator. Therefor, we have used an linearized conservative implicit form [4] which can be written in Delta form as :

$$\left[I + \frac{\Delta t}{\Delta\xi}\left(H_{\xi_{i+1/2,j}} - H_{\xi_{i-1/2,j}}\right) + \frac{\Delta t}{\Delta\eta}\left(H_{\xi_{i,j+1/2}} - H_{\xi_{i,j-1/2}}\right)\right] \times \Delta^n Q_{ij} \qquad (4)$$

$$= -\frac{\Delta t}{\Delta\xi}\left(\overline{F}^n_{i+1/2,j} - \overline{F}^n_{i-1/2,j}\right) + \frac{\Delta t}{\Delta\eta}\left(\overline{G}^n_{i,j+1/2} - \overline{G}^n_{i,j-1/2}\right)$$

with : $H_{\xi_{i+1/2,j}} = \frac{1}{2}\left(\widehat{A}^n_{i+1,j} + \Omega^n_{\xi_{i+1/2,j}}\right)$ and $\Omega_{\xi_{i+1/2,j}} = \left(R_\xi [D] R_\xi^{-1}\right)_{i+1/2,j}$.
$D_{i+1/2,j}$ is a diagonal matrix, its elements are :
$D_{i+1/2,j} = \left(g^l_{\xi_{i+1,j}} + g^l_{\xi_{i,j}}\right)/\alpha^l_{\xi_{i+1/2,j}} - \psi(a^l_\xi + \gamma^l_\xi)_{i+1/2,j}$.
Notice that this scheme is spacially second order accurate ; nevertheless, it may no longer be unconditionally TVD.

The implicit operator is factorized using an Alternating Direction Implicit (ADI) form [4]. Therefor, in each coordinate direction, a block tridiagonal matrix is solved using a classic algorithm.

In the present calculations, the Courant number (CFL) is prescribed. The time step ($\Delta t$) is then calculated using the minimum value on the grid of the local time step expressed by the use of the CFL and the spectral radius of the jacobian matrices of the Euler flux functions.

**Boundary conditions.** On the airfoil, a no-slip condition is applied and a wall temperature ($T_w$) is prescribed. At the inlet ($\eta = \eta_{max}$), we suppose that the flow is inviscid and the governing equations are projected along the charcateristic directions. Therefor, if the characteristic speeds are positive we employ an upwind discretization of the derivative in the $\eta$ direction ; while if the characteristic speeds are negative, we impose that there is no variation in time. At the oulet ($\xi = 0.$ or $\xi = \xi_{max}$), the $\xi$ derivatives are supposed to be equal to zero. let us add that all the boundary conditions are applied in the implicit part of the algorithm.

## Results and discussion

First, this numerical code was checked on the calculation of the flow in the vicinity of a NACA 0012 airfoil without incidence (without vortex interaction). The Mach number is equal to 0.6, the temperature is $300^o K$ and the Reynolds number, based on the chord length, is about 22000. The $CFL$ number is equal to 20. The pressure coefficient ($Cp$) distribution along the chord of the airfoil [Fig. 1] is in good agreement with the results obtained by Srinivasan and McCroskey [3] on a similar configuration. The plateau like-shape behaviour situated near the trailing edge of the airfoil is due to a slight recirculation zone [Fig. 1]. According to the distribution of the skin friction coefficient, the recirculation region starts at 0.65 chord length from the leading edge which corresponds to the region of the plateau like-shape evolution on the $Cp$ distribution.

Secondly, the interaction between a vortex and a NACA 0012 airfoil has been undertaken. The dimensionless strength of the vortex is 0.2 ; it is situated initialy at the following location : $x_v/C = -0.6$ ; $y_v/C = -0.25$ (where $C$ is the chord length and $x = 0.$, $y = 0$ refers to the leading edge of the airfoil). The Mach and Reynolds numbers are respectively equal to 0.6 and 22000., the $CFL$ number is 20. The calculation starts from the steady state solution using the previous results. At the initial state, as the imposed strength of the vortex is positive, the vortex induces a negative pitching moment on the airfoil. Nevertheless, the vortex is imposed initially too close to the leading edge of the airfoil therefore, as we can see on the time variation of the lift coefficient ($C_z$) [Fig. 3], a positive lift force is induced initially on the airfoil. When the vortex is close to the leading edge of the airfoil $x_v/C = 0.$ [Fig. 2], the distribution of the pressure coefficient ($C_p$) on the upper side of the airfoil stays comparable to the undisturbed distribution (that means the distribution obtained without vortex interaction) ; while the $C_p$ distribution on the lower side is greatly affected [Fig. 2]. Then, the vertical force on the airfoil induces by the pressure distribution is positive. However, when the vortex is located at about the mid-chord of the airfoil ($x_v/C = 0.5$) [Fig. 2], the sign of the lift force due to the pressure distribution has changed and became negative. When we look at the time variation of

$C_z$ [Fig. 3], we see that the change in the $C_z$ sign occurs when the vortex is located at about $x/C = 0.3$ which corresponds roughly at the position of the maximum thickness of the airfoil. At the vortex location $x_v/C = 0.5$, the $C_p$ distribution on the lower side of the airfoil has almost recorvered the undisturbed distribution, except in the vicinity of the trailing edge. On the upper side, the pressure distribution seem to be greatly affected everywhere along the chord length. When the vortex is situated at $x_v/C = 1.$, corresponding to the trailing edge of the airfoil, the pressure distribution on the lower side is only affected close to the trailing edge ; while, on the lower side, the $C_p$ distribution is modified all along the chord length, mainly at the trailing edge, where unsteady, extended recirculation regions occur. Time-steps later, when the vortex is located one and half chord downstream of the airfoil, the pressure distribution on the upper side of the profile tends to recover the undisturbed distribution [Fig. 2]. That time, the lift coefficient ($C_z$) oscillates and tends to recover its undisturbed value ($C_z = 0.$) [Fig. 3].

At last, we could see on the isovalues of $V$ [Fig. 4] that, when the vortex is close to the leading edge of the airfoil, alternate vorticies occur in the vicinity of the trailing edge of the airfoil, therefore a vortex wake is created. Then, the more the interacting vortex moves downstream, the more the vortex-wake grows [Fig. 4]. Nevertheless, one chord downstream of the trailing edge [Fig. 4], the vortex-wake seems to be dissipated. This might be explained by a too important numerical diffusion due to a too coarse grid in the $\xi$ direction one chord downstream of the airfoil. On the static pressure field [Fig. 5], as the vortex is imposed too close to the leading edge of the airfoil, pressure waves are visible which grow from the wall of the airfoil to the upper boundary of the computation domain. Let us mention that no reflexion of the pressure wave is observed using the appropriate boundary condition on the external boundary. When the cocentrated vortex is situated dowstream of the airfoil, pressure waves are then created from each side of the vortex wake [Fig. 5]. These waves might be generated by the alternate vorticies. To get better results one chord downstream of the airfoil than those obtained, finer grid added to a subgrid scale modelling should be used.

## Conclusion

The numerical simulation of the compressible flow passed a NACA 0012 airfoil has been undertaken using the Navier-Stokes and total energy equations. The integration of these governing equations is based on a Roe's approximate Riemann solver through the use of an implicit upwind TVD scheme, devolopped by Harten [1] and Yee [4]. The resolution of the set of the governing equations is based on an Alternating Direction Implicit (ADI) formulation. To check the numerical code, two configurations has been investigated. First, this numerical code was checked on the calculation of the compressible flow in the vicinity of a NACA 0012 airfoil without incidence. The Mach and Reynolds numbers are respectively equal to 0.6 and 22000., the $CFL$ number is 20. A good agreement with numerical results obtained by Srinivasan and McCroskey [3], is achieved on the distribution of the pressure coefficient ($Cp$) along the chord of the airfoil. Secondly, the interaction between a vortex and a NACA 0012 airfoil has been investigated. The initial conditions are those obtained in the first calculation. At the initial state, we have imposed a positive strength vortex ($\Gamma = 0.2$). The influence of the isolated vortex on the distribution of $Cp$ along the chord of the NACA 0012 airfoil is then reviewed. When the concentrated votex is close to the leading edge of the airfoil, alternate vorticies occur in the vicinity of the trailing edge of the airfoil, therefore a vortex wake is created. The more the interacting vortex moves downstream, the more the vortex-wake grows. Therefor, due to these alternate vorticies, pressure waves have been created from each side of the vortex wake. Nevertheless, one chord downstream of the trailing edge, the vortex-wake seems to be dissipated. This might be explained by a too important numerical diffusion due to a too coarse grid in the $\xi$ direction one chord downstream of the airfoil. To get better results for this configuration, finer grid added to a subgrid scale modelling should be used. This approach is actually undertaken using a three different subgrid scale models.

# References


[1] **A. Harten** *High Resolution Schemes for Hyperbolic Conservation Laws.* Journal of Computational Physics Vol. 49, pp. 357-393, (1983).

[2] **P.L. Roe** *Approximate Riemann Solvers, Parameter Vectors and Difference Schemes.* Journal of Computational Physics Vol. 43, pp. 357-372, (1981).

[3] **G.R. Srinivasan - W.J. McCroskey** *Numerical Simulations of Unsteady Airfoil-Vortex Interactions.* Vertica Vol. 11, N° 2, pp. 3-28, (1987).

[4] **H.C. Yee - A. Harten** *Implicit TVD Schemes for Hyperbolic Conservation Laws in Curvilinear Coordinates.* AIAA Journal Vol. 25, N° 2, pp. 266-274, (1987).

[5] **H.C. Yee** *Construction of Explicit and Implicit Symmetric TVD Schemes and Their Applications.* Journal of Computational Physics Vol. 68, pp. 151-179, (1987).

[6] **H.C. Yee** *Numerical Experiments With a Symmetric High Resolution Shock-Capturing Scheme.* NASA TM-88325, (1986).

[7] **B. Van Leer** *Towards the Ultimate Finite Difference Scheme II : Monotonicity and Conservation combined in a second order scheme.* Journal of Computational Physics Vol. 14, pp. 361-370, (1974).

[8] **C. Tenaud - T.P. Loc** *Numerical Simulation of Unsteady Compressible Viscous Flow Around NACA 0012 Airfoil.* Workshop on Numerical Methods for the Navier-Stokes equations, Heidelberg, Oct. 1993.


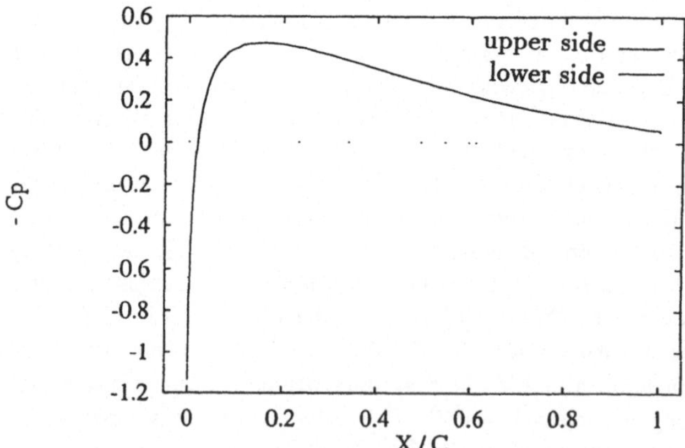


Figure 1: Distribution of the pressure coefficient along a NACA 0012 airfoil without vortex interaction ($M_\infty = 0.63$ ; $Re_C = 22000$.)

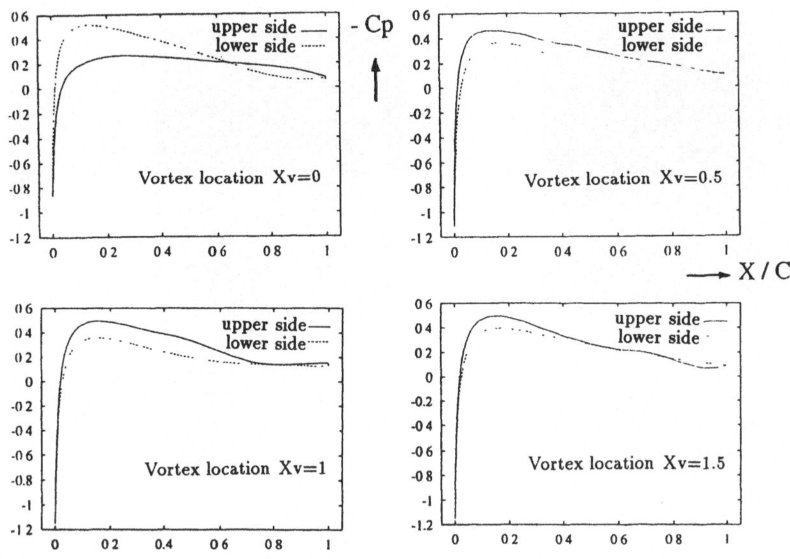

Figure 2: Distribution of the pressure coefficient along a NACA 0012 airfoil with vortex interaction ($M_\infty = 0.63$ ; $Re_C = 22000.$ ; $\Gamma = 0.2$)

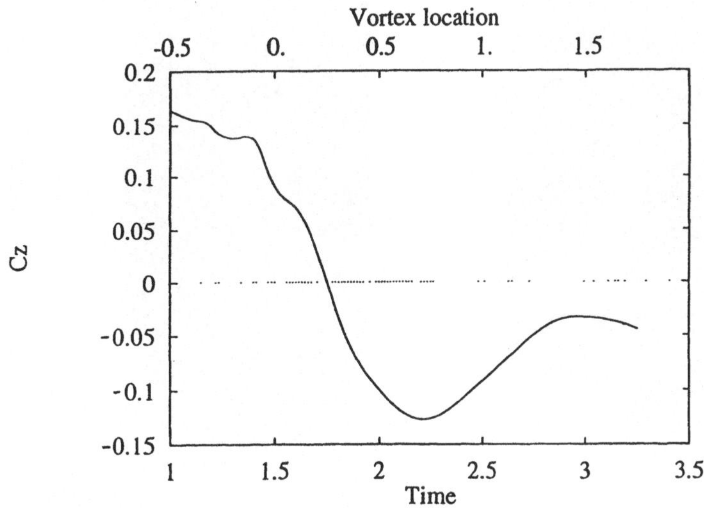

Figure 3: Time variation of the lift coefficient $C_z$ during vortex-NACA 0012 airfoil interaction ($M_\infty = 0.63$ ; $Re_C = 22000.$ ; $\Gamma = 0.2$)

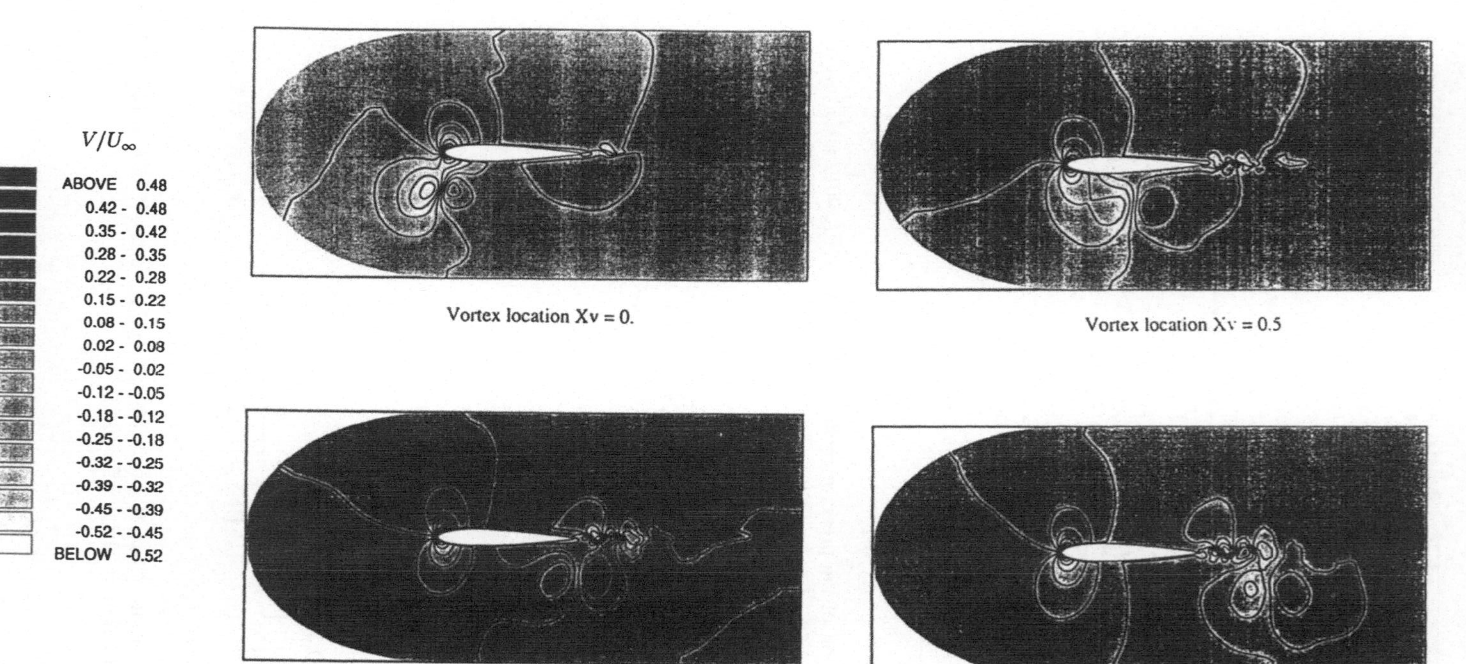



Figure 4: vortex-NACA 0012 airfoil interaction ($M_\infty = 0.63$ ; $Re_C = 22000.$ ; $\Gamma = 0.2$) : vertical velocities field

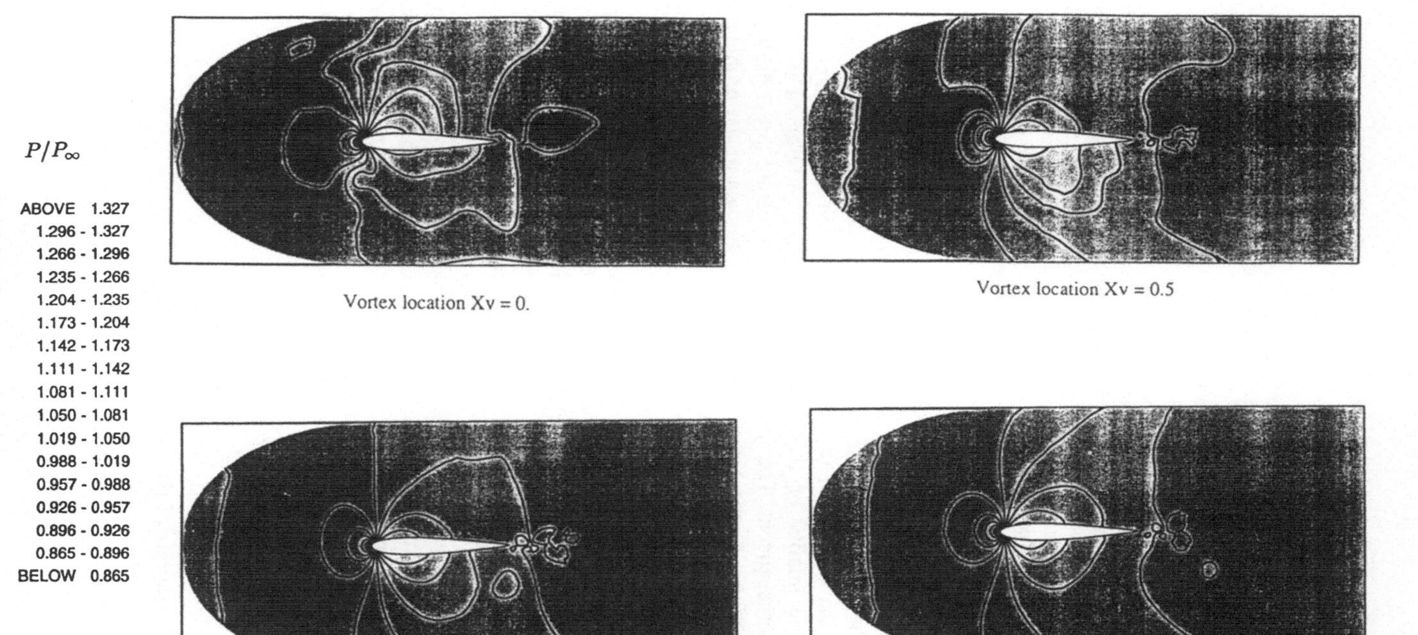



Figure 5: vortex-NACA 0012 airfoil interaction ($M_\infty = 0.63$ ; $Re_C = 22000.$ ; $\Gamma = 0.2$) : static pressure field

# A POSTERIORI ERROR ESTIMATES
# FOR NON-LINEAR PROBLEMS

*R. Verfürth*

Fakultät für Mathematik, Ruhr-Universität Bochum, Universitätsstr. 150, D-44780 Bochum, Germany


**Abstract**

We give a general framework for deriving a posteriori error estimates for approximate solutions of non-linear problems. In a first step it is proven that the error of the approximate solution can be bounded from below and from above by an appropriate norm of its residual. In a second step this norm of the residual is bounded from above and from below by a similar norm of a suitable finite dimensional approximation of the residual. This quantity can easily be evaluated and for many practical applications sharp explicit upper and lower bounds are readily obtained. The general results are applied to finite element discretizations of scalar quasilinear elliptic partial differential equations of 2nd order and yield residual a posteriori error estimates which can easily be computed from the given data of the problem and the computed numerical solution and which give global lower bounds on the error of the numerical solution.


## 1. Introduction

The efficiency of a numerical method for the solution of partial differential equations strongly depends on the choice of an "optimal" discretization, the use of a fast and efficient algorithm for the solution of the discrete problem, and a simple, but reliable method for judging the quality of the obtained numerical solution. These three objectives are often interdependent. The first and last one are related to the problem of a posteriori error estimation, i.e., of extracting from the given data of the problem and the computed numerical solution reliable bounds on the error of the numerical solution. Of course, the computation of the a posteriori error estimates should be much less costly than the solution of the original discrete problem.

Within the framework of finite element methods various strategies of a posteriori error estimation have been devised during the last 15-20 years (cf., e.g., [1,2,11,12] and the literature cited there). They often depend on a particular class of problems and discretizations. A close inspection, however, reveals that they have certain principles in common. It is the aim of this paper to give a rather general framework that allows to construct a posteriori error estimators and to prove that they yield upper and lower bounds on the error. In this general context we are satisfied with proving that the upper and lower bounds differ by a multiplicative constant which is independent of the mesh-size. We neither intend to derive optimal estimates for this constant nor to prove efficiency of the error estimators, i.e., that the ratio of the true and the estimated error asymptotically tends to 1. This latter question is addressed for linear problems in e.g. [1-6,9,10].

We consider non-linear equations of the form

$$F(u) = 0 \tag{1.1}$$

and corresponding discretizations of the form

$$F_h(u_h) = 0. \tag{1.2}$$

Here, $F \in C^1(X, Y^*)$ and $F_h \in C(X_h, Y_h^*)$, $X_h \subset X$ and $Y_h \subset Y$ are finite dimensional subspaces of the Banach spaces $X$ and $Y$, and $*$ denotes the dual of a Banach space.

If $u_0 \in X$ is a solution of Equation (1.1) such that $DF(u_0)$ is an isomorphism of $X$ onto $Y^*$ and $DF$ is Lipschitz continuous at $u_0$, we prove in Proposition 2.1 that

$$\underline{c}\|F(u)\|_{Y*} \le \|u - u_0\|_X \le \bar{c}\|F(u)\|_{Y*} \tag{1.3}$$

holds for all u in a suitable neighborhood of $u_0$. The constants $\underline{c}$ and $\bar{c}$ depend on $DF(u_0)$ and $DF(u_0)^{-1}$.

In Section 3, we estimate the residual $\|F(u_h)\|_{Y*}$ where $u_h$ is an approximate solution of Equation (1.2). To this end we introduce a restriction operator $R_h : Y \to Y_h$, a finite dimensional subspace $\tilde{Y}_h \subset Y$, and an approximation $\tilde{F}_h : X_h \to Y^*$ of F at $u_h$ which are coupled via Inequality (3.1). For practical applications, the construction of $R_h$ and $\tilde{F}_h$ is rather straightforward. Usually, $\tilde{F}_h(u_h)$ is obtained by locally projecting $F(u_h)$ onto suitable finite dimensional spaces. This corresponds to the well-known technique of locally freezing the coefficients of a differential operator. We then prove in Proposition 3.1 that, up to multiplicative constants and additive correction terms, $\|F(u_h)\|_{Y*}$ is bounded from below and from above by $\|\tilde{F}_h(u_h)\|_{\tilde{Y}_h^*}$. The latter can be evaluated quite easily since its computation is equivalent to a finite dimensional maximization problem. Moreover, sharp explicit bounds on $\|\tilde{F}_h(u_h)\|_{\tilde{Y}_h^*}$ are readily obtained for many pratical applications.

The construction of $\tilde{Y}_h$ is considerably simplified by the following auxiliary result (cf. Lemma 4.1)

$$0 \le \alpha \le \inf_{u \in V_S \setminus \{0\}} \sup_{u \in V_S \setminus \{0\}} \frac{\int_S u \psi_S v}{\|u\|_{L^p(S)} \|v\|_{L^q(S)}} \le 1. \tag{1.4}$$

Here, $1 \le p \le \infty$, $\frac{1}{p} + \frac{1}{q} = 1$, S is either a simplex in $IR^n$ or a face of such a simplex, $V_S$ is a finite dimensional space of functions defined on S, and $\psi_S$ is a cut-off function. It is important to note that the constant $\alpha$ is independent of S. Thanks to Inequality (1.4) one can show that for finite element methods $\tilde{Y}_h$ can be chosen as the space of all linear combinations of functions $\psi_S v$ where $v \in V_S$ and S varies through all elements and their faces.

In Section 5 we apply the general results of the previous sections to finite element approximations of scalar quasilinear elliptic partial differential equations of 2nd order. We obtain upper and lower bounds for the finite element error in terms of a residual a posteriori error estimator. This error estimator essentially consists of the elementwise error of the finite element functions with respect to the strong form of the differential equation and of jumps across inter-element boundaries of that boundary operator which naturally links the strong and weak forms of the differential equation.

## 2. Error Estimation via the Residual

Let X, Y be two Banach Spaces with norms $\|.\|_X$ and $\|.\|_Y$. For any element $u \in X$ and any real number $R > 0$ set $B(u,R) := \{v \in X: \|u-v\|_x < R\}$. $\mathcal{L}(X,Y)$ and $Isom(X,Y) \subset \mathcal{L}(X,Y)$ denote the Banach space of continuous linear maps of X and Y equipped with the operator-norm $\|.\|_{\mathcal{L}(X,Y)}$ and the open subset of linear homeomorphisms of X onto Y. $Y^* := \mathcal{L}(Y,IR)$ and $<.,.>$ are the dual space of Y and the corresponding duality pairing. Finally, $A^* \in \mathcal{L}(Y^*,Y^*)$ denotes the adjoint of a given linear operator $A \in \mathcal{L}(Y,Y)$.

Let $F \in C^1(X,Y^*)$ be a given continuously differentiable funtion. The following proposition yields a posteriori error estimates for elements in a neighborhood of a solution of Equation (1.1).

**Proposition 2.1.** Let $u_0 \in X$ be a regular solution of Equation (1.1), i.e. $DF(u_0) \in$ Isom$(X,Y^*)$. Assume that $DF$ is Lipschitz continuous at $u_0$, i.e., there is an $R_0 > 0$ such that

$$\gamma := \sup_{u \in B(u_0, R_0)} \frac{\|DF(u)-DF(u_0)\|_{\mathcal{L}(X,Y^*)}}{\|u-u_0\|_X} < \infty .$$

Set

$$R := \min\{R_0, \ \gamma^{-1}\,\|DF(u_0)^{-1}\|^{-1}_{\mathcal{L}(Y^*,X)}, \ 2\gamma^{-1}\,\|DF(u_0)\|_{\mathcal{L}(X,Y^*)}\} .$$

Then the following error estimates hold for all $u \in B(u_0,R)$:

$$\frac{1}{2}\,\|DF(u_0)\|^{-1}_{\mathcal{L}(X,Y^*)}\|F(u)\|_{Y^*} \leq \|u-u_0\|_X \leq 2\|DF(u_0)^{-1}\|_{\mathcal{L}(Y^*,X)}\|F(u)\|_{Y^*} . \tag{2.1}$$

**Proof:** Let $u \in B(u_0,R)$. We then have

$$u-u_0 = DF(u_0)^{-1}\left\{ F(u) + \int_0^1 [DF(u_0)-DF(u_0+t(u-u_0))]\,(u-u_0)dt \right\}$$

and thus

$$\|u-u_0\|_X \leq \|DF(u_0)^{-1}\|_{\mathcal{L}(Y^*,X)} \left\{ \|F(u)\|_{Y^*} + \int_0^1 \|DF(u_0)-DF(u_0+t(u-u_0))\|_{\mathcal{L}(X,Y^*)}\|u-u_0\|_X dt \right\}$$

$$\leq \|DF(u_0)^{-1}\|_{\mathcal{L}(Y^*,X)} \left\{ \|F(u)\|_{Y^*} + \frac{1}{2}\gamma\|u-u_0\|_X^2 \right\}$$

$$\leq \|DF(u_0)^{-1}\|_{\mathcal{L}(Y^*,X)}\,\|F(u)\|_{Y^*} + \frac{1}{2}\,\|u-u_0\|_X .$$

This yields the second inequality in (2.1).
On the other hand, we have for all $\varphi \in Y$ with $\|\varphi\|_Y = 1$

$$\langle F(u), \varphi\rangle = \langle DF(u_0)(u-u_0),\varphi\rangle + \langle\int_0^1 [DF(u_0+t(u-u_0))-DF(u_0)](u-u_0)dt,\varphi\rangle \tag{2.2}$$

and thus

$$\|F(u)\|_{Y^*} \leq \|DF(u_0)\|_{\mathcal{L}(X,Y^*)}\,\|u-u_0\|_X + \int_0^1 \|DF(u_0+t(u-u_0))-DF(u_0)\|_{\mathcal{L}(X,Y^*)}\|u-u_0\|_X dt$$

$$\leq \|DF(u_0)\|_{\mathcal{L}(X,Y^*)}\|u-u_0\|_X + \frac{1}{2}\gamma\|u-u_0\|_X^2$$

$$\leq 2\|DF(u_0)\|_{\mathcal{L}(X,Y^*)}\|u-u_0\|_X .$$

This proves the first inequality in (2.1). ♦

**Remark 2.2.** In the examples of Section 5, $X$ and $Y$ are closed subspaces of suitable Sobolev spaces of functions defined on an open set $\Omega \subset \mathrm{IR}^n$. When considering in Equation (2.2) only functions $\varphi$ with support in a given open subset $\omega \subset \Omega$, one then often obtains lower bounds for $u - u_0$ restricted to $\omega$. ♦

**Remark 2.3.** The conditions about $F$ can be weakened such that Hölder continuous and monotone operators, which are not necessarily differentiable, can be handled. The above results can also be extended to branches of solutions including singular points such as simple limit or bifurcation points (cf. [13]).  ◆

### 3. Estimation of the Residual

Let $X_h \subset X$ and $Y_h \subset Y$ be finite dimensional subspaces and $F_h \in C(X_h, Y_h^*)$ be an approximation of F. We want to estimate $\|F(u_h)\|_{Y*}$ where $u_h \in X_h$ is an approximate solution of Equation (1.2). In doing this $c, c_0, c_1, \dots$ denote various constants which are independent of h.

**Proposition 3.1.** Let $u_h \in X_h$ be an approximate solution of Equation (1.2), i.e., $\|F_h(u_h)\|_{Y_h*}$ is "small". Assume that there are a restriction operator $R_h \in \mathcal{L}(Y, Y_h)$, a finite dimensional subspace $\tilde{Y}_h \subset Y$, and an approximation $\tilde{F}_h : X_h \to Y^*$ of F at $u_h$ such that

$$\|(Id_Y - R_h)^* \tilde{F}_h(u_h)\|_{Y*} \le c_0 \|\tilde{F}_h(u_h)\|_{\tilde{Y}_h^*}. \tag{3.1}$$

Then the following estimates hold

$$\|F(u_h)\|_{Y*} \le c_0 \|\tilde{F}_h(u_h)\|_{\tilde{Y}_h^*} + \|(Id_Y - R_h)^*[F(u_h) - \tilde{F}_h(u_h)]\|_{Y*}$$
$$+ \|R_h\|_{\mathcal{L}(Y,Y_h)} \|F(u_h) - F_h(u_h)\|_{Y_h*} + \|R_h\|_{\mathcal{L}(Y,Y_h)} \|F(u_h)\|_{Y_h*} \tag{3.2}$$

and

$$\|\tilde{F}_h(u_h)\|_{\tilde{Y}_h^*} \le \|F(u_h)\|_{\tilde{Y}_h^*} + \|F(u_h) - \tilde{F}_h(u_h)\|_{\tilde{Y}_h^*}. \tag{3.3}$$

**Remark 3.2.** In the examples of section 5, $X_h$ and $Y_h$ are suitable finite element spaces. The choice of $R_h$ then is quite natural. $\tilde{F}_h(u_h)$ is obtained by projecting $F(u_h)$ element-wise onto suitable finite dimensional spaces. This construction is also rather canonical. The main difficulty is to find a space $\tilde{Y}_h$ such that Inequality (3.1) is satisfied. This task is simplified by the auxiliary results of Section 4. The second terms on the right-hand sides of Equations (3.2) and (3.3) measure the quality of the approximation $\tilde{F}_h(u_h)$ to $F(u_h)$. Usually they are higher order terms when compared with $\|\tilde{F}_h(u_h)\|_{\tilde{Y}_h^*}$. The term $\|F(u_h) - F_h(u_h)\|_{Y_h^*}$ is the consistency error of the discretization. The term $\|F_h(u_h)\|_{Y_h^*}$ measures the residual of the algebraic equation (1.2) and can easily be evaluated.  ◆

**Proof of Proposition 3.1:** Consider an arbitrary element $\varphi \in Y$ with $\|\varphi\|_Y = 1$. We then have

$$\langle F(u_h), \varphi \rangle = \langle \tilde{F}_h(u_h), \varphi - R_h\varphi \rangle + \langle F(u_h) - \tilde{F}_h(u_h), \varphi - R_h\varphi \rangle$$
$$+ \langle (F(u_h) - F_h(u_h), R_h\varphi \rangle + \langle F_h(u_h), R_h\varphi \rangle$$
$$\le \|(Id_Y - R_h)^* \tilde{F}_h(u_h)\|_{Y*} + \|(Id_Y - R_h)^*[F(u_h) - \tilde{F}_h(u_h)]\|_{Y*}$$
$$+ \|R_h\|_{\mathcal{L}(Y,Y_h)} \|F(u_h) - F_h(u_h)\|_{Y_h*} + \|R_h\|_{\mathcal{L}(Y,Y_h)} \|F_h(u_h)\|_{Y_h*}.$$

Together with Inequality (3.1), this proves Estimate (3.2).
Estimate (3.3) follows from the triangle inequality.  ◆

When combining Propositions 2.1 and 3.1 we obtain a residual a posteriori error estimator. In a similar way one may obtain a framework for a posteriori error estimators which are based on the solution of auxiliary local problems (cf. [13]).

## 4. Auxiliary Results

Let $\Omega$ be a bounded, connected, open domain in $IR^n$, $n \geq 2$, with polyhedral boundary $\Gamma$. For any open subset $\omega \subset \Omega$ with Lipschitz boundary $\gamma$, we denote by $W^{k,s}(\omega)$, $k \in IN$, $1 \leq s \leq \infty$, $L^s(\omega) := W^{0,s}(\omega)$, and $L^s(\gamma)$ the usual Sobolev- and Lebesgue-spaces equipped with the standard norms $\|.\|_{k,s;\omega} := \|.\|_{W^{k,s}(\omega)}$ and $\|.\|_{s;\gamma} := \|.\|_{L^s(\gamma)}$. If $\omega = \Omega$, we omit the index $\omega$.

Let $\mathcal{T}_h$, $h > 0$, be a family of partitions of $\Omega$ into n-simplices which satisfies the following conditions:

(1) Any two simplices in $\mathcal{T}_h$ are either disjoint or share a complete smooth submanifold of their boundaries.

(2) The ratio $h_T/\rho_T$ is bounded from above independently of $T \in \mathcal{T}_h$ and $h > 0$.

Here, $h_T, \rho_T$, and $h_E$ denote the diameter of $T \in \mathcal{T}_h$, the diameter of the largest ball inscribed into $T$, and the diameter of a face $E$ of $T$. Note, that condition (2) allows the use of locally refined meshes and that it implies that the ratio $h_T/h_E$ is for all $T \in \mathcal{T}_h$ and all faces $E$ of $T$ bounded from above and from below by constants which are independent of $h$, $T$, and $E$.

Denote by $\mathcal{E}_h$ the set of all faces of all $T \in \mathcal{T}_h$. $\mathcal{E}_h$ may be decomposed as $\mathcal{E}_h = \mathcal{E}_{h,\Omega} \cup \mathcal{E}_{h,\Gamma}$, $\mathcal{E}_{h,\Omega} \cap \mathcal{E}_{h,\Gamma} = 0$, where $\mathcal{E}_{h,\Gamma}$ denotes the set of all faces lying on $\Gamma$. Given an $E \in \mathcal{E}_h$, we denote by $\omega_E$ the union of all simplices in $\mathcal{T}_h$ having $E$ as a face. Similarly $\omega_T, T \in \mathcal{T}_h$, is the union of all simplices sharing a face with $T$. For any $E \in \mathcal{E}_h$ and any piecewise continuous function $\varphi$, we denote by $[\varphi]_E$ the jump of $\varphi$ across $E$ in a fixed direction. Here, $\varphi$ is continued by 0 outside $\Omega$ and the direction is given by the exterior normal of $\Gamma$ if $E \in \mathcal{E}_{h,\Gamma}$.

For $k \in IN$, we define

$$S_h^{k,-1} := \{\varphi : \Omega \to IR : \varphi|_T \in \Pi_k \,\forall\, T \in \mathcal{T}_h\} \ , \ S_h^{k,0} := S_h^{k,-1} \cap C(\overline{\Omega}).$$

Here, $\Pi_k$, $k \geq 0$, is the space of polynomials of degree at most k. Moreover, we denote by $\pi_{k,S}$, $S \in \mathcal{T}_h \cup \mathcal{E}_h$, the $L^2$-projection of $L^1(S)$ onto $\Pi_{k|S}$.

Using standard scaling arguments for finite elements, we conclude from [8] that there is an "interpolation" operator $I_h : L^1(\Omega) \to S_h^{1,0}$ which satisfies the following error estimates for all $T \in \mathcal{T}_h$, $E \in \mathcal{E}_h$, and $1 \leq q \leq \infty$:

$$\|\varphi - I_h\varphi\|_{k,q;T} \leq c_1 h_T^{1-k} \|\varphi\|_{1,q;\tilde{\omega}_T} \quad \forall\ 0 \leq k \leq 1,\ \varphi \in W^{1,q}(\tilde{\omega}_T), \tag{4.1}$$

$$\|\varphi - I_h\varphi\|_{q;E} \leq c_2 h_E^{1-1/q} \|\varphi\|_{1,q;\tilde{\omega}_E} \quad \forall\ \varphi \in W^{1,q}(\tilde{\omega}_E), \tag{4.2}$$

where $\tilde{\omega}_T$ and $\tilde{\omega}_E$ denote the union of all elements having a non-empty intersection with $T$ and $E$, respectively. Here and in what follows we adopt the usual convention that $1/\infty := 0$.

Denote by $\hat{T} := \{\hat{x} \in IR^n : \sum_{i=1}^{n} \hat{x}_i \leq 1, \hat{x}_j \geq 0, 1 \leq j \leq n\}$ the reference n-simplex. Set $\hat{E} := \hat{T} \cap \{\hat{x} \in IR^n : \hat{x}_n = 0\}$ and let $\hat{x}_T$ and $\hat{x}_E$ be the barycentres of $\hat{T}$ and $\hat{E}$, respectively. Let $\psi_T, \psi_E \in C^\infty(\hat{T}, IR)$ be functions such that

$$0 \leq \psi_T \leq 1, \qquad \psi_T(\hat{x}_T) = 1, \qquad \psi_T = 0 \text{ on } \partial\hat{T},$$

$$0 \le \psi_{\hat{E}} \le 1, \qquad \psi_{\hat{E}}(\hat{x}_{\hat{E}}) = 1, \qquad \psi_{\hat{E}} = 0 \text{ on } \partial\hat{T} \setminus \hat{E}.$$

We define a continuation operator $\hat{P} : L^\infty(\hat{E}) \to L^\infty(\hat{T})$ by

$$\hat{P}\hat{u}(\hat{x}_1,...,\hat{x}_n) := \hat{u}(\hat{x}_1,...,\hat{x}_{n-1}) \quad \forall \; \hat{x} \in \hat{T}, \hat{u} \in L^\infty(\hat{E}).$$

Finally, $V_{\hat{T}} \subset L^\infty(\hat{T})$ and $V_{\hat{E}} \subset L^\infty(\hat{E})$ are two arbitrary finite dimensional spaces which are kept fixed throughout this section.

Let $T \in \mathcal{T}_h$ be an arbitrary n-simplex and $E \subset \partial T$ be a face of T. There is an invertible affine mapping $F_T : \hat{T} \to T$, $\hat{x} \to x := F_T(\hat{x}) = b_T + B_T \hat{x}$ such that $\hat{T}$ is mapped onto T and $\hat{E}$ is mapped onto E. Set

$$\psi_T := \psi_{\hat{T}} \circ F_T^{-1}, \qquad\qquad \psi_E := \psi_{\hat{E}} \circ F_T^{-1},$$
$$V_T := \{\hat{u} \circ F_T^{-1} : \hat{u} \in V_{\hat{T}}\}, \qquad V_E := \{\hat{\sigma} \circ F_T^{-1} : \hat{\sigma} \in V_{\hat{E}}\},$$

and define the continuation operator $P : L^\infty(E) \to L^\infty(T)$ by

$$P\sigma := [\hat{P}\sigma \circ F_T] \circ F_T^{-1}.$$

In what follows, p, q are two fixed real numbers with $1 \le p \le \infty$ and $\frac{1}{p} + \frac{1}{q} = 1$ and $|||.|||$ denotes the spectral norm on $\mathbb{R}^{n \times n}$.

**Lemma 4.1.** There are constants $c_1,...,c_7$, which only depend on the functions $\psi_{\hat{T}}, \psi_{\hat{E}}$, the spaces $V_{\hat{T}}, V_{\hat{E}}$, the number p, and the ratio $h_T/\rho_T$, such that the following inequalities hold for all $u \in V_T$ and all $\sigma \in V_E$:

$$c_1\|u\|_{0,p;T} \le \sup_{v \in V_T} \frac{\int_T u\psi_T v}{\|v\|_{0,q;T}} \le \|u\|_{0,p;T} \tag{4.3}$$

$$c_2\|\sigma\|_{p;E} \le \sup_{\tau \in V_E} \frac{\int_E \sigma\psi_E \tau}{\|\tau\|_{q;E}} \le \|\sigma\|_{p;E}, \tag{4.4}$$

$$c_3 \, h_T^{-1}\|\psi_T u\|_{0,q;T} \le \|\nabla(\psi_T u)\|_{0,q;T} \le c_4 \, h_T^{-1}\|\psi_T u\|_{0,q;T}, \tag{4.5}$$

$$c_5 \, h_T^{-1}\|\psi_E P\sigma\|_{0,q;T} \le \|\nabla(\psi_E P\sigma)\|_{0,q;T} \le c_6 \, h_T^{-1}\|\psi_E P\sigma\|_{0,q;T}, \tag{4.6}$$

$$\|\psi_E P\sigma\|_{0,q;T} \le c_7 \, h_T^{1/q}\|\sigma\|_{q;E}. \tag{4.7}$$

**Proof:** The upper bounds of Equations (4.3), (4.4) immediately follow from Hölder's inequality and $0 \le \psi_T \le 1, 0 \le \psi_E \le 1$.

The lower bounds of Equations (4.3), (4.4) are proven be transforming all quantities to $\hat{T}$ and $\hat{E}$, by observing that the mappings

$$\hat{u} \to \sup_{\hat{v} \in V_{\hat{T}}} \frac{\int_{\hat{T}} \hat{u}\psi_{\hat{T}} \hat{v}}{\|\hat{v}\|_{0,q;\hat{T}}} \quad \text{and} \quad \hat{\sigma} \to \sup_{\hat{\tau} \in V_{\hat{E}}} \frac{\int_{\hat{E}} \hat{\sigma}\psi_{\hat{E}} \hat{\tau}}{\|\hat{\tau}\|_{q;\hat{E}}}$$

define norms on $V_T$ and $V_E$, resp., and using the equivalence of norms on finite dimensional spaces.

Estimates (4.5)-(4.7) are similarly proven by transforming all quantities to $\hat{T}$ and $\hat{E}$ and using the equivalence of norms on finite dimensional spaces.

## 5. Scalar Quasilinear Elliptic Equations of 2nd Order

Consider the boundary value problem

$$-\nabla \cdot \underline{a}(x,u, \nabla u) = b(x,u, \nabla u) \ \text{ in } \Omega$$

$$u = 0 \ \text{ on } \Gamma \tag{5.1}$$

where $b \in C(\Omega \times \mathrm{IR} \times \mathrm{IR}^n, \mathrm{IR})$ and $\underline{a} \in C^1(\Omega \times \mathrm{IR} \times \mathrm{IR}^n, \mathrm{IR}^n)$ are such that the matrix $A(x,y,z) := (\frac{1}{2}(\partial_{z_j}a_i(x,y,z) + \partial_{z_i}a_j(x,y,z))_{1\leq i,j\leq n}$ is positive definite for all $x \in \Omega$, $y \in \mathrm{IR}$, $z \in \mathrm{IR}^n$.

Under suitable growth-conditions on $\underline{a}$, b, and their derivatives there are real numbers $1 < r, q < \infty$ such that the weak formulation of Problem (5.1) fits into the framework of Section 2 with

$$X := \{u \in W^{1,r}(\Omega) : u = 0 \text{ on } \Gamma\} \ , \ \|.\|_X := \|.\|_{1,r},$$
$$Y := \{\varphi \in W^{1,q}(\Omega) := 0 \text{ on } \Gamma\} \ , \ \|.\|_Y := \|.\|_{1,q},$$
$$<F(u),\varphi> := \int_\Omega \underline{a}(x,u,\nabla u)\nabla\varphi - \int_\Omega b(x,u,\nabla u)\varphi .$$

Denote by $p := \dfrac{q}{q-1}$ the dual exponent of q. Note, that $DF(u) \in \mathrm{Isom}(X,Y^*)$ if the linear boundary value problem

$$- \nabla \cdot (A(x,u,\nabla u)\nabla v) - \nabla \cdot (\partial_y\underline{a}(x,u,\nabla u)v) - \nabla_z b(x,u,\nabla u) \cdot \nabla v - \partial_y b(x,u,\nabla u)v = f \text{ in } \Omega$$
$$v = 0 \text{ on } \Gamma$$

admits for each $f \in Y^*$ a unique weak solution $v \in X$ which depends continuously on f. A particular example falling into the present category is the subsonic flow of an irrotational, ideal, compressible gas:

$$\underline{a}(x,u,\nabla u) := \left[1 - \frac{\gamma-1}{2} \|\nabla u\|^2\right]^{1/(\gamma-1)} \nabla u, \ \gamma > 1,$$
$$b(x,u,\nabla u) := f(x) \in L^p(\Omega),$$
$$r := q := \frac{2\gamma}{\gamma-1} .$$

We do not specify the discretization of Problem (5.1) in detail. We only assume that $X_h \subset X \cap W^{1,\infty}(\Omega)$ and $Y_h \subset Y \cap W^{1,\infty}(\Omega)$ are finite element spaces corresponding to $\mathcal{T}_h$ consisting of affine equivalent elements in the sense of [7] and that $S_h^{1,0} \cap Y \subset Y_h$.

In order to construct $R_h$, $\tilde{F}_h$ and $\tilde{Y}_h$, we define two integers $k,\ell$ and approximations $\underline{a}_h$ of $\underline{a}$ and $b_h$ of b as follows

$$\underline{a}_h(x,u_h,\nabla u_h) := \begin{cases} \underline{a}(x,u_h,\nabla u_h), & \text{if } \underline{a}(x,v_h,\nabla v_h)\in S_h^{k,-1} \ \forall \ v_h \in X_h \\[2ex] \sum_{T \in \mathcal{T}_h} \pi_{1,T}\underline{a}(x,u_h,\nabla u_h), k:=1, & \text{otherwise} \end{cases}$$

$$b_h(x,u_h,\nabla u_h) := \begin{cases} b(x,u_h,\nabla u_h), & \text{if } b(x,v_h,\nabla v_h) \in S_h^{\ell-1} \ \forall \ v_h \in X_h \\ \displaystyle\sum_{T \in \mathcal{T}_h} \pi_{0,T} b(x,u_h,\nabla u_h), \ \ell := 0, & \text{otherwise} \end{cases}$$

Here, $u_h \in X_h$ is arbitrary. Now, $\tilde{F}_h$ is defined in the same way as F with $\underline{a}$ and b replaced by $\underline{a}_h$ and $b_h$, respectively, $R_h := I_h$, and

$$\tilde{Y}_h := \text{span}\{\psi_T v, \psi_E P\sigma : v \in \Pi_{m|T}, \sigma \in \Pi_{k|E}, T \in \mathcal{T}_h, E \in \mathcal{E}_{h,\Omega}\},$$

where $m := \max\{k-1,\ell\}$.

Put for abbreviation for all $T \in \mathcal{T}_h$

$$\varepsilon_T := \{h_T^p \|- \nabla \cdot (\underline{a}(\cdot,u_h,\nabla u_h) - \underline{a}_h(\cdot,u_h,\nabla u_h)) - (b(\cdot,u_h,\nabla u_h) - b_h(\cdot,u_h,\nabla u_h))\|_{0,p;T}^p$$
$$+ \sum_{E \subset \partial T \backslash \Gamma} h_E \|[n \cdot (\underline{a}(\cdot,u_h,\nabla u_h) - \underline{a}_h(\cdot,u_h,\nabla u_h))]_E\|_{p;E}^p \}^{1/p} \tag{5.2}$$

$$\eta_T := \{h_T^p \|- \nabla \cdot \underline{a}_h(\cdot,u_h,\nabla u_h) - b_h(\cdot,u_h,\nabla u_h)\|_{0,p;T}^p$$
$$+ \sum_{E \subset \partial T \backslash \Gamma} h_E \|[n \cdot \underline{a}_h(\cdot,u_h,\nabla u_h)]_E\|_{p;E}^p \}^{1/p}. \tag{5.3}$$

The quantity $\varepsilon_T$ obviously measures the quality of the approximation of $\underline{a}$ and b by $\underline{a}_h$ and $b_h$, respectively and can be estimated explicity. Note that

$$\varepsilon_T = h_T \|f - \pi_{0,T} f\|_{0,p;T} \ \forall \ T \in \mathcal{T}_h,$$

if $X_h \subset S_h^{1,0}$ in the above example.

Using integration by parts elementwise, we obtain for all $\varphi \in Y$

$$\langle F(u_h),\varphi\rangle = \sum_{T \in \mathcal{T}_h} \int_T \{-\nabla \cdot \underline{a}(x,u_h,\nabla u_h) - b(x,u_h,\nabla u_h)\}\varphi + \sum_{E \in \mathcal{E}_{h,\Omega}} \int_E [n \cdot \underline{a}(x,u_h,\nabla u_h)]_E \varphi, \tag{5.4}$$

and

$$\langle \tilde{F}_h(u_h),\varphi\rangle = \sum_{T \in \mathcal{T}_h} \int_T \{-\nabla \cdot \underline{a}_h(x,u_h,\nabla u_h) - b_h(x,u_h,\nabla u_h)\}\varphi$$
$$+ \sum_{E \in \mathcal{E}_{h,\Omega}} \int_E [n \cdot \underline{a}_h(x,u_h,\nabla u_h)]_E \varphi. \tag{5.5}$$

Lemma 4.1, Inequalities (4.1), (4.2), the definition of $\tilde{Y}_h$, and Equalities (5.4), (5.5) then imply that

$$\|(\text{Id}_Y - R_h)^*[F(u_h) - \tilde{F}_h(u_h)]\|_{Y^*} \le c\Big\{\sum_{T \in \mathcal{T}_h} \varepsilon_T^p\Big\}^{1/p} \tag{5.6}$$

and

$$\|F(u_h) - \tilde{F}_h(u_h)\|_{\tilde{Y}_h^*} \le c\Big\{\sum_{T \in \mathcal{T}_h} \varepsilon_T^p\Big\}^{1/p} \tag{5.7}$$

Similarly, we obtain

$$\|(\text{Id}_Y - R_h)^*\tilde{F}_h(u_h)]\|_{Y^*} \le c\Big\{\sum_{T \in \mathcal{T}_h} \eta_T^p\Big\}^{1/p} \tag{5.8}$$

and

$$\|\tilde{F}_h(u_h)\|_{Y_h^*} \le c\left\{\sum_{T\in \mathcal{T}_h} \eta_T^p\right\}^{1/p}. \tag{5.9}$$

In order to prove Inequality (3.1), consider an arbitrary simplex $T \in \mathcal{T}_h$ and an arbitrary face $E \in \mathcal{E}_{h,\Omega}$ of $T$ and denote by $\tilde{Y}_{h|\omega}$, $\omega \in \{T,\omega_E,\omega_T\}$, the set of all functions $\varphi \in \tilde{Y}_h$ with $\mathrm{supp}(\varphi) \subset \omega$. Lemma 4.1, Equation (5.5), and the definition of $\tilde{Y}_h$ then yield

$$
\begin{aligned}
& c_1 c_4^{-1} h_T \|{-\nabla} \cdot a_h(\cdot,u_h,\nabla u_h) - b_h(\cdot,u_h,\nabla u_h)\|_{0,p;T} \\
&\le \sup_{v\in \Pi_{m|T}\setminus\{0\}} \|\nabla(\psi_T v)\|_{0,q;T}^{-1} \int_T \{ -\nabla \cdot a_h(x,u_h,\nabla u_h)-b_h(x,u_h,\nabla u_h)\}\psi_T v \\
&= \sup_{v\in \Pi_{m|T}\setminus\{0\}} \|\nabla(\psi_T v)\|_{0,q;T}^{-1} \; <\tilde{F}_h(u_h),\psi_T v> \\
&\le \sup_{\substack{\varphi\in \tilde{Y}_{h|T} \\ \|\varphi\|_Y = 1}} <\tilde{F}_h(u_h),\varphi>
\end{aligned}
\tag{5.10}
$$

and, using Inequality (5.10),

$$
\begin{aligned}
& c_2\, c_6^{-1}\, c_7^{-1} h_E^{1/p} \|[n\cdot a_h(\cdot,u_h,\nabla u_h)]\|_{p;E} \\
&\le \sup_{\sigma\in \Pi_{k|E}\setminus\{0\}} c_6^{-1} c_7^{-1} h_E^{1/p} \|\sigma\|_{q;E}^{-1} \int_E [n\cdot a_h(x,u_h,\nabla u_h)]_E\psi_E P\sigma \\
&= \sup_{\sigma\in \Pi_{k|E}\setminus\{0\}} c_6^{-1} c_7^{-1} h_E^{1/p} \|\sigma\|_{q;E}^{-1}\left\{ \int_E \{<\tilde{F}_h(u_h),\psi_E P\sigma> \right. \\
&\qquad\qquad\qquad\qquad\qquad \left. -\int_{\omega_E} \{-\nabla\cdot a_h(x,u_h,\nabla u_h)-b_h(x,u_h,\nabla u_h)\}\psi_E P\sigma \right\} \\
&\le \sup_{\substack{\varphi\in \tilde{Y}_{h|\omega_E} \\ \|\varphi\|_Y = 1}} <\tilde{F}_h(u_h),\varphi> + c_6^{-1} h_E \|{-\nabla}\cdot a_h(\cdot,u_h,\nabla u_h) - b_h(\cdot,u_h,\nabla u_h)\|_{0,p;\omega} \\
&\le c \sup_{\substack{\varphi\in \tilde{Y}_{h|\omega_E} \\ \|\varphi\|_Y = 1}} <\tilde{F}_h(u_h),\varphi>.
\end{aligned}
\tag{5.11}
$$

Inequalities (5.10) and (5.11) imply that

$$\eta_T \le c \sup_{\substack{\varphi\in \tilde{Y}_{h|\omega_T} \\ \|\varphi\|_Y = 1}} <\tilde{F}_h(u_h),\varphi> \tag{5.12}$$

and

$$\left\{\sum_{T\in \mathcal{T}_h} \eta_T^p\right\}^{1/p} \le c\|\tilde{F}_h(u_h)\|_{Y_h^*}. \tag{5.13}$$

Inequalities (5.8) and (5.13) in particular prove Inequality (3.1).

Propositions 2.1 and 3.1 and Inequalities (5.6)-(5.9), (5.12), and (5.13) yield the following a posteriori error estimates for Problem (5.1).

**Proposition 5.1.** Let $u \in X$ be a weak solution of problem (5.1) which is regular in the sense of Proposition 2.1 and let $u_h \in X_h$ be an approximate solution of the corresponding discrete problem which is sufficiently close to u in the sense of Proposition 2.1. Then the following a posteriori error estimates hold

$$\|u-u_h\|_{1,r} \le c_1 \left\{ \sum_{T \in \mathcal{T}_h} \eta_T^p \right\}^{1/p} + c_2 \left\{ \sum_{T \in \mathcal{T}_h} \varepsilon_T^p \right\}^{1/p} + c_3 \|F(u_h) - F_h(u_h)\|_{Y_h*} + c_4 \|F_h(u_h)\|_{Y_h*}$$

and

$$\eta_T \le c_5 \|u-u_h\|_{1,r;\omega_T} + c_6 \left\{ \sum_{T' \subset \omega_T} \varepsilon_{T'}^p \right\}^{1/p} \forall\, T \in \mathcal{T}_h \,.$$

Here, $\varepsilon_T$ and $\eta_T$ are given by Equations (5.2) and (5.3) and $\|F(u_h)-F_h(u_h)\|_{Y_h*}$ and $\|F_h(u_h)\|_{Y_h*}$ are the consistency error of the discretization and the residual of the discrete problem, respectively. ◆

**Remark 5.2.** Proposition 5.1 can easily be extended to the case of Neumann boundary conditions. One only has to replace in Equations (5.2) and (5.3) $\Gamma$ by the part of the boundary on which Dirichlet boundary conditions are imposed. The first estimate of Proposition 5.1 also holds if $\eta_T$ is defined using the original functions $\underline{a}$ and b instead of the projected ones $\underline{a}_h$ and $b_h$. The $\varepsilon_T$-term then of course disappears. If the functions $\underline{a}$ and b are sufficiently smooth one may also use higher order approximations $\underline{a}_h$ and $b_h$ instead of the present low-order ones. ◆

## 6. Bibliography


1. I. Babuska, Feedback, adaptivity, and a posteriori estimates in finite elements: aims, theory, and experience, Accuracy Estimates and Adaptive Refinements in Finite Element Computation (I. Babuska et all., eds.), Wiley, New York, 1986, pp. 3-23.
2. I. Babuska, W. Gui, Basic principles of feedback and adaptive approaches in the finite element method, Comp. Meth. Appl. Mech. Engrg. 55 (1986), 27-42.
3. I. Babuska, W.C. Rheinboldt, Error estimates for adaptive finite element computations, SIAM J. Numer. Anal. 15 (1978), 736-754.
4. I. Babuska, W.C. Rheinboldt, A posteriori error estimates for the finite element method, Int. J. Numer. Meth. Engrg. 12 (1978), 1597-1615.
5. I. Babuska, R. Rodriguez, The problem of the selection of an a-posteriori error indicator based on smoothing techniques, Int. J. Numer. Math, Engrg. (to appear).
6. R.E. Bank, A Weiser, Some a posteriori error estimators for elliptic partial differential equations, Math. Comput. 44 (1985), 283-301.
7. Ph. G. Ciarlet, The Finite Element Method for Elliptic Problems, North Holland, 1978.
8. Ph. Clément, Approximation by finite element functions using local regularization, RAIRO Anal. Numér. 2 (1975), 77-84.
9. R. Duran, M.A. Muschietti, R. Rodriguez, On the asymptotic exactness of error estimators for linear triangular elements, Numer. Math. 59 (1991), 107-127.
10. R. Duran, R. Rodriguez, On the asymptotic exactness of Bank-Weiser's estimator, Numer. Math. 62 (1992), 297-303.
11. W.C. Rheinboldt, On a theory of mesh-refinement processes, SIAM J. Numer. Anal. 17 (1980), 766-778.
12. R. Verfürth, A posteriori error estimators and adaptive mesh-refinement techniques for the Navier-Stokes equations, In: Incompressible CFD-Trends and Advances (M.D. Gunzburger, R.A. Nicolaides, eds.), Cambridge University Press, 1993, pp. 447-477.
13. R. Verfürth, A posteriori error estimates for non-linear problems. Finite element discretizations of elliptic equations. Math. Comput. (to appear 1994).

COMPUTING INCOMPRESSIBLE FLOWS IN GENERAL DOMAINS

P. Wesseling, C.G.M. Kassels, C.W. Oosterlee,
A. Segal, C. Vuik, S. Zeng, M. Zijlema

Delft University of Technology
Department of Technical Mathematics and Informatics
Mekelweg 4, 2628 CD Delft
The Netherlands


## ABSTRACT

We describe a generalization of the classical staggered grid discretization of the incompressible
Navier-Stokes equations to general coordinates, using the coordinate-invariant tensor formulation of the equations of motion. Provided a certain choice is made for the dependent variables
and for the computation of the geometric quantities, satisfactory accuracy is obtained on fairly
non-uniform grids. The pressure-correction method is used to solve the time-dependent equations with preconditioned GMRES and multigrid methods. Turbulent and three-dimensional
applications are presented.


## INTRODUCTION

For the numerical discretization of systems of partial differential equations, one may use
structured or unstructured grids, and a staggered (different unknowns in different grid
points) or a collocated (different unknowns in the same grid point) grid point arrangement. At the moment it is not clear which options are best, and all approaches are
being vigorously pursued. The generation of unstructured grids in complicated domains
is far easier than that of structured grids. But with structured grids data structures are
simpler and solution time is less, at least on present day vector machines. Furthermore,
multigrid, preconditioning and matrix-vector multiplications are easier to implement
and requires less computing time per iteration. Whether a staggered or a collocated
grid is better depends on the system to be solved. For the incompressible Navier-Stokes
equations, the staggered MAC (marker-and-cell) placement of variables ([5]), given in
figure 1, is more attractive than a collocated grid, because on the latter one has to cope
with spurious pressure oscillations, and the physical boundary conditions do not suf-
fice. But in the compressible case this disadvantage of collocated schemes disappears.

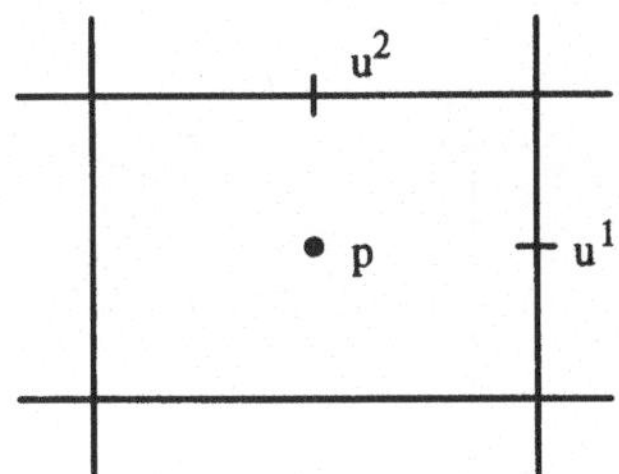


Figure 1: Staggered placement of unknowns according to [5].

The various schemes have been studied in the past in Cartesian coordinates, of course. Collocated schemes are much easier to generalize from Cartesian to general coordinates than staggered schemes. This is probably why collocated schemes are more preponderant than staggered schemes at present; for a brief review of some recent publications on both approaches, see [15]. In addition, the papers [18], [19] and [24] (on staggered grids) are of substantial interest. The theoretical basis of the staggered MAC scheme for the incompressible Navier-Stokes equations has recently been strengthened in [6]. A comparison of a collocated and a staggered scheme is given in [1].

Staggered schemes in general coordinates not only require more programming effort than collocated schemes, but can be inaccurate on non-smooth grids, unless derived carefully, as we will see.

In Delft we pursue the development of codes for the numerical solution of the incompressible Navier-Stokes equations in general three-dimensional geometries by using structured boundary-fitted staggered grids and domain decomposition. A list of our recent publications is [10], [29], [12], [30], [23], [13], [14], [11], [28], [37], [36] . In this paper we present some recent developments.


## DISCRETIZATION IN GENERAL COORDINATES ON A STAGGERED GRID

Let $x = x(\xi)$ be a boundary-fitted coordinate mapping, with $x$ Cartesian coordinates in physical space. The mapping is one-to-one, transforming the physical domain $\Omega$ in $x$-space to a square $G$ (restricting ourselves to two dimensions for convenience; a three-dimensional application will be presented later) in $\xi$-space. $G$ is subdivided uniformly in cells. The coordinate mapping is assumed to be generated numerically, and $x = x(\xi)$ is known only in the vertices of the grid cells (cf. fig. 2). In order to avoid the inaccuracies alluded to before, it is important to specify $x = x(\xi)$ everywhere, and to make approximations accordingly. The simplest way to extend the definition of $x = x(\xi)$ to all of $G$ is by bilinear interpolation in each cell. This allows abrupt changes of meshsize in $\Omega$, which is what one wants in practice. As a consequence, the sides of the cells in $\Omega$ are straight, as shown in figure 2.

The Navier-Stokes equations in general coordinates are best described by means of tensor notation. For an introduction to tensor analysis, see [2], [21], [25]; for more details, see [11], [14] and [23].

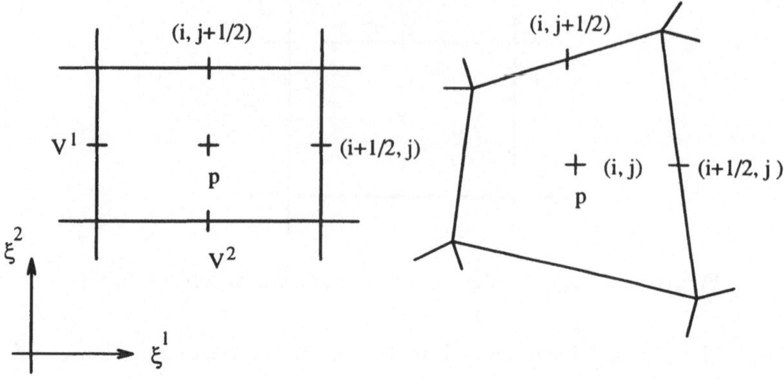



Figure 2: Cell in computational domain $G$ and its image in the physical domain $\Omega$

The contravariant and covariant basevectors are defined by, respectively,

$$\boldsymbol{a}^{(\alpha)} = grad\ \xi^\alpha\ , \quad \boldsymbol{a}_{(\alpha)} = \frac{\partial \boldsymbol{x}}{\partial \xi^\alpha}\ . \tag{1}$$

We obtain $\boldsymbol{a}_{(\alpha)}$ simply by numerical differentiation, which is exact, since $\boldsymbol{x} = \boldsymbol{x}(\boldsymbol{\xi})$ is piecewise bilinear.

Note that $\boldsymbol{a}_{(\alpha)}$ is discontinuous across cell boundaries $\xi^\alpha = constant$, but continuous across all boundaries $\xi^\beta = constant$, $\beta \neq \alpha$. We find $\boldsymbol{a}^{(\alpha)}$ from the relations

$$\boldsymbol{a}^{(1)} = \frac{1}{\sqrt{g}} \left( \begin{array}{c} a^2_{(2)} \\ -a^1_{(2)} \end{array} \right)\ , \quad \boldsymbol{a}^{(2)} = \frac{1}{\sqrt{g}} \left( \begin{array}{c} -a^2_{(1)} \\ a^1_{(1)} \end{array} \right) \tag{2}$$

where $\sqrt{g}$ is the Jacobian of $\boldsymbol{x} = \boldsymbol{x}(\boldsymbol{\xi})$, given by

$$\sqrt{g} = a^1_{(1)}a^2_{(2)} - a^2_{(1)}a^1_{(2)}\ . \tag{3}$$

Note that $\sqrt{g}$ and $\boldsymbol{a}^{(\alpha)}$ are discontinuous at all cell boundaries. Further geometric quantities that will be required are the metric tensor:

$$g^{\alpha\beta} = \boldsymbol{a}^{(\alpha)} \cdot \boldsymbol{a}^{(\beta)} \tag{4}$$

and the Christoffel symbol:

$$\{ {}^{\ \alpha}_{\beta\gamma} \} = \boldsymbol{a}^{(\alpha)} \cdot \frac{\partial \boldsymbol{a}_{(\gamma)}}{\partial \xi^\beta} = \frac{\partial \xi^\alpha}{\partial x^\delta} \cdot \frac{\partial^2 x^\delta}{\partial \xi^\beta \partial \xi^\gamma}\ . \tag{5}$$

Notice that $g^{\alpha\beta}$ is discontinuous at cell boundaries, and that some components of $\{ {}^{\ \alpha}_{\beta\gamma} \}$ are zero in the interior of cells, but have a delta-function infinity at cell boundaries. The contravariant components $U^\alpha$ of a vector field $\boldsymbol{u}$ are defined by

$$U^\alpha = \boldsymbol{a}^{(\alpha)} \cdot \boldsymbol{u}\ . \tag{6}$$

Invariant (or tensor) components will be denoted by capital letters, Cartesian components by lower case letters, bold letters refer to vectors themselves.

Since $\boldsymbol{a}^{(\alpha)}$ is normal to $\xi^{\alpha} = constant$, on a staggered grid we want to store $U^{\alpha}$ in the center of cell segments $\xi^{\alpha} = constant$, in order to generalize the MAC scheme to general coordinates. However, here $\boldsymbol{a}^{(\alpha)}$ is discontinuous, and $U^{\alpha}$ is not well-defined. Instead, as contravariant representation of $\boldsymbol{u}$ we will use

$$V^{\alpha} = \sqrt{g}U^{\alpha} \tag{7}$$

which is continuous at the points where we want to locate $V^{\alpha}$ on a staggered grid, namely $V^1$ in $(i + 1/2, j)$ and $V^2$ in $(i, j + 1/2)$, cf. figure 2. An additional argument to prefer the use of $V^{\alpha}$ over that of $U^{\alpha}$ has been put forward in [23].
It may be shown that $V^{\alpha}$ has a physical interpretation as mass-flux; for example, $V^1\Delta\xi^2$ is the mass flux through a vertical cell boundary of size $\Delta\xi^2$ in $G$.

The tensor formulation of the incompressible Navier-Stokes equations in general coordinates is given by

$$U^{\alpha}_{,\alpha} = 0 \tag{8}$$

$$\frac{\partial U^{\alpha}}{\partial t} + T^{\alpha\beta}_{,\beta} = 0 \tag{9}$$

with the tensor $T^{\alpha\beta}$ defined by

$$T^{\alpha\beta} = -Re^{-1}(g^{\alpha\gamma}U^{\beta}_{,\gamma} + g^{\gamma\beta}U^{\alpha}_{,\gamma}) + g^{\alpha\beta}p + U^{\alpha}U^{\beta}. \tag{10}$$

Here $p$ is the pressure, and $Re$ is the Reynolds number, which is not constant if a turbulent flow is modeled.

We have

$$U^{\alpha}_{,\alpha} = \frac{1}{\sqrt{g}}\frac{\partial V^{\alpha}}{\partial \xi^{\alpha}} = 0 \; . \tag{11}$$

This is discretised by finite volume integration over the cell depicted in figure 2 according to

$$\int_{G_{ij}} U^{\alpha}_{,\alpha}\sqrt{g}d\xi^1 d\xi^2 = \int_{G_{ij}} \frac{\partial V^{\alpha}}{\partial \xi^{\alpha}}d\xi^1 d\xi^2 \cong \Delta\xi^2 V^1|_{i-1/2,j}^{i+1/2,j} + \Delta\xi^1 V^2|_{i,j-1/2}^{i,j+1/2} = 0 \; . \tag{12}$$

Next, consider (9). We have

$$T^{\alpha\beta}_{,\beta} = \frac{1}{\sqrt{g}}\frac{\partial\sqrt{g}T^{\alpha\beta}}{\partial\xi^{\beta}} + \{^{\alpha}_{\beta\gamma}\}T^{\beta\gamma} \tag{13}$$

where the Christoffel symbol defined in (5) appears. In our discretization of the momentum equation (9) we let ourselves guided by the requirement, that in the case of the identity mapping $\boldsymbol{x} = \boldsymbol{\xi}$ the classical MAC scheme of [5] should be recovered. This ensures that we will have second order accuracy when the mapping is sufficiently smooth. Assuming this, we neglect the lack of smoothness of the geometrical quantities that was discussed before, and make approximations by simple finite differences and average when needed. For example,

$$\{^{1}_{11}\}_{i+1/2,j} \cong \boldsymbol{a}^{(1)}_{i+1/2,j} \cdot \boldsymbol{a}_{(1)}|_{ij}^{i+1,j}/\Delta\xi^1 \tag{14}$$

$$V^2_{i+1/2,j+1/2} \cong \frac{1}{2}(V^2_{i,j+1/2} + V^2_{i+1,j+1/2}) \; . \tag{15}$$

Approximations that take (lack of) smoothness into account more properly will be developed in the near future. But as will be illustrated, satisfactory results are obtained with the above procedure on reasonably smooth grids.

For $\alpha = 1$, equation (9) is integrated over a shifted cell centered at the $V^1$-point $(i+1/2, j)$ with the surrounding $V^2$-points as vertices. We have

$$\int_{G_{i+1/2,j}} \frac{\partial U^1}{\partial t} \sqrt{g} d\xi^1 d\xi^2 \cong dV^1_{i+1/2,j}/dt \ . \tag{16}$$

Using (13) we obtain

$$\int_{G_{i+1/2,j}} T^{1\beta}_{,\beta} \sqrt{g} d\xi^1 d\xi^2 \cong \Delta\xi^2 (\sqrt{g} T^{11})|^{i+1,j}_{ij}$$

$$+ \Delta\xi^1 (\sqrt{g} T^{12})|^{i+1/2,j+1/2}_{i+1/2,j-1/2} + \Delta\xi^1 \Delta\xi^2 (\sqrt{g} \{ {}^{\ 1}_{\beta\gamma} \} T^{\beta\gamma}))_{i+1/2,j} \ . \tag{17}$$

$T^{\alpha\beta}$ is further approximated using (10), evaluating $U^\beta_{,\gamma}$ by averaging over suitable volumes, similar to what is done in [17] in the Cartesian case. This results in the 19-point stencil presented in figure 3 (after elimination of four $V^2$-points by using (12)).  The

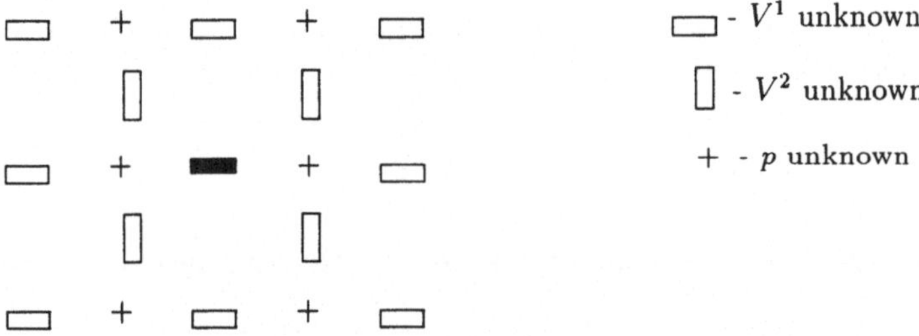

Figure 3: Stencil for the $V^1$-momentum equation

implementation of the boundary conditions is described in [23].

The following numerical experiments are instructive. First we compute the Poiseuille flow in a channel using a non-uniform grid (cf. figure 4). The flow direction is from bottom to top. The grid has been chosen in this way solely for testing purposes. The exact isobars are straight as in figure 4(3); the isobars of figure 4(2) are completely wrong.

Figure 4(2) is obtained with the discretization just described. This bad result can be explained as follows. For the pressure term equation (17) gives, with $g^{12} = 0$ for this orthogonal case,

$$\int_{G_{i+1/2,j}} (g^{1\beta} p)_{,\beta} \sqrt{g} d\xi^1 d\xi^2 \cong \Delta\xi^2 (\sqrt{g} g^{11} p)|^{i+1,j}_{ij} + \Delta\xi^1 \Delta\xi^2 (\sqrt{g} \{ {}^{\ 1}_{11} \} g^{11} p)_{i+1/2,j} \ . \tag{18}$$

Because the first term is a difference of a step-function when $i + 1/2$ is at the location of the grid discontinuity and the second formally implies taking the derivative of a step

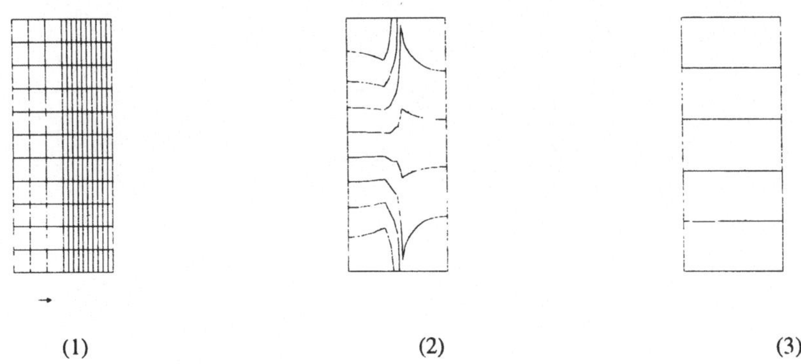

(1)                 (2)                 (3)

Figure 4: Grid (1) and isobars (2), (3) for Poiseuille flow; (2): with equation (18); (3): with (19).

function, inaccuracies are to be expected on a nonsmooth grid like this. However, we may also proceed as follows. Since $g^{\alpha\beta}_{,\gamma} = 0$ we can replace (18) by (in the general non-orthogonal case)

$$\int\limits_{G_{i+1/2,j}} g^{1\beta}\frac{\partial p}{\partial \xi^\beta}\sqrt{g}d\xi^1 d\xi^2 \cong \Delta\xi^2(\sqrt{g}g^{11})_{i+1/2,j}p|_{i,j}^{i+1,j} + \Delta\xi^1(\sqrt{g}g^{12})_{i+1/2,j}p|_{i+1/2,j-1/2}^{i+1/2,j+1/2}. \quad (19)$$

This gives the result of figure 4(3). However, we do not claim that this modification eliminates all errors arising from lack of grid smoothness.

Next, we compute the flow in the $L$-shaped channel depicted in figure 5. Again, the grid is deliberately chosen non-smooth. Figure 6 shows the resulting velocity vectors using $U^\alpha$-unknowns and using $V^\alpha$-unknowns (on a refined version of the grid of figure 5). The solution presented in the right half of figure 6 is known to be correct. The case for using $V^\alpha$ as unknowns is clear.

For time discretization we use a linear combination of the forward and backward Euler method. Gathering the pressure unknowns in an algebraic vector $\boldsymbol{P}$, the $V^\alpha$-unknowns in a vector $\boldsymbol{V}_I$ and the $V^\alpha$-values prescribed by Dirichlet boundary conditions in a vector $\boldsymbol{V}_B$, a time-step can be described by

$$\begin{aligned}
(\boldsymbol{V}_I^{n+1} - \boldsymbol{V}_I^n)/\Delta t &= \theta\boldsymbol{f}(\boldsymbol{V}_I^{n+1}, \boldsymbol{V}_B^{n+1}) \\
&\quad + (1-\theta)\boldsymbol{f}(\boldsymbol{V}_I^n, \boldsymbol{V}_B^n) + \theta G\boldsymbol{P}^{n+1} + (1-\theta)G\boldsymbol{P}^n \quad (20) \\
D_I\boldsymbol{V}_I^{n+1} + D_B\boldsymbol{V}_B^{n+1} &= 0. \quad (21)
\end{aligned}$$

Here the nonlinear operator $\boldsymbol{f}$ and the linear operator $G$ come from the discretization of (13), and (21) is equivalent to (12).

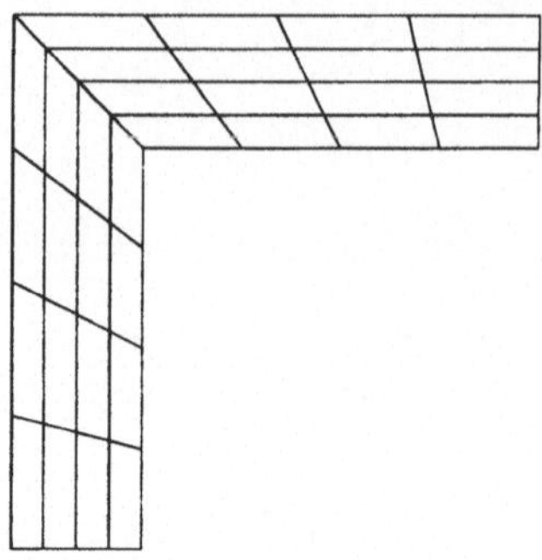

Figure 5: *L*-shaped channel with grid.

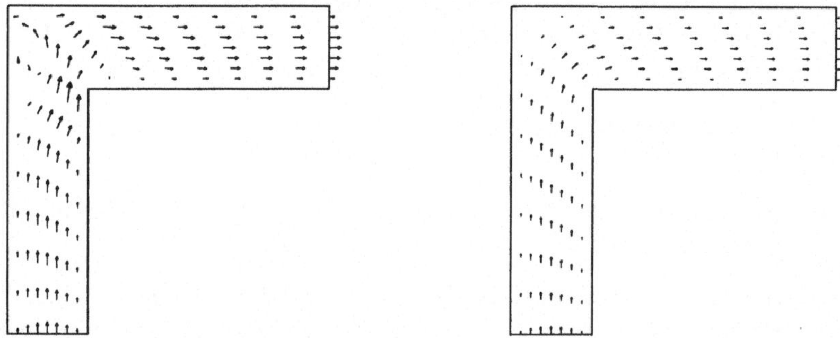

Figure 6: Velocity vectors in *L*-chaped channel, for Reynolds number = 10. Left: with $U^\alpha$-unknowns; right: with $V^\alpha$-unknowns.

## SOLUTION METHODS

We solve (20), (21) with the pressure-correction method ([5], further developed and studied in [3], [16], [7], [4], [6], [27]), and frequently used for computing time-dependent flows. Using Newton-linearization of $\boldsymbol{f}$, the pressure-correction method gives

$$(\boldsymbol{V}_I^* - \boldsymbol{V}_I^n)/\Delta t + \theta Q_1 \boldsymbol{V}_I^* = \boldsymbol{f}(\boldsymbol{V}_I^n, \boldsymbol{V}_B^n) + Gp^n \tag{22}$$

$$(\boldsymbol{V}_I^{n+1} - \boldsymbol{V}_I^*)/\Delta t = \theta G(p^{n+1} - p^n) \tag{23}$$

$$D_I \boldsymbol{V}^{n+1} = -D_B \boldsymbol{V}_B^{n+1} \tag{24}$$

where the linear operator $Q_1$ depends on $\boldsymbol{V}_I^n, \boldsymbol{V}_B^n$ and $\boldsymbol{V}_B^{n+1}$. Combination of (23) and (24) gives the pressure equation:

$$D_I Gp^{n+1} = D_I Gp^n - \frac{1}{\theta \Delta t}(D_B \boldsymbol{V}_B^{n+1} + D_I \boldsymbol{V}_I^*) . \tag{25}$$

The systems (22) and (25) are both non-symmetric in general coordinates. For such problems GMRES-type and multigrid methods are robust and efficient, and these are the methods that we use. We usually take $\theta = 1$ for stationary problems.

The GMRES method is described in [20], and its application to our problems is more fully described in [28]. Let the pressure equation be denoted by

$$Py = b. \tag{26}$$

Using the GMRES(m) method ([20]), the vector $z_m$ is found in $m$ iterations satisfying

$$z_m = arg\min\{\|b - P(y_0 + z)\|_2 : z \in K^m(P; r_0)\| \tag{27}$$

where $r_0 = b - Py_0$ and the Krylov subspace $K^m(P; r_0)$ is defined by

$$K^m(P; r_0) = Span\ \{r_0, Pr_0, ..., P^{m-1}r_0\} \tag{28}$$

where we take $y_0 = (0, ..., 0)^T$. After obtaining $z_m$ the method restarts using $x_0 := z_m$. For implementation details, see [20] and [26]. It is found that preconditioning is required to obtain satisfactory convergence. After some experiments it was found that ILU preconditioning with MILU(0.95) was suitable; for details see [28]. For a typical problem on a grid with $16 \times 24 = 1024$ cells the results of table 1 are obtained on an Convex C240 with full GMRES (no restart, $m$ not fixed in advance and MILU(0.95) preconditioning).

Table 1: Statistics of preconditioned GMRES for the pressure equation.

| Matrix × vector | Vector updates | Inner products | Memory vectors | Preconditioning matrix × vector | CPU $s$ |
|---|---|---|---|---|---|
| 31 | 481 | 481 | 31 | 31 | 0.6 |

The Krylov subspace dimension $m$ equals the number of matrix-vector products. We can afford the memory required by a Krylov subspace of dimension as high as 31 because the memory required by the matrix solver for the momentum equation (22) (which can be overwritten) is sufficient to accommodate the pressure equation solution strategy just described. The termination criterion is $\|r_m\|_2/\|r_0\|_2 < 10^{-6}$.

For the momentum equation (22) GMRES convergence without preconditioning is satisfactory, thanks to the enhancement of the main diagonal by the time-derivative. It is essential to use diagonal scaling to make the main diagonal elements of comparable size.

Table 2: Statistics of GMRES(5) for the momentum equation.

| Matrix × vector | Vector updates | Inner products | Memory vectors | CPU $s$ |
|---|---|---|---|---|
| 68 | 144 | 144 | 5 | 0.6 |

Numerical experiments indicate that a low value of $m$ ($m = 5$) is best for computing time on Convex. For details, see [28]. Table 2 gives some statistics for the same case as table 1.

Denoting the momentum system as $My = b$, the termination criterion is $\|r_k\|_2/\|b\|_2 \leq 10^{-5}$ with $k$ the number of GMRES iterations carried out. Unlike the Cartesian case, the computing time is not dominated by the time required for the pressure equation.

We are also developing multigrid solution methods. Our experiences can be summarized by saying that standard multigrid works. Of course, the smoothing method must be sufficiently robust to handle stretched grids and mixed derivatives. Smoothers of collective block Gauss-Seidel and ILU type combined with an $r$-transformation (i.e. distributive iteration or postconditioning) (as in [31], [32], [33], [34], [35]) have been found to be satisfactory. For the stationary case (not to be further discussed here) the nonlinear multigrid method has been studied in [12], [13], [14], [11], [36]. Extension to the nonstationary case has been considered in [11], [13], [14].

A disadvantage of nonlinear multigrid is that discretization and solution are interwined. In many applications this does not matter, but in our case discretization is a complicated affair, and code-development is a multi-person job. Therefore software modularity is at a premium.

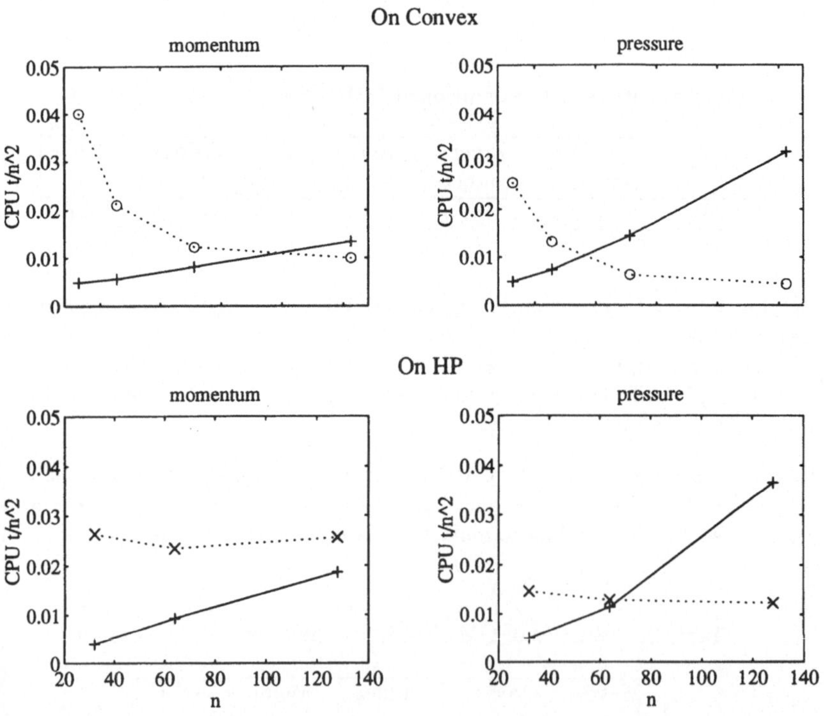



Figure 7: Soluton time per grid point on a vector computer (Convex) and a scalar computer (HP workstation) as function of grid-size on a ($n \times n$) grid. —: GMRES; ⋯: multigrid.

The discretization and solution phases can be separated by using linear multigrid inside an outer Newton iteration method. If one uses Galerkin coarse grid approximation in multigrid, solution code becomes independent of discretization code, so that modifications of the discretization do not propagate into the multigrid code. This linear multigrid approach has been investigated in [37].

We find, not unexpectedly, that Krylov subspace methods are faster than multigrid if the grid size is not too large, and vice-versa. The break-even point is problem-dependent and dependent on the type of computer (scalar or vector), because of short vector lengths occurring on coarse grids in multigrid. This is illustrated in figure 7 for a typical case. This figure gives the number of seconds per grid point required to execute 40 time-steps. The termination criterion in each time-step is $\|r\|/\|r_0\| < tol$, with $r$ the residual and $r_0$ the residual at the start of the first time-step; $tol = 10^{-4}$ for momentum and $tol = 10^{-6}$ for pressure. Significant reduction of computing times could be realized by relating iterative termination criteria to spatial discretization accuracy, and by improvements in the code. Figure 7 is intended merely to give a qualitative impression.

## TURBULENT FLOW

Turbulence is modeled by the standard high Reynolds number $k-\varepsilon$ model. The viscosity coefficient is given by

$$\mu = c_\mu k^2/\varepsilon \tag{29}$$

with $c_\mu = 0.09$, and the turbulent kinetic energy $k$ and the turbulent energy dissipation rate $\varepsilon$ satisfying, in general coordinates,

$$\frac{\partial k}{\partial t} + (kU^\alpha)_{,\alpha} - (\frac{\mu_t}{\sigma_k}g^{\alpha\beta}k_{,\alpha})_{,\beta} = P - \varepsilon \tag{30}$$

$$\frac{\partial \varepsilon}{\partial t} + (\varepsilon U^\alpha)_{,\alpha} - (\frac{\mu_t}{\sigma_\varepsilon}g^{\alpha\beta}\varepsilon_{,\alpha})_{,\beta} = \frac{\varepsilon}{k}(c_{1\varepsilon}P - c_{2\varepsilon}\varepsilon) \tag{31}$$

with $\sigma_k = 1$, $\sigma_\varepsilon = 1.3$, $c_{1\varepsilon} = 1.44$ and $c_{2\varepsilon} = 1.92$. The values of these constants and $c_\mu$ are recommended in [8]. Furthermore,

$$P = 2\mu_t S^{\alpha\beta} S_{\alpha\beta} \tag{32}$$

in the rate of production of turbulent kinetic energy. Furthermore, $S^{\alpha\beta}$ and $S_{\alpha\beta}$ are the contravariant and covariant mean rate of strain tensors, given by

$$S^{\alpha\beta} = \frac{1}{2}(g^{\alpha\gamma}U^\beta_{,\gamma} + g^{\beta\gamma}U^\alpha_{,\gamma}), \quad S_{\alpha\beta} = \frac{1}{2}(U_{\alpha,\beta} + U_{\beta,\alpha}). \tag{33}$$

The unknowns $k$ and $\varepsilon$ are located in the pressure points in the staggered grid. Time discretization takes place with the implicit Euler method. Space discretization is done as before, except that upwind discretization is applied to the convection terms, to ensure positivity of $k$ and $\varepsilon$. For the same reason, care has to be taken in linearising the right-hand-sides. This is done as follows:

$$P^{n+1} - \varepsilon^{n+1} \cong P^{n+1} + \varepsilon^n - 2\frac{\varepsilon^n}{k^n}k^{n+1} \tag{34}$$

$$\frac{\varepsilon}{k}(c_{1\varepsilon}P - c_{2\varepsilon}\varepsilon)^{n+1} \cong \frac{\varepsilon^n}{k^n}(c_{1\varepsilon}P^{n+1} + c_{2\varepsilon}\varepsilon^n) - 2c_{2\varepsilon}\frac{\varepsilon^n}{k^n}\varepsilon^{n+1}. \tag{35}$$

The momentum equations are solved first with $\mu_t = \mu_t^n$. It may be shown that (30)-(35) give positive $k^{n+1}$ and $\varepsilon^{n+1}$, if $k^n$ and $\varepsilon^n$ are positive.

For the boundary conditions for the momentum equations the wall function method is applied, similar to what is proposed in [8]. For $k$ a homogeneous Neumann condition is applied, and in the center of cells adjacent to a solid wall we use

$$\varepsilon = c_\mu^{3/4} k^{3/2} / (\kappa Y) \tag{36}$$

with $Y$ the distance to the wall, following [8].

## NUMERICAL EXPERIMENTS

First we give an illustration of not smooth enough and smooth enough grids for our method. Figure 8 gives an example. The top wall moves from right to left. The resulting

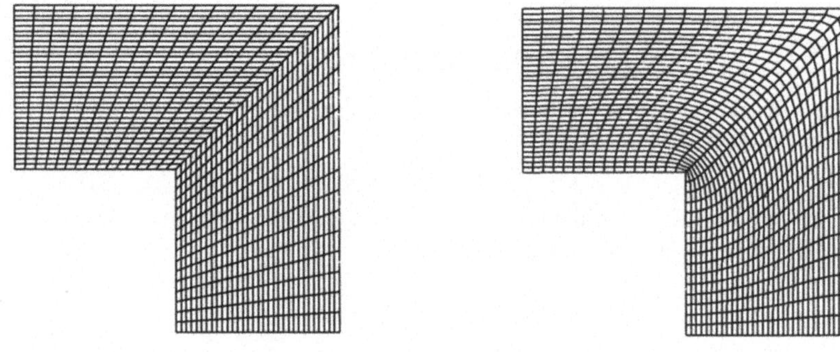

Figure 8: $L$-shaped cavity with non-smooth and smooth grids.

streamlines are given in figure 9 for Reynolds number = 1000.

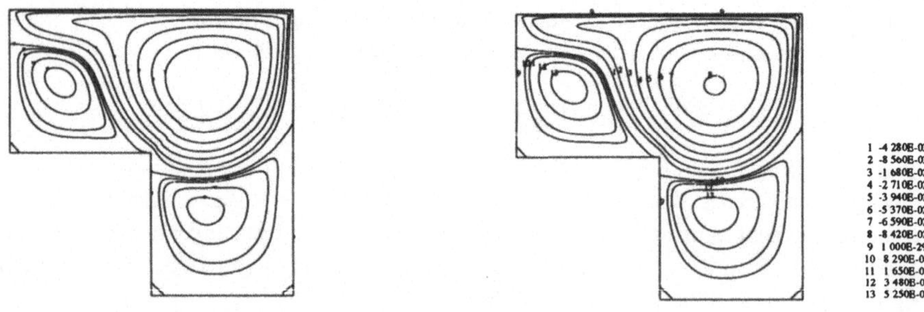



Figure 9: Streamline patterns obtained with the grids of 8.

Although still rather satisfactory, the streamline pattern obtained with the non-smooth grid shows traces of the kinks in the gridlines. The smooth grid was generated with the LiSS package ([9]). Further details are given in [15]. The following example concerns turbulent flow across a staggered tube bundle, as sketched in fig. 10. This was a test

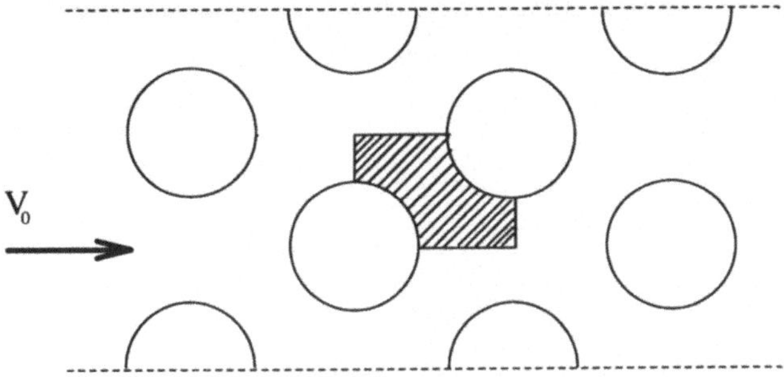


Figure 10: Staggered tube bank geometry

problem the Second ERCOFTAC-IAHR Workshop on Refined Flow Modelling, Manchester, 1993. Because of symmetry, the solution needs to be computed in the hatched region only. For details about how the symmetry conditions are handled in the solver, see [22]. The grid (again generated with LiSS ([9])) is shown in fig. 11. Figure 12 shows

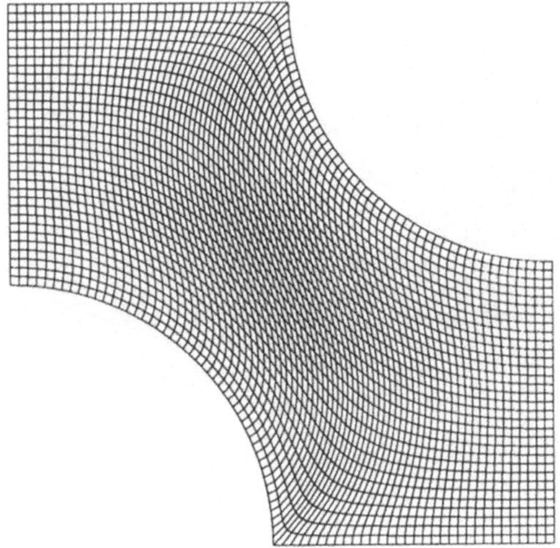

Figure 11: Computational grid

a typical comparison with experiment.

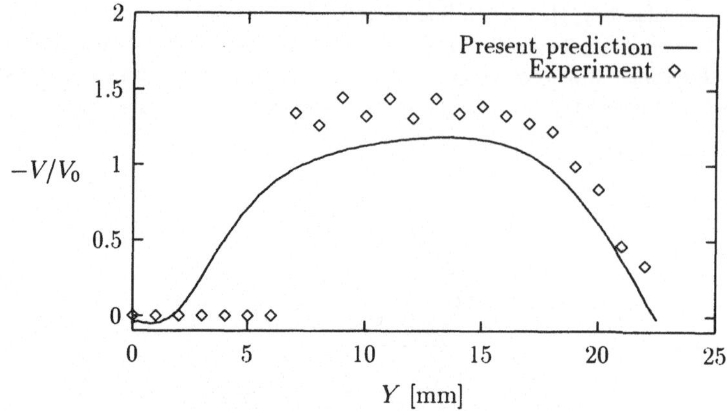

Figure 12: Vertical velocity profile at $X = 11$ mm

The present method was found to compare well with other methods presented at the workshop. Our final example is 3D flow over a backward facing step. Figure 13 shows the grid, and figure 14 two streamlines.

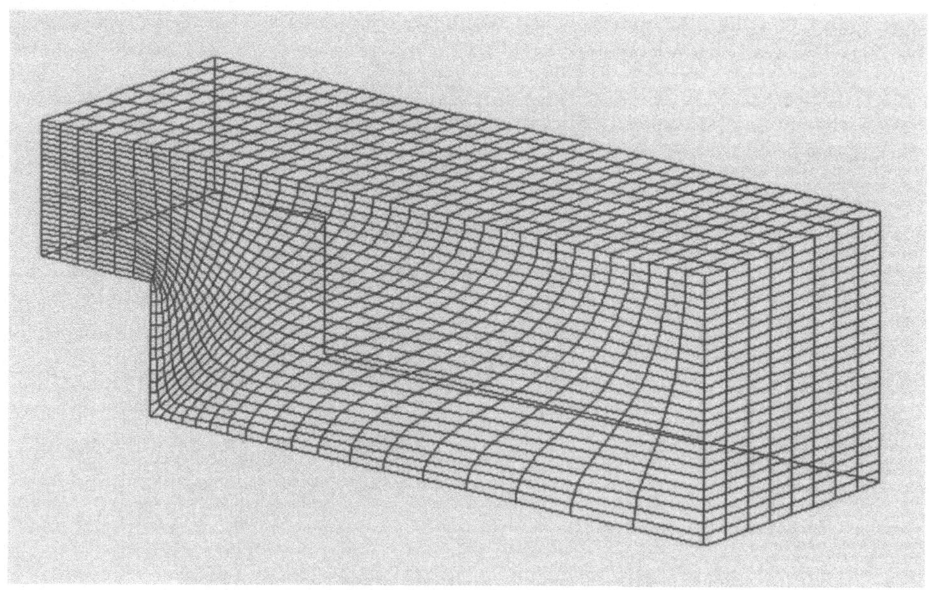

Figure 13: Backward facing step with $24 \times 38 \times 8$ grid

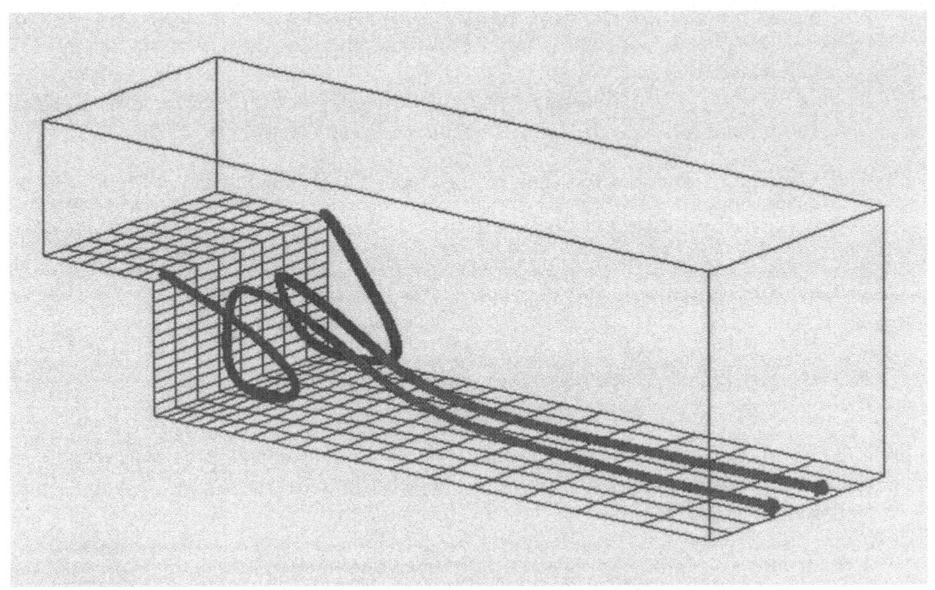

Figure 14: Two streamlines

Unlike the 2D case, a closed recirculation zone seems to be absent. The Reynolds number is 100. Starting from rest, the residual of the stationary equations is reduced by a factor $10^{-5}$ in 40 time steps with $\Delta t = 0.25$, using a multigrid solver with one $F$-cycle with one pre- and post-smoothing of ILU type per time step, giving a mesh-independent reduction factor of 0.11 of better. Further details may be found in [37].

## FINAL REMARKS

We have found that a generalization of the classical staggered-grid MAC scheme ([5]) to general coordinates with sufficient discretization accuracy may be based on the co-ordinate invariant tensor formulation of the equations of motion, provided a judicious implementation of the geometric quantities is employed and the contravariant flux components are used as unknowns. In general coordinates the fast solution methods (such as fast Poisson solvers for the pressure based on the fast Fourier transform) available in the Cartesian case cannot be applied. But using GMRES or multigrid, the computing load remains bearable. Turbulent and 3D applications are feasible.

## REFERENCES

[1] H.I. Andersson, J.I. Billdal, P. Eliasson, and A. Rizzi. Staggered and non-staggered finite-volume methods for nonsteady viscous flows: a comparative study. In K.W.

Morton, editor, *Lecture Notes in Physics 371*, pages 172–176. Twelfth International Conference on Numerical Methods in Fluid Dynamics, Springer, Berlin, 1990.

[2] R. Aris. *Vectors, tensors and the basic equations of fluid mechanics.* Prentice-Hall, Inc., Englewood Cliffs, N.J., 1962. Reprinted, Dover, New York, 1989.

[3] A.J. Chorin. Numerical solution of the Navier-Stokes equations. *Math. Comp.*, 22:745–762, 1968.

[4] M.P. Gresho and R.L. Sani. On pressure boundary conditions for the incompressible Navier-Stokes equations. *Int. J. Numer. Fluids*, 7:1111–1145, 1987.

[5] F.H. Harlow and J.E. Welch. Numerical calculation of time-dependent viscous incompressible flow of fluid with a free surface. *The Physics of Fluids*, 8:2182–2189, 1965.

[6] T.Y. Hou and B.T.R. Welton. Second-order convergence of a projection scheme for the incompressible Navier-Stokes equations with boundaries. *SIAM J. Num. Anal.*, 30:609–629, 1993.

[7] J. Kim and P. Moin. Application of a fractional-step method to incompressible Navier-Stokes equations. *J. Comp. Phys.*, 59:308–323, 1985.

[8] B.E. Launder and D.B. Spalding. The numerical computation of turbulent flows. *Comp. Methods Appl. Mech. Eng.*, 3:269–289, 1974.

[9] G. Lonsdale and K. Stüben. The LiSS package. Arbeitspapiere der GMD 524, Gesellschaft für Mathematics und Datenverarbeitung mbH, Sank Augustin, Germany, 1991.

[10] A.E. Mynett, P. Wesseling, A. Segal, and C.G.M. Kassels. The ISNaS incompressible Navier-Stokes solver: invariant discretization. *Applied Scientific Research*, 48:175–191, 1991.

[11] C.W. Oosterlee. *Robust multigrid methods for the steady and unsteady incompressible Navier-Stokes equations in general coordinates.* PhD thesis, Delft University of Technology, The Netherlands, 1993.

[12] C.W. Oosterlee and P. Wesseling. A multigrid method for an invariant formulation of the incompressible Navier-Stokes equations in general co-ordinates. *Communications in Applied Numerical Methods*, 8:721–734, 1992.

[13] C.W. Oosterlee and P. Wesseling. Multigrid schemes for time-dependent incompressible Navier-Stokes equations. *Impact Comp. Science Engng.*, 5:153–175, 1993.

[14] C.W. Oosterlee and P. Wesseling. A robust multigrid method for a discretization of the incompressible Navier-Stokes equations in general coordinates. *Impact Comp. Science Engng.*, 5:128–151, 1993.

[15] C.W. Oosterlee, P. Wesseling, A. Segal, and E. Brakkee. Benchmark solutions for the incompressible Navier-Stokes equations in general co-ordinates on staggered grids. *Int. J. Num. Meth. Fluids*, 17:301–321, 1993.

[16] S.V. Patankar and D.B. Spalding. A calculation procedure for heat and mass transfer in three-dimensional parabolic flows. *Int. J. Heat Mass Transfer*, 15:1787–1806, 1972.

[17] R. Peyret and T.D. Taylor. *Computational Methods for Fluid Flow.* Springer, Berlin, 1983.

[18] M. Rosenfeld. Validation of numerical simulation of incompressible pulsatile flow in a constricted channel. *Computers and Fluids*, 22:139–156, 1993.

[19] M. Rosenfeld and D. Kwak. Time-dependent solution of viscous incompressible flows in moving coordinates. *Int. J. Numer. Meth. Fluids*, 13:1311–1328, 1991.

[20] Y. Saad and M.H. Schultz. GMRES: a generalized minimal residual algorithm for solving non-symmetric linear systems. *SIAM J. Sci. Stat. Comp.*, 7:856–869, 1986.

[21] L.I. Sedov. *A course in continuum mechanics, Vol. I. Basic equations and analytical techniques.* Wolters-Noordhoff Publishing, Groningen, The Netherlands, 1971.

[22] A. Segal, C. Vuik, and C.G.M. Kassels. On the implementation of symmetric and antisymmetric periodic boundary conditions for incompressible flow. Report 93-61, Faculty of Technical Mathematics and Informatics, Delft University of Technology, Delft, 1993. To appear in Int. J. Numer. Meth. Fluid Dyn.

[23] A. Segal, P. Wesseling, J. Van Kan, C.W. Oosterlee, and K. Kassels. Invariant discretization of the incompressible Navier-Stokes equations in boundary fitted co-ordinates. *Int. J. Num. Meth. Fluids*, 15:411–426, 1992.

[24] B.R. Shin, T. Ikohagi, and H. Daiguji. An implicit finite-difference scheme for solving the unsteady 3-D incompressible Navier-Stokes equations. In Ch. Hirsch, J. Périaux, and W. Kordulla, editors, *Computational Fluid Dynamics (1992). Vol. 1*, pages 457–464, Amsterdam, 1992. Elsevier.

[25] I.S. Sokolnikoff. *Tensor analysis.* John Wiley & Sons, Inc., Englewood Cliffs, N.J., 1964.

[26] H.A. Van der Vorst. The convergence behaviour of some iterative solution methods. In P. Grüber, J. Périaux, and R.P. Shaw, editors, *Proc. 5th Int. Symp. on Numerical Methods in Engineering. Vol. 1*, pages 61–72, Berlin, 1989. Springer.

[27] J.J.I.M. Van Kan. A second-order accurate pressure correction method for viscous incompressible flow. *SIAM J. Sci. Stat. Comp.*, 7:870–891, 1986.

[28] C. Vuik. Solution of the discretized incompressible Navier-Stokes equations with the GMRES method. *Int. J. for Num. Meth. Fluids*, 16:507–523, 1993.

[29] P. Wesseling. Large scale modeling in computational fluid dynamics. In E.F. Deprettere and A.-J. van der Veen, editors, *Algorithms and parallel VLSI architectures, Volume A: Tutorials*, pages 277–308, Amsterdam, 1991. Elsevier.

[30] P. Wesseling, A. Segal, J.J.I.M. van Kan, C.W. Oosterlee, and C.G.M. Kassels. Finite volume discretization of the incompressible Navier-Stokes equations in general coordinates on staggered grids. *Comp. Fluid Dynamics Journal*, 1:27–33, 1992.

[31] G. Wittum. Linear iterations as smoothers in multigrid methods: Theory with applications to incomplete decompositions. *Impact of Comp. Science Engng.*, 1:180–215, 1989.

[32] G. Wittum. Multi-grid methods for Stokes and Navier-Stokes equations with transforming smoothers: Algorithms and numerical results. *Numer. Math.*, 54:543–563, 1989.

[33] G. Wittum. On the convergence of multi-grid methods with transforming smoothers. *Num. Math.*, 57:15–38, 1990.

[34] G. Wittum. R-transforming smoothers for the incompressible Navier- Stokes equations. In W. Hackbusch and R. Rannacher, editors, *Numerical treatment of the Navier-Stokes equations*, pages 153–162, Braunschweig, 1990. Vieweg. Notes on Numerical Fluid Mechanics 30.

[35] G. Wittum. The use of fast solvers in computational fluid dynamics. In P. Wesseling, editor, *Proceedings of the Eighth GAMM-Conference on Numerical Methods in Fluid Mechanics*, pages 574–581, Braunschweig, 1990. Vieweg. Notes on Numerical Fluid Mechanics 29.

[36] S. Zeng and P. Wesseling. Numerical study of a multigrid method with four smoothing methods for the incompressible navier-stokes equations in general coordinates. In N. Duane Melson, T.A.Manteuffel, and S.F.McCormick, editors, *Sixth Copper Mountain Conference on Multigrid Methods. NASA Conference Publication 3224*, pages 691–708. NASA, Hampton VA., 1993.

[37] S. Zeng and P. Wesseling. Multigrid solution of the incompressible Navier-Stokes equations in general coordinates. *SIAM J. Num. Anal.*, 1994. To appear.

# LIST OF PARTICIPANTS

**Andrich, D.**, Dr., Inst. f. Schiffstechnik, Albert-Einstein-Str. 2, D-18059 Rostock, FAX: 0831-4405253

**Babovsky, H.**, PD Dr., IBM Sc. Center, Vangerowstr. 18, D-69115 Heidelberg, TEL.: 06221-404353

**Bader, G.**, Prof. Dr., Institut für Mathematik, Postfach 101344, D-03013 Cottbus, TEL.: 0355-692444, bader@math.uni-cottbus.de

**Bastian, P.**, Dipl.-Math., IWR, Im Neuenheimer Feld 368, D-69120 Heidelberg, bastian@iwr.uni-heidelberg.de

**Becker, R.**, Dipl.-Math., Inst. f. Ang. Mathematik, Im Neuenheimer Feld 294, D-69120 Heidelberg, roland@gaia.iwr.uni-heidelberg.de

**Benz, E.**, Inst. f. Therm. Strömungsmaschinen, Kaiserstr. 12, D-76128 Karlsruhe, FAX: 0721-8082767

**Berzins, M.**, School of Computer Studies, GB-Leeds LS29JT, FAX: 0044-532-335468, martin@scs.leeds.ac.uk

**Bikker, S.**, Dipl.Ing., Inst. f. Strahlantriebe, Templergraben 55, D-52062 Aachen, FAX: 0241-28226

**Birken, K.**, Rechenzentrum der Uni Stuttgart, Allmandring 30, D-70550 Stuttgart, FAX: 0711-6788363

**Blum, H.**, Prof. Dr., FB Mathematik, Campus Nord, Vogelpothsweg 87, D-44227 Dortmund, FAX: 0231-7555307, blum@euler.mathematik.uni-dortmund.de

**Braess, D.**, Prof. Dr., Inst. f. Mathematik, Universitätsstr. 150, D-44801 Bochum, FAX: 0234-7094103

**Daniels, H.**, Dr., IBM Sc. Center, Vangerowstr. 18, D-69115 Heidelberg, FAX: 06221-593500

**Dick, E.**, Dr., Dept. of Mach., U Gent, Sint Pietersnieuwstraat 41, B-9000 Gent, FAX: 0032-91-643586

**Drikakis, D.**, Dr., Lehrst. f. Strömungsmechanik, Cauerstr.4, D-91058 Erlangen, FAX: 09131-859503

**Eggers, D.**, Institut für Mathematik, Postfach 101344, D-03013 Cottbus, eggers@math.uni-cottbus.de

**Feistauer, M.**, Prof., Fac. of Math. and Phys., Charles Univ., Sokolovská 83, CR-18600 Praha 8, FEIST%CSPGUK11.BITNET

**Fellehner, S.**, Inst. f. Ang. Math., U Hamburg, Bundesstraße 55, D-20146 Hamburg, FAX: 040-41235117

**Fontaine, J.**, Dr., LIMSI/CNRS, BP 133, F-91403 Orsay Cedex, FAX: 0033-1-69858088, fontaine@limsi.fr

**Franca, L. P.**, Prof., Lab. Nacional de Comp. Cientifica, Rua Lauro Muller 455, 22290 Rio de Janeiro, franca@titan.Colorado.EDU

**Führer, Chr.**, Dipl.-Math., Inst. f. Ang. Mathematik, Im Neuenheimer Feld 294, D-69120 Heidelberg, fuehrer@gaia.iwr.uni-heidelberg.de

**Gehrke, E.**, Dipl.-Math., Inst. f. Ang. Mathematik, Im Neuenheimer Feld 294, D-69120 Heidelberg, FAX: 06221-565634, eckard@gaia.iwr.uni-heidelberg.de

**Gerz, T.**, Dr., DLR, Inst. f. Physik d. Atmosphäre, D-82234 Oberpfaffenhofen, FAX: 08153-281841

**Glowinski, R.**, Prof., Dept. of Math., 4800 Calhoun, USA-Houston, TX 77204-3476, FAX:001-713-7494626

**Gresho, P. M.**, Dr., Lawrence Livermore Lab., PO Box 5501, USA-Livermore, CA 94551, FAX:001-510-4225844

**Greza, H.**, Dipl.-Ing., Inst. f. Strahlantriebe, Templergraben 55, D-52062 Aachen, FAX: 0241-28226

**Grimmer, A.**, Dipl.-Math., Inst. f. Ang. Mathematik, Im Neuenheimer Feld 294, D-69120 Heidelberg

**Groh, U.**, Dr., TU Chemnitz, FB Mathematik, D-09009 Chemnitz, FAX: 0371-5612657

**Hannemann, V.**, Dr., DLR, Inst. f. Strömungsmechanik, Bunsenstr. 10, D-37073 Göttingen, FAX: 0551-7092446

**Hauser, T.** , Lehrst. f. Fluidmechanik, TU München, Arcisstr. 21, D-80333 München, FAX: 089-21052505, hauser@lsm.mw.tu-muenchen.de

**Hebeker, F.K.**, Prof. Dr., IBM Sc. Center, Vangerowstr. 18, D-69115 Heidelberg, FAX: 06221-593500, HEBEKER%DHDIBM1.BITNET

**Heinrichs, W.**, Dr., Inst. f. Ang. Math., Universitätsstr. 1, D-40225 Düsseldorf, FAX: 0211-3113117

**Hemforth, F.**, Dipl.-Math., Inst. f. Math., Ruhruniversität, Universitätsstr. 150, D-44801 Bochum

**Heywood, J.**, Prof., Dept. of Math., Univ. of British Columbia, CAN-Vancouver, BC V6T 1Y4, heywood@math.ubc.ca

**Hujeirat, A.**, Dipl.-Math., Inst. f. Ang. Math., Im Neuenheimer Feld 294, D-69120 Heidelberg, FAX: 06221-565634, hujeirat@gaia.iwr.uni-heidelberg.de

**Hupertz, B.**, Dipl.-Ing., Abt. Klimawindkanal, Volkswagen AG, D-38436 Wolfsburg

**Jäger, J.**, Dipl.-Math., IBM Sc. Center, Vangerowstraße 18, D-69115 Heidelberg, FAX: 06221-593500, JJAEGER%DHDIBM1.BITNET

**Jäger, W.**, Prof. Dr., Inst. f. Ang. Math., Im Neuenheimer Feld 294, D-69120 Heidelberg, TEL.: 06221-565780, jaeger@iwr1.iwr.uni-heidelberg.de

**Janßen, R.**, Akad. Rat Dr., IBM Sc. Center, Vangerowstr. 18, D-69115 Heidelberg, FAX: 06221-593500, JANSSEN@DHDIBM1.BITNET

**Johannsen, K.**, Inst. f. Ang. Mathematik, Im Neuenheimer Feld 294, D-69120 Heidelberg, johannse@iwr1.iwr.uni-heidelberg.de

**Johnson, C.**, Prof., Math. Dept., Chalmers University, S-41296 Göteborg, claes@math.chalmers.se

**Kallinderis, Y.**, Prof., Dept. of Aerospace Engng., Univ. of Texas, USA-Austin, TX 78712, kallind@cfdlab.ae.utexas.edu

**Kanschat, G.**, Dipl.-Math., Inst. f. Ang. Mathematik, Im Neuenheimer Feld 294, D-69120 Heidelberg, Guido.Kanschat@iwr.uni-heidelberg.de

**Kerschl, P.**, Lehrstuhl für Fluidmechanikund Prozeßautomation, TU München, D-85350 Freising

**Keyes, D.**, Prof., Dept. of Mechanical Engng., Yale Univ., USA-New Haven, CT 06520, keyes@icase.edu

**Kilian, S.**, Dipl.-Math., Inst. f. Ang. Mathematik, Im Neuenheimer Feld 294, D-69120 Heidelberg, Susanne.Kilian@iwr.uni-heidelberg.de

**Klingenberg, Chr.**, PD Dr., Inst. f. Ang. Mathematik, Im Neuenheimer Feld 294, D-69120 Heidelberg, TEL.: 06221-563223, B07%DHDURZ2.BITNET

**Kreiss, H.**, Prof., Firestone Laboratory, Appl. Math., 101-50 CALTECH, USA-Pasadena, CA 91125

**Kröner, D.**, Prof. Dr., Inst. f. Ang. Math., Hermann-Herder-Str. 10, D-79104 Freiburg, FAX: 0761-2033066, dietmar@titan.mathematik.uni-freiburg.de

**Kurreck, M.** , Inst. f. Therm. Strömungsmaschinen, Kaiserstr. 12, D-76128 Karlsruhe, FAX: 0721-699222

**Lechner, R.**,  Lehrst. f. Fluidmechanik, TU München, Arcisstr. 21, D-80333 München, FAX: 089-21052505

**Lilek, Z.**, Inst. f. Schiffbau, Lämmersieth 90, D-22305 Hamburg, FAX: 040-29843199

**Linde, J.**, Dipl.-Math., Campus Nord, Vogelpothsweg 87,  D-44227 Dortmund, FAX: 0231-7555307, linde@euler.mathematik

**Maaß, C.**, DLR, Inst. f. Physik d. Atmosphäre, D-82234 Oberpfaffenhofen, FAX: 08153-281841

**Malcherek, A.**, Inst. f. Strömungsmechanik, Appelstraße 9a, D-30167 Hannover, FAX: 0511-762-3777

**Marion, M.**, Prof., Ecole Centrale de Lyon, Depart. de Mathem.-Im Neuenheimer Feldormat.-Syst., F-69131 Ecully cedex, marion@cc.ec-lyon.fr

**Mayerle, R.**, Inst. f. Strömungsmech., Appelstr. 9a, D-30167 Hannover, FAX: 0511-7623777

**Meinel, S.**, Dr.,Projektgruppe SPC, TU Chemnitz, Reichenhainer Str. 88, D-09126 Chemnitz, FAX:0371-5614748

**Michl, T.**, Inst. f. Aerodynamik, Pfaffenwaldring 21, D-70550 Stuttgart, FAX: 0711-6853438

**Morgan, K.**, Prof., Dept. of Civil Engng., University College, Singleton Park, GB-Swansea SA2 8PP, K.Morgan@swansea.ac.uk

**Morton, K.W.**, Prof., Oxford Univ., Computing Laboratory, Wolfson Building, Parks road, GB-Oxford OX1 3QD

**Munz, C.-D.**, Dr., KfK INR, Postfach 3640, D-76021 Karlsruhe, MAIL: INR338@DKAKFK3.BITNET

**Nau, M.**, Dipl.-Ing. Inst. f. Techn. Verbrennung, U Stuttgart, Pfaffenwaldring 12, D-70569 Stuttgart

**Neises, J.**, Siemens Nixdorf Sc. Computing, Godesberger Allee 83, D-53175 Bonn, FAX: 0228-9588502

**Neuss, N.**, Dipl.-Math., Inst. f. Ang. Mathematik, Im Neuenheimer Feld 294, D-69120 Heidelberg, neuss@iwr1.iwr.uni-heidelberg.de

**Nirschl, H.**, Dipl.-Ing., TU München, Lehrst. f. Fluidmechanik, D-85350 Freising, FAX: 08161-714510

**Oswald, H.**, Inst. f. Ang. Mathematik, Im Neuenheimer Feld 294, D-69120 Heidelberg, oswald@gaia.iwr.uni-heidelberg.de

**Peters, A.**, Dr., IBM Sc. Center, Vangerowstraße 18, D-69115 Heidelberg, APETERS%DHDIBM1.BITNET

**Pinelli, A.**, Dr., v. Karman Inst. for Fluid Dynamics, Chaussee de Waterloo 72, B-1640 Rhode-St.-Genese

**Pironneau, O.**, Prof., INRIA Domaine de Voluceau, Roquencourt B.P. 105, F-78153 Le Chesnay Cedex, pironnea@menusin.inria.fr

**Pospiech, Chr.**, Dr., IBM Sc. Center, Vangerowstraße 18, D-69115 Heidelberg, POSPIECH%DHDIBM1.BITNET

**Prohl, A.**, Dipl.-Math.,Inst. f. Ang. Mathematik, Im Neuenheimer Feld 294, D-69120 Heidelberg, prohl@gaia.iwr.uni-heidelberg.de

**Quarteroni, A.**, Prof., Dipartimento di Matematica, Politecnico di Milano, I-20133 Milano

**Rannacher, R.**, Prof. Dr., Inst. f. Ang. Mathematik, Im Neuenheimer Feld 294, D-69120 Heidelberg, TEL.: 06221-564873, rannacher@gaia.iwr.uni-heidelberg.de

**Rautmann, R.**, Prof. Dr., FB Math./Im Neuenheimer Feldormatik, UGH Paderborn, Warburger Str. 100, D-33098 Paderborn

**Reichert, A.**, Dipl.-Ing., FB7-FG3, UGH Duisburg, Lotharstraße 1, D-47057 Duisburg, FAX: 0203-3793052

**Reichert, H.**, Dipl.-Math., Inst. f. Ang. Math., Im Neuenheimer Feld 294, D-69120 Heidelberg, FAX: 06221-565331, reichert@iwr1.iwr.uni-heidelberg.de

**Rexroth, C.-H.**, Inst. f. Therm. Strömungsmaschinen, Kaiserstr. 12, D-76128 Karlsruhe

**Riedel, U.**, Dr., IWR, Im Neuenheimer Feld 368, D-69120 Heidelberg, riedel@iwr1.iwr.uni-heidelberg.de

**Risch, U.**, Inst. f. Analysis und Numerik, TU Magdeburg, PF 4120, D-39016 Magdeburg

**Rivkind, V.**, Prof., Inst. of Math., St. Petersburg State University, Bibliotechnaja sqn. 2, 197022 St. Petersburg

**Roos, H.-G.**, Prof. Dr., FB Mathematik, Mommsenstr. 13, D-01069 Dresden, FAX: 0351-4634268

**Sarazin, R.**, Dipl.-Math., Inst. f. Math., Universitätsstr. 150, D-44801 Bochum

**Schenk, K.**, Dr., Inst. f. Math., Karl-Marx-Str. 17, D-03044 Cottbus, FAX: 0355-692402, schenk@math.uni-cottbus.de

**Schieweck, F.**, Dr., Fakultät f. Math., PSF 4120, D-39016 Magdeburg, FAX: 0391-55922758, schiewec@DMDTU11.BITNET

**Schmachtel, R.**, Inst. f. Ang. Mathematik, Im Neuenheimer Feld 294, D-69120 Heidelberg, schmachtel@gaia.iwr.uni-heidelberg.de

**Schneider, R.**, KfK INR, Postfach 3640, D-76021 Karlsruhe, TEL.: 07247-822449

**Schreck, E.**, Lehrstuhl f. Strömungsmechanik, Cauerstr. 4, D-91058 Erlangen

**Schreiber, P.**, Dipl.-Math., Inst. f. Ang. Mathematik, Im Neuenheimer Feld 294, D-69120 Heidelberg, schreib@gaia.iwr.uni-heidelberg.de

**Schäfer, I.**, Inst. f. Grundwasserwirtschaft, Mommsenstr. 13, D-01062 Dresden, FAX: 0351-2326118

**Schäfer, M.**, Dr., Lehrst. f. Strömungsmech., Caverstr. 4, D-91058 Erlangen, FAX: 09131-810450

**Schöll, E.**, Dipl.-Ing., Inst. f. Raumfahrtsysteme, Pfaffenwaldring 31, D-70550 Stuttgart, FAX: 0711-6852489, schoell@grandma.irs.uni-stuttgart.de

**Schüller, A.**, Dr., GMD, Postfach 1316, D-53731 St. Augustin

**Segatz, J.**, Dipl.-Ing., IWR, Im Neuenheimer Feld 368, D-69120 Heidelberg,TEL.: 06221-564981, segatz@cerberus.iwr.uni-heidelberg.de

**Silvester, D. J.**, Dr., Math. Dept., UMIST, PO Box 88, GB-Manchester M60IQD, FAX: 0044-61-200-3669, djd@courant.ma.umist.ac.uk

**Stolcis, L.**, Dr., CRS4, Via Nazario Sauro 10, I-09123 Cagliari, MAIL: Stolcis@crs4.it

**Strietzel, M.**, Dipl.-Math., DLR, Inst. f. Physik d. Atmosphäre, D-82234 Oberpfaffenhofen, FAX: 08153-281841

**Suttmeier, F.-T.**, Dipl.-Math., Inst. f. Ang. Mathematik, Im Neuenheimer Feld 294, D-69120 Heidelberg, Franz-Theo.Suttmeier@iwr.uni-heidelberg.de

**Szepessy, A.**, Dr., INADA, Royal Institute of Technology, S-10044 Stockholm, szepessy@nada.kth.se

**Ta Phuoc Loc**, Dr., LIMSI-CNRS, BP 133, F-91403 Orsay Cedex, FAX: 0033-1-69858088

**Tenaud, Chr.**, LIMSI, BP.133, F-91403 Orsay Cedex, tenaud@limsi.fr

**Thevenin, D.**, IWR Heidelberg, Im Neuenheimer Feld 368, D-69120 Heidelberg, TEL.: 06221-564984

**Tobiska, L.**, Prof. Dr., FB Mathematik, Universitätsplatz 2, D-39106 Magdeburg, tobiska@dmdtu11

**Turek, St.**, Dr., Inst. f. Ang. Math., Im Neuenheimer Feld 293, D-69120 Heidelberg, ture@gaia.iwr.uni-heidelberg.de

**Valli, A.**, Prof., Dipertimento di Matematica, Università degli Studi, I-38050 Povo (Trento), FAX:0039-461881624

**Verfürth, R.**, Prof. Dr., Math. Inst., Universitätsstr. 150, D-44801 Bochum

**Warnatz, J.**, Prof. Dr., Inst. f. techn. Verbrennung, Pfaffenwaldring 12, D-70569 Stuttgart, TEL.: 0711-6855653

**Weidner, J.**, Dr., IBM Sc. Center, Vangerowstr. 18, D-69115 Heidelberg

**Weinbrecht, R.**, Inst. f. Ang. Mathematik, Im Neuenheimer Feld 294, D-69120 Heidelberg

**Wesseling, P.**, Prof., Faculty of Tech. Math., University of Technology, Julianalaan 132, NL-2628 BL Delft

**Wittum, G.**, Prof. Dr., Inst. f. Ang. Math., Im Neuenheimer Feld 294, D-69120 Heidelberg, FAX: 06221-565331

**Wölfert, A.**, Dipl.-Ing. Inst. f. Techn. Verbrennung, U Stuttgart, Pfaffenwaldring 12, D-70569 Stuttgart

**Zhou, G.**, Dr., Inst. f. Ang. Mathematik, Im Neuenheimer Feld 294, D-69120 Heidelberg, zhou@gaia.iwr.uni-heidelberg.de

**Zulehner, W.**, Dr., Inst. f. Math., Univ. Linz, A-4040 Linz, zulehner@miraculix.numa.uni-linz.ac.at

**Brief Instruction for Authors**

Manuscripts should have well over 100 pages. As they will be reproduced photomechanically they should be typed with utmost care on special stationary which will be supplied on request.
In print, the size will be reduced linearly to approximately 75 per cent. Figures and diagrams should be lettered accordingly so as to produce letters not smaller than 2 mm in print. The same is valid for handwritten formulae. Manuscripts (in English) or proposals should be sent to the general editor, Prof. Dr. E. H. Hirschel, Herzog-Heinrich-Weg 6, D-85604 Zorneding.